Confluence, Tech Comm, Chocolate

A wiki as platform extraordinaire for technical communication

Sarah Maddox

Confluence, Tech Comm, Chocolate

Graphics credits

Illustrations: Ryan Maddox

The cover design, book design, and the XML Press logo are copyright © 2012 XML Press. The cover illustration and interior illustrations are copyright © 2012 Ryan Maddox.

Disclaimer

Trademarks

XML Press
Laguna Hills, California
http://xmlpress.net

First Edition
ISBN: 978-1-937434-00-7

Table of Contents

Introduction

Hallo and welcome to

Confluence, Tech Comm, Chocolate

I've noticed that the word "chocolate" creeps into many a technical writer's daily parlance. People are talking about wikis too. This book is about both. It is also about technical communication and the people who devote their energies and skills to that field.

Part I introduces the book, wikis in general, and Confluence in particular.

1

Welcome to this book

The aim of this book is to get everyone fired up and skilled up.

Fired up? Yes. About wikis, technical communication and chocolate. By the time you have read a few chapters and paged through the rest to see what is coming up, I hope that you will be as passionate about wikis as I have become over the last four years of working on one. Web-based documentation is the face of any company. Potential customers drop in unexpectedly via Google searches. Existing customers come looking for information. All of them gain an instant impression of the company, good or bad. Technical communication, both internal and external, is core to an organization's health and growth. This book shows how a wiki is the platform for developing and presenting that information.

If you are not already a technical writer, you will soon wish you were. And if you don't already have a technical writer at hand, you will go right out and get one. (In this book, the terms technical writer and technical communicator are used interchangeably unless otherwise stated.) Technical writers are skilled, knowledgeable, and enthusiastic people. The work they do is central to our understanding of the ever more technical world we inhabit.

Skilled up? The book is an in-depth guide to technical communication on a wiki. It focuses on Confluence because that is the wiki I know best. By the end of the book you will know everything I know about developing technical documentation on Confluence. And you will be able to bake the best chocolate cake in the known universe.

Getting passionate about technical communication and wikis

I have been a technical communicator for fourteen years. Before that, I was an analyst programmer. That role also included writing documents and communicating requirements to the project stakeholders. I have used many different technical writing tools. They are all excellent, interesting, and fun.

Until four years ago, my only wiki experience was to edit a couple of pages on *Wikipedia*, just to see what it was like. Then I started working full time on wiki-based documentation. I loved it from the start. It is very different from other technical writing platforms, as you will see in this book. The tools and techniques are different, and even more significantly the interaction with readers and other stakeholders is different. Yet all of a technical writer's skills in writing and information design are just as vital on a wiki.

A wiki as a platform of innovation

Wikis have been around since 1995, yet they are still under active development. Competition in the wiki world is fierce. The stakes are high. New technologies offer online collaboration, content management, and social media, and they all overlap with a wiki's offering. Wiki developers are forward thinking, open to new ideas, and very involved in the burgeoning world of online platforms and tools. Ward Cunningham, inventor of the first wiki, is still coming up with new ideas and pushing wikis to reach new heights of innovation. (See Chapter 23.)

Confluence wiki is made by Atlassian, a software development company that embodies this spirit of innovation and eagerness to embrace new ideas. I work at Atlassian. No two working days are the same. New ideas bounce around all day long. People try things just to see if they work. The company's products are designed to encourage customers to experiment, too. In this book I hope to show you how to stretch your wiki and watch it bounce.

The value of documentation

The purpose of technical documentation is to help people use a product or perform a procedure. It tells them what the product can do for them, how to get started, and how to work magic with the product. It tells them why the procedure is necessary and how to perform it efficiently and correctly.

People come to the product documentation when they have questions. Particularly after they have bought the product, they are more likely to come to the user's guide than to the website. Even before making a purchase, people will find their way to the documentation. They get there by clicking a link on the website in search of more detailed information or by searching the web for a specific term and following the trail of search results. Technical documentation is, by its very nature, optimized for high search engine ranking. It is concise, full of keywords, well structured, authoritative, and comprehensive. More about this in Chapter 9.

The result? Documentation is the face of the company. It contributes to the reputation and perceived character of the product and of the organization. Companies are becoming increasingly aware of the value of product documentation in this role. Chapter 17 discusses this, and describes the art of engaging readers by giving them a place to exchange information. As technical communicators, we want to make sure readers keep coming back to the documentation when they have questions. There are so many other sources of information out there on the web. What if a reader finds a less authoritative source before they find ours, with perhaps even incorrect information about our products? We want to help them avoid that. The best way is to make sure they find the help we provide and remember where they found it.

Internal procedures and information, such as those housed on an intranet, are no less important. They help to form every employee's sense of the nature of the organization and their own place in it. Knowledge is power. People who have access to clear, correct, and up-to-date information can devote their energies to building the company and its products.

What about customers who have gained plenty of experience in using the product and now have ideas and tips of their own to share? What about employees who have essential information to add to the procedures, both internal and external? It is excellent if they add the information in a spot where others can find it. Attached to the related documentation, perhaps. People will do that, given the chance.

A wiki is designed for just such collaboration. Technical documentation can be a hub of activity. People come to read the documentation, of course, as in any set of online manuals. But they also come to swap ideas and solutions, to find out if other people have encountered the same issue that they have, and simply to see what everyone else is saying about a new feature in the product or a new quirk in the procedures. They will keep coming back, to keep the conversation going.

The skill, inventiveness and enthusiasm of technical communicators

Technical writers know the value of collaboration. We know that a document is never finished, and that the secret of near-perfection is peer review. Collaboration is in our bones. It is the guts of what we do. It is our heart and soul. Wikis are designed for collaboration, and especially for technical teams. It is as if wikis were made just for us.

Technical writers are bursting with ideas. We are knowledgeable and passionate about our field. Every time someone mentions wikis, other people chip in with suggestions about what works well and what wikis could do better to suit our needs. Witness the ongoing discussion on *Technical Writing World* in a post titled *What do you think about using a wiki for technical documentation?*[1] (You will find the references in the "Endnotes" section at the end of this and every chapter.) The discussion started in May 2011 and is still attracting comments at the time of writing this chapter (October 2011). People are asking in-depth questions, giving detailed answers from their own experience, and contributing innovative ideas.

Combine all this skill and passion with the innovative approach that characterizes wiki development, and we have a winner. Let's help build our own platform! In Chapter 23 we will take a look at the opportunities and ideas floating around, and see how we can hitch a ride on the wings of a wiki.

The evolving nature of a tech comm team

Technical communicators are highly adaptable people. We have to be, because a large part of our role is to help other people adapt to changing environments. Our own role is constantly changing in response to outside pressures and to our own striving for excellence. This has always been the case, but perhaps it is especially so right now.

Technology allows us to have more direct contact with our readers than ever before. Customers and community authors are willing and able to contribute to the documentation. The products that we document, especially software products, are more and more integrated with other products and services, presenting interesting challenges for the documentation. We cope with ever faster development cycles, enabled by rapid technological growth and new processes such as agile methodologies. Emerging roles arise and abut onto ours: content curators, content managers, content strategists, developers of elearning material, online content marketers, and more.

Tech comm teams are therefore managing a continuously evolving sense of their own purpose and identity. Their audience has become more than just people who read the documentation. Readers are now primarily users of the documentation, then participants in it and even co-creators of it. Subject matter experts, whether inside the organization or even outside it, contribute directly to the content. Technical writers provide the framework, the core content, the advice, and the curatorship.

To survive and thrive in such an environment, we need the right tools and the techniques to use the tools well. That is why this book came into being. It introduces a wiki as the ideal platform for the job and shows you how to use it. The book is more than just a "how to" guide. It is full of useful techniques, hints, and ideas. You will see how to set up a basic documentation space and develop content. Then pick and choose from the ideas to make your wiki fly. Pull live information from websites like Twitter,

Flickr, SlideShare, and YouTube. Display a list of fixed issues directly from your organization's issue tracker. Have readers interact with the documentation. Go where no page has gone before.

Now to a thorny problem. And a suggestion for tackling it. First, a question: Is technical communication undervalued? Some people say yes, and many technical communicators feel undervalued, uncertain of their job security, and powerless to change their situation. I have seen and heard people discussing this problem online and at conferences. The suggestion: Get naked! Given the right tools, technical writers can make their product and themselves highly visible. Both will shine. Chapter 17 has a section titled "The naked technical writer" (p. 330). Nudge, nudge. Say no more.

How to use this book

This is a quick guide to the contents of the book, the versions of Confluence to which it applies, and the references to the Confluence documentation used in the book.

What is in the book, and for whom?

If you are a technical communicator, you will want to read the book from cover to cover. The book works as a step-by-step guide to getting a documentation wiki up and running. If you already have a wiki documentation site, you can skim through the first few chapters, enjoy the tidbits of supplementary information, and hop in when you find something that you do not already know. **Hint:** Be sure to find the first part of the recipe for Kay's chocolate cake, in Chapter 5.

Are you considering a wiki, but have not yet chosen one? Part I describes wikis in general and gives examples of wiki-based technical documentation. The first chapter in part II contains guidelines on choosing a wiki and planning the documentation suite. Part III is about life on a wiki, from a technical communicator's point of view and also from the point of view of other wiki inhabitants: product managers, support engineers, developers and customers. Part IV is full of ideas on how to use the unique capabilities of a wiki and combine them with other tools to develop shiny, useful, and fit-to-audience documentation.

If you are setting up an intranet site to contain employee information and procedures, start with chapters 2 and 3 for an overview of wikis and Confluence. Chapter 4 helps you to plot your requirements against the features of a wiki. Then get hold of Confluence in Chapter 5 and build your documentation space. Chapters 6 and 7 are a quick-start guide to the wiki editor, specifically for building structured information. In the chapters that follow, see how to create procedures that are easy to read and remember. Use templates for a consistent structure, which is essential for quick absorption and retrieval.

With judicious content reuse, you can make sure that often-used pieces of information are consistently presented too. Help readers find the information they need by applying the principles of findability and searchability. Do your procedures require a workflow, moving from draft through review to approval? Chapter 10 is all about workflow on a wiki. People who rely on procedural documentation also need to know when something changes. With a wiki, readers can request and manage their own change notifications. See Chapter 14.

Part IV is all about bringing your procedures to life, making them memorable, and encouraging people to feel a sense of ownership and pride in the documentation. This is a good way to ensure that staff members will find, remember, and follow the procedures and regulations. Read about the benefits of engaging internal readers, and how to make it happen. Explore the world of collaboration. The life of a page does not end at publication. See how to manage updates by other people, and choose the collaboration model that suits your organization.

Are you a product manager or business owner? See the strategic, organizational and team benefits that a wiki brings as a collaborative platform for document development. Technical communication will become your not-so-secret weapon against competitors. The book gives a solid grounding in what it takes to design and build a wiki documentation site. Part I introduces wikis and Confluence, and gives a foretaste of why a wiki is the platform extraordinaire for technical communication. In part II, chapters 4 and 5 offer a guide to planning and designing wiki documentation. Are you interested in attracting readers and customers to your site? Chapter 9 describes search engine optimization (SEO) and other ways of helping people find their way to the documentation. The rest of part II is an in-depth guide to developing documentation. See the opportunities for maximizing efficiency by content reuse, the wiki's capabilities for rich content, the integration with other services and systems, and more. Part III is a practical guide to working on a wiki, including best practices, team work, and wikis in an agile environment. Product managers feature in the chapter on "a day in the life." Part IV is for you too. Now that your technical communication team has a wiki, how can they build the best documentation in the world? Perhaps they can become the happiest, most motivated technical communicators, too.

Are you a developer, looking to provide the documentation for a project of your own? Follow the guide in part II from start to finish, and you will be ready for launch. Then dip into part IV to put the icing on the cake. Working in an agile team? Chapter 15 is about a wiki in an agile environment. Keen to contribute to the development of wikis? Jump right into Chapter 23, which is all about driving wiki innovation. Oh, and take a look especially at Chapter 13, where a developer walks into a wiki inhabited by technical writers.

Confluence, Tech Comm, Chocolate online

Part V of the book is an invitation to everyone to come and talk about wikis, technical communication, and chocolate, on the *Confluence, Tech Comm, Chocolate* wiki at `http://wikitechcomm.onconfluence.com`.

Introducing Ganache

Now is a good time to meet Ganache. She is our hero throughout the book. She is a technical writer at a fictitious company called *Choc Ink*. Ganache handles the technical communication for the company's chocolate assessment tool, *ChocTactics*, and for a new product called *ChocSpice*. She is also writing a guide to using Twitter, for use by her colleagues at *Choc Ink*, as well as the world at large.

Follow Ganache as she sets up a technical documentation wiki. Meet the people she works with. See how the wiki forms the bedrock on which they all thrive. Ganache is my *nom de wiki*. I invented her to avoid the boredom of having to say "I" all the time and of seeing my name on all the screenshots. As well as following in her footsteps in the book, you can visit her and her Twitter guide on the *Confluence, Tech Comm, Chocolate* wiki.

Confluence versions

The content of this book applies to versions 3.5, 4.0 and 4.1 of Confluence. As the book is going to press, Confluence 4.1 has just been released. I have tested the guidelines against both versions 3.5 and 4.0. There are no changes in version 4.1 that significantly affect the content of the book.

Confluence 4 presents a big change from Confluence 3.5, in that the editor and the wiki storage format were both completely rewritten. Chapter 3 describes these changes in detail, and the "how to" sections of the book cover both versions. You can apply most of the ideas and concepts you will learn in this book to any version of Confluence, and often to other wikis too.

Atlassian follows a rapid development cycle, and it is likely that another version of Confluence will be released soon. It will include some improvements to the editor and a couple of new wiki features. For information about the latest releases, please refer to the Confluence release notes at `http://confluence.atlassian.com/display/DOC/Release+Notes`.

References to the Confluence documentation

The latest Confluence documentation is at `http://confluence.atlassian.com/display/DOC`. In this book, references to the Confluence documentation refer always to the DOC space, as shown in the above URL. The documentation for each version is held in a separate space. The DOC space always holds the most recent version, which at time of writing is Confluence 4.1. The earlier versions are in spaces with a key of CONFxx, where xx is the version number. For example, the Confluence 3.5 documentation is in CONF35 and the Confluence 4.0 documentation is in CONF40. To find the version-specific guides, just replace the key DOC with the version-specific key in the URL. For example, the most recent *Confluence User's Guide* is at `http://confluence.atlassian.com/display/DOC/-Confluence+User's+Guide`. The guide for Confluence 3.5 is at `http://confluence.atlassian.com/display/CONF35/Confluence+User's+Guide`.

How we produced this book

In the spirit of eating your own dog food, we developed this book on a Confluence wiki. ("Eating your own dog food" is a phrase used by software developers. It means using your own products, so that you can experience them in the same way that your customers do. A slightly posher version of the phrase is "drinking your own champagne.") This book is, after all, a type of technical documentation. This is a good test case of one of the scenarios we discuss in Chapter 11, where a team develops content on a wiki then publishes it to another medium. It is also useful to see that the content produced on a wiki, including diagrams and images, is of a quality suitable for print publication.

We also conducted the technical review on the wiki. This meant that the reviewers, author, editor, and publisher could work together on the content in a single location. There was no need to email documents to and fro, with the nightmare of version management and content merging that can ensue. The reviewers read the content online, added their comments to the pages, and conversed with each other and the publishing team via comments too.

Once we were all happy with the content, we exported it to DocBook XML for processing by the publisher and printer. We used the *Scroll Wiki DocBook Exporter*, a Confluence plugin. There is more about this plugin, and exporting to other formats, in Chapter 11.

I used the *Gliffy Confluence plugin* to draw the diagrams used in the book. With Gliffy installed on your site, you can draw diagrams directly in the web browser. There is no need for any other tool. When you save the diagram, it is displayed on the wiki page as a .png image. Chapter 6 tells you all about Gliffy and other wiki drawing tools.

Disclaimer

Although I work at Atlassian, this book is entirely my own work and any opinions expressed are my own. Atlassian does not necessarily agree with or endorse anything written in this book.

Acknowledgments

My husband, Peter, has been a champion of this book since its first conception. It is thanks to him that I have managed to write the book while holding down a full time, exciting, and demanding day job. Peter also has a good eye for design, and helped greatly with the cover layout for the book. Our two sons, Duncan and Ryan, have given tireless and uncomplaining support, have suffered (if that is the word) a plethora of pizza dinners, and have endured a dearth of home-baked chocolate cake while the book was under development. My mother, Kay Espley, arrived on her first visit ever to Australia while I was in the throes of composition, and gave many a quiet word of encouragement. Thank you, family.

Thank you to Atlassian, for the free Confluence site that I have used to write the book, and for the four years of fun as an Atlassian employee. Being part of the Atlassian technical writing team is a privilege and a pleasure. The technical writers and other Atlassians have helped build up the store of knowledge that I have tapped to write this book. The Confluence product team develop, promote and support the versatile, appealing and feature-rich wiki that we use as a tool. They are not afraid to push the boundaries, nor to change their minds when necessary. Thank you all.

A grin to Andrew Lui (@drewylui) for patiently composing Tweets with the tag `#techcomm2` so that I could take the screenshots for chapters 5 and 21. Thanks also to Andreas van Rienen for giving me permission to include his Tweets in the screenshots of the Twitter widget in Chapter 21. A number of bloggers and writers have given me permission to quote from their posts, and I have referred to a number of other posts in the book too. Those people's names are in the references section at the end of each chapter and in the index. Thank you to every one of you.

A big vote of appreciation to the plugin vendors who have so generously granted me a free license for the plugins used in the writing of this book and in the associated website:

- *Gliffy Confluence plugin.* Thank you to Debi Kohlhardt, Chris Kohlhardt and all the Gliffy team at `http://www.gliffy.com/`.

- *Scroll Wiki DocBook Exporter plugin*. Thank you to Stefan Kleineikenscheidt, Tobias Anstett, and the rest of the team at K15t Software (`http://www.k15t.com`) for the Scroll Wiki plugins, and especially for the Scroll Wiki DocBook Exporter. Kudos to Tobias for his work on enhancing the exporter to meet the requirements for publishing this book.

Thank you to Richard Hamilton at XML Press, for his professional, expert, and easy style of publishing. He made the process of writing this book a dream.

I love the illustrations in this book. Congratulations and thank you to the artist, Ryan Maddox, for interpreting the wishes of the author (me) and the publisher (Richard) so well and for creating such appealing, gorgeous pictures. They go a long way to defining Ganache's character and the tone of the book.

For two weeks in December 2011, the *Confluence, Tech Comm, Chocolate* wiki was buzzing. Comments flew hither and thither. Opinions jostled up against each other. Perfection in wording and concepts drew nigh. This was the technical review phase of the book. Thank you to all the technical reviewers, for their generous and invaluable input:

- Alan J. Porter – Writer, speaker, consultant

- Anne Gentle – Technical Writer for OpenStack

- Ellen Feaheny – CEO of AppFusions

- Mark Fidelman – Chief Social Strategist at harmon.ie

- Robert Rhyne Armstrong – Director of Documentation for RouteMatch, Inc

- Sherif Mansour – Technical Product Manager at Atlassian

The worldwide community of technical communicators is full of generous people who contribute their ideas and time to the furtherance of our field. Thanks to everyone who attends conferences, writes blog posts and drops comments on other blogs, participates in social sites, and shares their knowledge in so many ways. You are all technical communicators extraordinaire.

Confluence quick start

This book begins with a gentle introduction to wikis, then leads you through the points to consider when choosing a wiki and planning the wiki documentation. Some readers may want to get started on Confluence more quickly than that. Here's how.

You can sign up for a hosted Confluence site (Confluence OnDemand) or choose to download and install Confluence on your own computer. Atlassian offers a thirty-day free trial for either option. Details and signup options are at `http://www.atlassian.com/software/confluence/try/`. Then jump to Chapter 5 for help with getting a documentation space up and running.

Have fun in this book!

The book reveals the passion of a technical writer. Some of it is my personal opinion, and all of it reflects my personal experiences on Confluence. It speaks to technical writers personally, from someone who knows their needs and wishes. It also speaks to business owners who are looking for the ideal solution for their documentation needs. I hope you enjoy it.

References

For an excellent thesis on crowd sourcing and wiki philosophy on a grand economic scale, read *Wikinomics, How Mass Collaboration Changes Everything*, a book by Don Tapscott and Anthony D. Williams. (First published in the United States of America in 2006 by Portfolio. First published in Great Britain in 2007 by Atlantic Books.) The book is primarily about developing systems and products rather than documentation. But its principles apply rather beautifully to our needs and concerns in the tech comm sphere.

Endnotes

[1] *What do you think about using a wiki for technical documentation?* a discussion on *Technical Writing World*, started in May 2011 and ongoing at time of writing this chapter (October 2011): `http://technicalwritingworld.com/forum/topics/what-do-you-think-about-using`

2

Introducing the wiki

In times of yore, web pages used to just put themselves up in front of you and that was that. Then along came the first wiki, and our lives changed for ever. Suddenly we could click a button and change the content of the web page. Even more revolutionary was that we could click "Save" and have our changes visible to the rest of the world. Immediately.

On 25 March 1995 the very first wiki site was born. That was Ward Cunningham's *WikiWikiWeb*, at address `http://c2.com/cgi/wiki`. This protowiki has other names too, including "Wards Wiki" and just plain "Wiki."

Why did Ward call his site a wiki? The word comes from the Hawaiian "wiki-wiki," which means quick. Ward's own words say it best: "I chose wiki-wiki as an alliterative substitute for quick and thereby avoided naming this stuff quick-web." (From the *WikiHistory* page on *WikiWikiWeb*.)[1]

Let's talk about wikis

This chapter is an introduction to wikis. It has the following sections:

- **What is a wiki?** A simple explanation of what a wiki is and how it looks to a person browsing a wiki page.

- **What do people mean when they say "wiki?"** They may mean the brand of wiki software, also called the wiki engine, or the wiki platform, or a specific site hosted on a wiki.

- **The major wiki brands.** Selected wiki software brands that are used for technical documentation.

- **A wiki by any other name...** The debate about whether a particular piece of software is a wiki or not.

- **Some well known wiki sites.** Famous websites that use a wiki as their platform.

- **Technical documentation wiki sites.** A focus on some documentation sites that use a wiki as their platform.

- **Do they all look the same?** Analyzing the appearance of those documentation wikis, what they have in common and how they display the characteristics of a wiki.

- **Plugins, add-ons and extensions.** Plugins and other extensions of wiki functionality, and the developers who build them.

- **Wiki and tech comm.** Why a wiki is the tool technical communicators dream of.

- **What you now know about wikis.** A summary of this chapter.

- **References.** The tools, websites, blog posts, and other references mentioned in this chapter.

What is a wiki?

A wiki adds an edit button to a web page. At its simplest, a wiki is a piece of server-based software that makes it possible for people to edit a web page and save their changes to the server. When other people request the page, the wiki serves them the updated content.

Let's look at it from the user's point of view. This is the first time we see Ganache, our intrepid hero, in action. Ganache is a wiki whisperer. She knows wikis inside out and will lead you right up to that wiki so that you can whisper in its ear.

✎ Edit   In this first scenario, Ganache is using a web browser to view a web page that is hosted on a wiki. The wiki has added that famous "Edit" button to the web page. Ganache spots an annoying typo on the page. She clicks the edit button. The page changes to edit mode and Ganache fixes the error.

She clicks "Save." The wiki applies the changes on the server side, updating the wiki database, and reloads the page in view mode. Ganache now sees the updated version of the page. So does everyone else.

Wikis have moved beyond the basic edit-update-save functionality. Today's wikis provide a number of additional features. Some have become sophisticated collaboration platforms. For example, a wiki will show you a history of updates made to a page, or an activity stream describing what everyone is up to on the wiki. A wiki may offer social features such as status updates and the ability to follow specific users. You can fill your wiki pages with rich and interactive content such as images, videos, and live data from other websites. At the backend, wikis interface with user management software such as LDAP and with enterprise database systems.

Wikis are mainstream tools and have been so for years.

What do people mean when they say "wiki?"

Usually when people talk about a wiki, they mean a specific website that is running on wiki software and therefore offers page editing and other wiki functionality. For example, people would refer to *Wikipedia* as a wiki and *WoWWiki* as another wiki. In this context, it does not matter whether the two sites are running on the same brand of wiki software or not.

When you start talking to wiki developers, or when you start thinking about setting up your own wiki site, you will find yourself talking about the wiki in terms of the brand or the software that delivers the functionality: Confluence wiki, MediaWiki, MindTouch, and so on. People may also talk about the wiki engine or the wiki platform when they want to distinguish between the different wiki software brands.

Of Confluence, impudence and imprudence

Looking at the capitalization of the wiki names in this chapter, it occurred to me that Confluence is the odd one out. Perhaps it should be spelled "ConFluence" to match the pattern of the other wiki names.

Just joking. But that does lead me to the all-important subject of pronunciation. When pronouncing the word Confluence, put the emphasis on the first syllable, not the second. Think about "impudence" rather than "imprudence."

The major wiki brands

There are many brands of wiki. At the time of writing this section of the book, *WikiMatrix* reported 133 different brands.[2] Here are a few that are used for technical documentation:

- **DokuWiki.** *WikiMatrix* itself runs on DokuWiki, a popular, lightweight wiki that stores its data as text files in the file system. DokuWiki is distributed as part of Debian, a free operating system based on Linux. As its name implies, DokuWiki is designed primarily for developing documentation. Its intended market is technical teams and small companies.

- **Confluence.** This book is primarily about Confluence, a wiki developed and sold by Atlassian Pty Ltd. Confluence is well on the way to becoming a full collaboration platform. The first part of this book highlights the features that are useful for technical documentation. Then we look at ways of using the collaborative features to optimize your organization's documentation processes. Getting even more adventurous, we dip into the social and interactive opportunities that the wiki provides. There is an overview of Confluence later in this chapter, giving you an idea of what the wiki contains. Then you can jump in and create your first documentation space.

- **MediaWiki.** This is probably the best known wiki because *Wikipedia* runs on MediaWiki software. MediaWiki is free. It is also open source, which means that any developer can contribute code to the MediaWiki source code. MediaWiki development is therefore a community project itself. MediaWiki is a popular choice as a platform for technical documentation.

- **MindTouch.** MindTouch is a wiki and collaboration platform. It offers add-ons and integration with other tools so that you can build a knowledge base, intranet, help center, or documentation suite based on your organization's requirements. MindTouch wiki was formerly known as Deki Wiki.

- **SharePoint.** MicroSoft SharePoint is a content management system and web portal, with wiki pages as one of the content types available. This is a simple implementation of a wiki, in that it offers instant editing of web pages. The rich-text editing and content integration capabilities are impressive. But the wiki itself, if you divorce it from the rest of SharePoint, is not rich in functionality. Instead, the wiki is an integral part of the SharePoint platform. Power users of SharePoint can manipulate a list of wiki pages in much the same way as they use other SharePoint components. Many people see SharePoint in their day-to-day office life and wonder what the difference is between a SharePoint site and a wiki. To help answer that question, Chapter 16 includes a comparison of SharePoint and Confluence.

Chapter 4 lists the questions to ask when choosing a wiki, in particular a wiki for technical communication purposes.

A wiki by any other name…

A few platforms have moved away from the name wiki. Some have come back to it again. There is a feeling that the term is too simplistic for what wiki software can provide, because it originally meant just the simple edit-update-save scenario. The name may also deter people because it implies a free-for-all on their content.

Anyone can edit anything, any time, not so? In fact, many wikis offer excellent control over who can do what. Another mark against the term wiki is that it brings to mind a spaghetti bowl of unstructured content. *Wikipedia* is great, as an amorphous repository of the world's wisdom. It is not so good as a showcase for an organization's intellectual property and corporate image. The chapters of this book address these points.

At times Confluence prefers to call itself something other than a wiki, although Atlassian has yet to find a better name. The term "collaboration platform" does not have quite the same ring to it. The default welcome message that you see on the dashboard after installing Confluence no longer includes the word wiki. At the time that this book is going to press, the product page on the Atlassian website[4] introduces Confluence with the title "Collaboration and Wiki Tools" but does not mention the w word anywhere else.

Chocolate by any other name…

Try Googling for "chocolate." The World Wide Web is chocolate coated. But where did chocolate get its name?

Web pages say it comes from the Aztec "xocolatl," meaning bitter water. Or perhaps the original was "cacahuatl." Or the Mayan "chokol" and "atl."

This beautiful script says it all:[3]

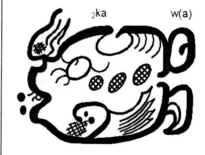

Over the last few years, MindTouch has suffered and overcome the same dilemma. At the moment, their product page says that MindTouch is the "world's most popular wiki and collaboration tool."[5]

Is Drupal a wiki? It is listed as such on *WikiMatrix*, but the Drupal website[6] describes it as "an open source content management platform." The word wiki does not appear on the home page of the website.

I like the word wiki. It has history, it is succinct, and it is cute. It evokes a tradition of openness and inclusiveness that the tool itself was designed to promote. Long live the wiki.

Some well known wiki sites

Some of the world's best known websites run on wikis.

- **Wikipedia.** At `http://www.wikipedia.org/`, this is probably the most famous wiki in the world. It runs on MediaWiki software. It is a crowd-sourced encyclopedia of the world's knowledge. Shifting, changing, controversial, inconsistent, at times unreliable? Perhaps. Nevertheless, it has become the first port of call for most people when looking for an introduction to a subject or topic.

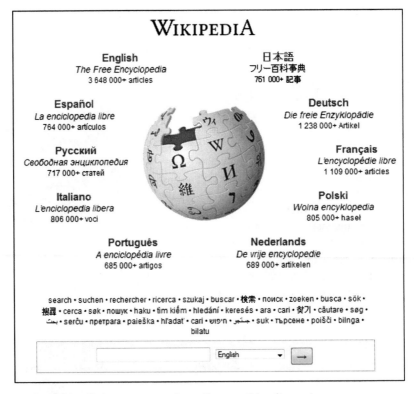

Figure 2.1. The Wikipedia home page at http://www.wikipedia.org/

In comparison to more traditional encyclopedias, *Wikipedia*'s pages can be more up-to-date and more descriptive, at very little cost. Anyone can edit a *Wikipedia* page. You do not even need to log in. If you update a page without logging in, *Wikipedia* will record your IP address (a string of numbers that identifies your computer's location on the Internet) and display it to the world. On the other hand, you can sign up for a username at *Wikipedia* and log in to make your changes. In that case, *Wikipedia* will show only your username and not your IP address as the source of any changes you make. A team of editors monitors all updates to *Wikipedia* pages. These people are volunteers who will reverse any malicious or obviously incorrect changes.

- **WikiLeaks.** WikiLeaks is a not-for-profit media organization dedicated to making it easy for whistle blowers to publish information they think needs exposing. The organization has been much in the news recently, due partially to the arrest in 2010 of its editor in chief, Julian Assange, and also to his strong defense of the right to freedom of speech. The *WikiLeaks* site at `http://-wikileaks.org/` publishes official secrets gleaned from whistle blowers all over the world.

  Is *WikiLeaks* a wiki? This is a very interesting question. From what I can see, it is based on MediaWiki software with some customizations. That makes it a wiki. The permissions are set so that the general public cannot edit the pages. That is a very viable wiki permissions model. I can understand why the *WikiLeaks* team have chosen to restrict editing and to ask people to submit content to the editorial team via other means. On the other hand, this has led to a bit of an outcry amongst people who feel that a wiki should be open to editing by all. A blog post titled *WikiLeaks: No Wiki, Just Leaks* gives a good overview of the controversy.[7] Read the comments for the alternative view.

- **WOWWiki.** The World of Warcraft wiki at `http://www.wowwiki.com/Portal:Main` runs on Wikia, which is also based on MediaWiki. This is a huge wiki. When I last looked, it had 90 960 pages. *WOWWiki* is the original wiki created for and written by everyone who plays the World of Warcraft game. The wiki was started in 2004 and has been hosted on Wikia since 2007. There was a split in 2010, when a change to Wikia's formatting caused some problems with the World of Warcraft pages. The *WOWWiki* administrators created a new wiki site, *Wowpedia* at `http://-www.wowpedia.org/`.

- **Chocolate Wiki.** If you are serious about technical communication, pay a visit to the *Chocolate Wiki* at `http://sites.google.com/site/chocolatewiki`. To be scrupulously fair to the title of this section, the *Chocolate Wiki* is probably not a well known wiki. It is not even an active wiki. But I like it for its no-nonsense introduction, complete with a proper definition of audience, needs and technological assessments, all wrapped up in a pretty mauve package and devoted to the serious subject of chocolate! Take a look at the recipe for making truffles. Surely that page embodies technical writing at its best. The *Chocolate Wiki* is hosted on Google Sites, part of the Google Apps suite. You can create your own wiki on Google Sites without having to install the wiki software.

Technical documentation wiki sites

As you delve further into this book, you will start seeing the possibilities that a wiki offers for technical communication. In fact, I hope you will get so excited that you will download a wiki and start playing with it immediately. There is a growing number of wiki documentation sites to prove the point. Let's take a look at just a few.

- **JIRA documentation.** Since this book focuses on Confluence wiki for documentation, it is only proper that our first port of call should be the documentation for Atlassian's software products. Atlassian is the developer of Confluence wiki. Do they use their wiki to develop and publish their own product documentation? Yes, I can personally and unequivocally attest to this fact, since I am one of the technical writers at Atlassian.

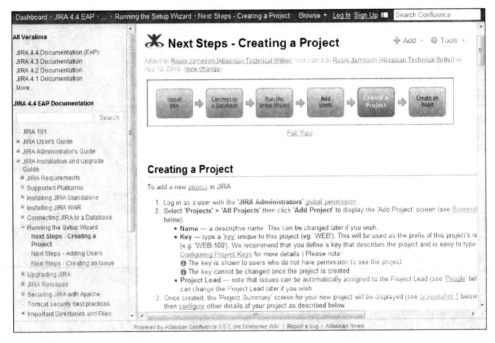

Figure 2.2. The JIRA installation guide

Confluence is one of the company's two flagship products. The other is JIRA, an issue tracking and project management application. Atlassian also develops a number of technical tools for use by development teams. All the documentation is developed and published on Confluence. The

JIRA documentation at `http://confluence.atlassian.com/display/JIRA` is a good example. It has the recognizable structure of large documentation suite: installation guide, administrator's guide, user's guide and various odds and ends.

- **OpenCloud Dev Portal.** The OpenCloud developer portal at `https://developer.open-cloud.com` is an attractive and information-rich documentation site for developers. Even without logging in, the reader has access to much news and information. To use the full functionality of the site, you can sign up for a username and then log in. This site runs on Confluence.

- **IBM developerWorks Wikis.** Another Confluence site hosts the IBM *developerWorks Wikis* at `http://www.ibm.com/developerworks/wikis`. This is a very active set of wiki spaces, designed to encourage developers to share information and ideas. The site goes beyond the usual edit-update-save freedom, in that it allows anyone to create a new space. This is equivalent to creating a wiki, because in Confluence a space is largely autonomous. Read the next few chapters of this book to see how.

- **ExactTarget Documentation Wiki.** ExactTarget offers interactive marketing services. Their documentation wiki at `http://wiki.memberlandingpages.com/` runs on MindTouch. This site has an appealing front page with big buttons in the "Don't Panic" style. Click a link to go to a documentation page, and you will see a well-structured documentation site. There is an enticing feedback form at the bottom of each page, asking "Was This Page Helpful?" and inviting you to "Send Adam Feedback." It is nice to see a real person's name mentioned on a form. Thanks to Adam Evans for letting me know about this site.

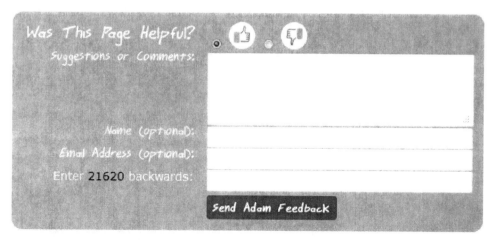

Figure 2.3. The feedback form on the ExactTarget Documentation Wiki

■ **splunk>docs.** Cool name! The Splunk documentation site at `http://www.splunk.com/base/-Documentation` is a very neat documentation site running on MediaWiki. This is the official product documentation for Splunk, a software package that indexes and reports on all the data generated by computers, applications and services. I like the dropdown list at the top of the page, where you can select the product version. Readers can generate their own PDF version of each manual. Every page asks, "Was this topic useful?" I was lucky enough to attend a walkthrough of this site. If you are interested in learning more about the way the team customized MediaWiki to suit their requirements, take a look at my blog post.[8]

■ **Mozilla Developer Network.** This is another MindTouch site, at `https://developer.mozilla.org/en`. Interestingly, this site does not offer a table of contents in a traditional left-hand navigation panel. Each page has breadcrumbs at the top – a trail of links leading you back to the home page and showing you where you are in the page hierarchy. An example looks like this: `MDC > Docs > CSS > Using CSS transforms`. Each page also offers a table of contents showing the main headings in that page. The lack of navigation panel is more in line with the traditional wiki style, where people find content via search rather than by navigating hierarchically through the site. For me, coming into the site blindly and trying to see what it was about, there was a moment of disorientation. "Where's my left-hand nav?" I quickly grew to like the clean look of the pages, but I do miss the ease of access that a table of contents provides.

What about **FLOSS Manuals**? I heartily recommend that you take a look at the *FLOSS Manuals* site, at `http://flossmanuals.net/`. It is an initiative to help people write free documentation for open source software. You can simply join up and start writing. The community also organizes regular book sprints, where people get together for a few days to focus on a particular project. Originally *FLOSS Manuals* ran on TWiki. At the time when I am writing this chapter, the original manuals are still available at `http://www.flossmanuals.net`. (Note the difference between the two web addresses in this paragraph. The second one has the "www" in front.) But *FLOSS Manuals* have recently moved to a new collaborative platform called Booki. Since Booki is not a wiki, I have not included *FLOSS Manuals* in my list of wiki documentation sites. But the site does deserve a very honorable mention. And it does offer an edit button!

There are many more technical documentation sites that use wikis as their platform. Anne Gentle has written two blog posts collecting a list of technical documentation wiki sites: *More technical documentation wikis*,[9] and *Even more technical documentation wikis*.[10] Anne's blog is an excellent read for anyone interesting in technical communication. She is also passionately involved in developing *FLOSS Manuals* and has worked on the Booki platform.

Do they all look the same?

Looking at the above documentation wikis, your first impression is that they look different. They have different themes or skins, and in particular the home pages have very different characters. They do all share a fresh and appealing look, which is why I chose them.

Delving deeper into the hierarchy of pages, you soon notice that most of them follow the familiar documentation format. They have a navigation pane, or table of contents, on the left and the page content on the right. In fact, they are very similar to the traditional tripane help system.

You may find yourself asking, *Where is the famous edit button?* How will our heroic Ganache fix that typo? In most of the above documentation wikis, the administrators have restricted the editing permissions so that only known people can update the content. (We will look at this phenomenon more closely in Chapter 5.) It is nice to see that the *Mozilla Developer Network* does display its edit button proudly!

Is a wiki just a glorified online help system? No! There is an entire section of this book dedicated to making your wiki fly.

Plugins, add-ons and extensions

Buzzing around most wikis you will find a community of developers writing plugins, add-ons and extensions to the wiki code. This is one of the strengths of wikis in general, and it is a strength that we technical communicators can use to our advantage. A number of wikis are built on open source software. Others, like Confluence, are not themselves open source but do support and encourage active contributions to open source development. Even where the wiki is not open source, the wiki developers offer APIs (application programming interfaces) and plugin frameworks, so that external developers can add new features and extend existing features of the wiki.

Wikis are still young enough, despite having been around for decades already, and still geeky enough, despite offering rich text editors, to attract the brightest and most passionate community enthusiasts.

A plugin is a small piece of software that you can install onto your wiki site. It is similar to the add-ons that you can install into your web browser. Once installed, the plugin adds extra features to the core wiki. It may add new options to the menus, or add a macro for use on a page, or change the look and feel with a new skin or theme. For example, the out-of-the-box wiki platform may provide a basic workflow solution. Plugin developers can add more sophisticated workflow, based on the specific needs of technical documentation. Some plugins are supplied by the company that develops the wiki itself. Other plugins are written and supported by community developers. This is the so-called "ecosystem," thriving alongside the wiki developers in a mutually dependent and mutually rewarding relationship.

Other developers build standalone applications, or extensions of existing applications, that interface with the wiki. These are not plugins, but separate applications that use the remote API to exchange data with the wiki. Thinking about Confluence wiki in particular, here are some examples. We will go into more detail about these tools in Chapter 11:

- **Universal Wiki Converter.** Often called UWC, a migration tool that assists in converting different flavors of wiki formatted pages to Confluence pages.

- **WebWorks ePublisher.** A set of tools for converting documents from a number of formats (Adobe FrameMaker, Microsoft Word, DITA XML) into other formats (Confluence wiki, PDF, HTML and others).

- **DITA2wiki.** An open source tool that converts DITA XML documents to Confluence format and uploads them into the wiki via a set of Ant commands.

Why is this vibrant development community a good thing for us technical communicators? We can tap into the enthusiasm, share our domain knowledge and skills, and encourage developers to add the features that we need.

Have you noticed that, very often, the features we technical communicators want are actually useful to other people too? When it comes down to it, there is not a huge difference between a documentation wiki and an intranet. Or a knowledge base. Features that seem at first glance to be documentation-specific are in fact handy in other use cases. Some of them fast become essential, such as improvements to the wiki search or metadata. In Chapter 23 we will examine more closely how technical communicators can influence, and maybe even drive, wiki development.

We have the knowledge and the skills. We have the words. We have the power!

Wiki and tech comm

What makes a wiki a good tool for technical communicators? Here are some of the features I have found most useful and most delightful. As a technical writer of many years' experience, I started working on a wiki for the first time about four years ago. It was like a breath of fresh air. I loved it from the start, and I still do.

- **Collaboration.** Both when drafting a document and when making revisions after publication, it is rewarding, efficient and effective to have all stakeholders working on the same document. In the busy period leading up to a product release, our delectable Ganache can work with the product managers and marketing team on crafting the perfect release notes. In those telling few days immediately after the release, the support team and developers can tweak the administrator's guide and add late-breaking items to the upgrade notes.

- **Customer feedback.** People can comment on the release notes, telling us what they really think of the release. And they do!

- **Keeping informed.** Who is doing what to the documentation? What are people saying about it, and about the products and the company? As a technical writer, it is essential to know what is happening on our pages. Wikis are good at keeping us informed. See Chapter 14 for the ins and outs of wiki notifications, feeds and activity streams.

- **Easy workflow.** To me as a technical writer new to the wiki, this was an eye opener. There is no need to wait while the content management system goes through its painful publication approval procedures. Your update is live immediately. It was a bit scary at times. My hand would hover over that save button for nanoseconds longer than necessary. Note that, if your environment does impose a strict level of control, you can add extended workflow functionality to the wiki.

- **Engaging our readers.** Why would we want to engage people in our documentation? To make our documentation easy to find, and to make people want to find it. We can write the most perfect technical documentation ever. We can use simple language, minimalism, progressive disclosure, you name it. We can be consistent, clear, concise, correct and complete until the cows come crawling home. None of that is any good if people do not find and then read our documentation.

- **Web 2.0 and social media.** It is a much-maligned term, but the power of Web 2.0 is still with us. With a wiki, we can bring Web 2.0 into the documentation, in a platform that offers the tools and benefits combined with security and a unified user experience. Wikis offer a number of new features, particularly in the area of integration with other applications and websites. Our release notes can display a "what's new" video, streamed from YouTube or another video hosting service. We might pull in a dynamic list of issues from our issue tracker showing real-time data, so that readers can click through to the details of the fixes. We can display presentations from SlideShare and "to do" lists from Remember The Milk. Suddenly you are not just a writer. You are a mashup artist too! (This is said tongue in cheek, of course. While writing this book, I do find that I miss the emoticons that we use so effectively in online writing.)

- **Documentation as a marketing asset.** Good technical documentation helps people. It is also an information-rich resource that promotes products, strengthens the corporate brand and draws customers to the company website. Add a wiki's power to engage readers and to provide up-to-date information. Now we have a platform that can provide exceptional strategic value to the organization. Chapter 17 is about engaging readers in the documentation, and how a wiki showcases the documentation as a corporate asset. Chapter 9 has guidelines on attracting readers by applying search engine optimization techniques.

- **Single authoring platform, available everywhere.** The wiki is one single platform for document development, and it is accessible from a web browser. You do not need any other tools. You can work from home, while traveling, even from your mobile phone.

- **Bottleneck breaker.** The ease of edit-update-save offers interesting possibilities for technical documentation. Perhaps our hero, Ganache, is a technical writer who needs to make a change to a published document. A software release has just gone out and the engineers have realized there is one vital element missing from the upgrade guide. Ganache adds the information on the spot and saves hundreds of customers the pain of a failed upgrade. Maybe Roald is a support engineer who wants to let customers know about a workaround for a known bug. He can add the description of the bug and the workaround directly to the documentation, without having to wait for a technical writer. Perhaps it is even a customer, Gianduja, who spots that bothersome typo. She can fix it as quickly as you can say "wiki." Technical writers no longer present a bottleneck in the stream of content updates.

On the other hand, the prospect of a quick edit-update-save may fill you with dread. What if Gianduja is a casual reader who thinks she is doing you a favor by correcting an error. What if she is wrong, or even malicious?

There are so many wikis, so many technical communicators, and so many writing environments that everyone's experience is bound to be different. Requirements are different too. A wiki is not always the best solution. But in this book I want to still some common fears about wikis. I will show you how you can take advantage of the quick turnaround time that a wiki offers in the publishing workflow, and also how you can protect your content using the permissions and monitoring tools built into the wiki. You will learn the nuts and bolts of wiki technical writing. You will take a peek into life on a wiki, from a technical communicator's point of view and from a reader's point of view. Then you will learn how to give your wiki wings. In short, I plan to share with you what my own experience with a wiki has taught me. I hope you enjoy the experience too!

What you now know about wikis

This chapter introduced the concept of a wiki as a tool that allows anyone to edit a web page, at any time and from anywhere. You wandered around some famous wiki sites, and then strolled through some less well known but very worthy documentation wikis. In your meanderings you noticed that the wiki documentation sites looked quite familiar, and in fact rather similar to other documentation platforms that you know so well. If you did not have time to do that while reading the chapter, I think you would have fun doing it now.

It also became apparent that the edit button is not always there. Wiki administrators can control who can do what, when and where.

You learned some more technical aspects of the wiki world, including a glimpse of the wiki brands, and the plugins and extensions that the developer community builds on top of the core wiki software. This flexibility and ongoing development offers opportunities for us as technical communicators to request the features we need as part of our wiki platform.

I had a go at convincing you that wikis are the ideal platform for technical communication. That was just a start. Read on.

References

These are the tools, websites, blog posts, and other references mentioned in this chapter.

Standalone platforms and tools

- *DITA2wiki*, an open source tool that converts DITA XML documents to Confluence format and uploads them into the wiki via a set of Ant commands: `http://sourceforge.net/projects/-dita2wiki/`

- *Universal Wiki Converter*, a standalone tool that assists in converting pages to Confluence format from other wiki formats: `https://studio.plugins.atlassian.com/wiki/display/UWC/-Universal+Wiki+Converter`

- *WebWorks ePublisher*, a set of tools for converting documents from a number of formats (Adobe FrameMaker, Microsoft Word, DITA XML) into other formats (Confluence wiki, PDF, HTML and others): `http://www.webworks.com/`

Endnotes

[1] The *WikiHistory* page on *WikiWikiWeb*: `http://c2.com/cgi/wiki?WikiHistory`

[2] *WikiMatrix*: `http://www.wikimatrix.org/`

[3] The Mayan word for cocoa on *Wikipedia*: `http://en.wikipedia.org/wiki/-File:Kakaw_%28Mayan_word%29.png`

[4] Product page on the Atlassian website: `http://www.atlassian.com/software`

[5] *MindTouch* website: `http://www.mindtouch.com/`

[6] *Drupal* website: `http://drupal.org/`

[7] *WikiLeaks: No Wiki, Just Leaks*, a blog post by William Beutler on *The Wikipedian*, July 2010: `http://thewikipedian.net/2010/07/31/wikileaks-no-wiki-just-leaks`

[8] *Wiki documentation – Splunk on MediaWiki*, a blog post by Sarah Maddox on *ffeathers*, June 2011: `http://ffeathers.wordpress.com/2011/06/18/wiki-documentation-splunk-on-mediawiki/`

[9] *More technical documentation wikis*, a blog post by Anne Gentle on *Just Write Click*, April 2009: `http://justwriteclick.com/2009/04/28/more-technical-documentation-wikis/`

[10] *Even more technical documentation wikis*, a blog post by Anne Gentle on *Just Write Click*, August 2010: `http://justwriteclick.com/2010/08/05/even-more-technical-documentation-wikis/`

3

Getting to know Confluence wiki

In the dark days when 2003 was old and 2004 was yet a distant dawn, a brand new wiki tumbled joyously into the world. This was the announcement of its birth:[1]

> We finally made it! Confluence 1.0b1!

Just a few days later, someone commented piteously on the above post:

> Is there actually a site, where those who do not know what confluence is all about, get explained what it is? I know that I'm part of a beta program, but am not sure what the I'm contributing to. [sic]

From such humble beginnings, Confluence has grown into the many-featured popular wiki that it is today. And yes, there is a site that tells us what it is all about – the Atlassian website is pretty good.[2] What's more, the site mentioned in the announcement is today the Atlassian documentation wiki.[3] And now there is a book too.

Let's talk about Confluence and its makers

In the last chapter you sampled a few wiki sites. Some of them were hosted on Confluence, some on other wikis. This chapter is all about Confluence:

- **Introducing Confluence and Atlassian.** A quick introduction to Confluence and the people who make it.

- **What is in the wiki?** A conceptual overview of the main content elements in Confluence: pages, spaces and blog posts. We also touch on the use of themes for customizing the appearance of the wiki, as well as RSS feeds, watches and notifications, the dashboard, and the administration console.

- **Wiki as subversive technology.** The wiki as destroyer of communication barriers and its relationship to more traditional documentation formats.

- **Why do they call Confluence an enterprise wiki?** The features that qualify Confluence for the name "enterprise wiki."

- **Confluence 4.** The big differences in this newest major update: the editor and the storage format.

- **Confluence is now your friend.** A summary of this chapter.

- **References.** The tools, websites, blog posts, and other references mentioned in this chapter.

Introducing Confluence and Atlassian

Confluence, or "Connie" as she is affectionately known amongst Atlassians, is one of the company's two flagship products. Over ten thousand organizations use Confluence in more than 100 countries. Atlassian offers a range of pricing options[4] based on the number of people who will use the wiki as registered users. Anonymous readers (people who have not logged in to the wiki) do not count towards the license fee – they do not cost you anything. Open source and not-for-profit organizations receive free licenses, as do those who use the wiki for classroom teaching.

Atlassian is a software development company based in Sydney, Australia, with offices in a number of locations around the world. The second largest office is in San Francisco, where most of the company's marketing activities occur. San Francisco also shares the development and support work, as do teams in the Netherlands, Brazil, Poland, Kuala Lumpur, and other locations. Mike Cannon-Brookes and Scott Farquhar are co-founders and CEOs of the company. They also did much of the early development of Confluence. If you ever find yourself walking through the hallowed halls of Atlassian, you may be lucky enough to hear the hushed tones of a developer describing the "founder code" that they recently came across in one of the source bases.

Chocolate grows on trees

Chocolate is made from cocoa beans that grow on the cocoa tree, genus *Theobroma*. The name Theobroma is derived from Greek, meaning food of the gods.

I work as a technical writer at Atlassian in Sydney. Confluence is where I spend most of my time. In reviewing the previous sentence, I would probably say that "hushed" and "hallowed" are not quite the right terms for the atmosphere at Atlassian. "Raucous" and "irreverent" are a better fit.

Donna McGahan, dubbed Atlassian's Support Diva, tells an engaging story about the early days in the San Francisco office.[5] She started work at Atlassian in November 2005, when the San Francisco office was new. Just "four desks, one phone, and a ratty Nerf soccer ball." The support team's mail server doubled as a footrest under an engineer's desk in Sydney. What drew Donna to Atlassian? The documentation, of course. She confesses, "I was already an Atlassian fan-girl but the fact that the company used Fight Club characters for user accounts in the documentation told me this nerd love ran deeper than most."[6]

The oldest public Confluence site is the Atlassian documentation and community wiki at `http://confluence.atlassian.com`. If, in those hallowed halls, you hear people speak about "CAC," then this is the site they are talking about. The oldest page that I can find is the home page of the Confluence User Community space.[7] The page was created on January 18, 2004, by Mike Cannon-Brookes. The content of the first version of the page was:

> This is the home page for the Discussion Space space.
>
> You can edit the content of this page using the *Edit* link on the right.

What could be clearer than that?

What is in the wiki?

Figure 3.1 gives a conceptual overview of the content elements of a Confluence wiki site. By content elements, I mean the bits and pieces that contain text, images, documents attached to pages, and other content that you can store and display in the wiki.

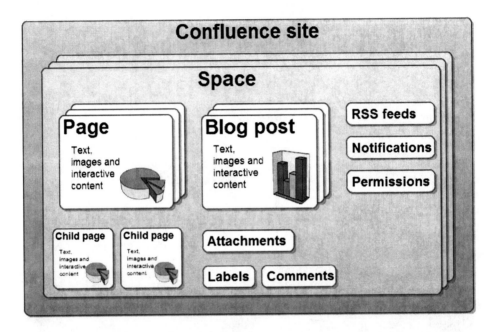

Figure 3.1. The content elements of a Confluence wiki site

Where are the people, you ask? People are a very important part of what happens in a wiki. There is a whole section about them in this book. For a quick overview, take a look at Figure 17.3 in Chapter 17.

Now, let's take a closer look at the content elements shown in the diagram. I will start with pages, because they the crux of it all. Spaces are collections of pages. Comments, attachments and labels need a page to hold them. Blog posts are special types of pages. And finally, RSS feeds, watches and notifications tell you what has happened to a page.

What follows is a quick, feature-based overview, to give you an idea of what makes up Confluence wiki. As you get deeper into the book, you will learn how to use each of these features in a technical communication environment. So do not despair if a particular feature seems a bit opaque at the moment. Sit back, have a hot chocolate, and imbibe the overview.

Pages

A page is just what you might expect it to be. It is where you write your words. When people read a wiki page, they see it in the form of a web page composed of HTML, CSS, and JavaScript. Like other web pages, Confluence pages are by default unrestricted in length and width.

As well as words, a page can display images and graphical layout elements including tables, panels, and heading levels. You can use text formatting such as bold text, italics, color, superscripts, and subscripts. Special pieces of code, called macros, embed multimedia files like audio and video onto the page. You can even pull in content from other websites and applications and display it live on your page. We will get into that later.

Figure 3.2 is a screenshot of a very simple page containing only text.

Figure 3.2. A simple Confluence page

At the top is the page title (in this case, "Home") and a few options that let the reader edit the page or perform other actions. Under the title is some information about the author and the date on which the page was last updated.

The body of the page contains the content.

A page has a number of other properties:

- **Labels.** You can add a label, or a tag, as a piece of metadata on a page. The Confluence search gives a higher weighting to pages where the label matches the search term. Using macros you can pull together all content tagged with a given label. Some plugins add more functionality that makes use of labels.

- **Child pages.** Each page can have any number of sub-pages, called child pages. When you add a new page, it is by default a child of the page you were on when you added it. Child pages can in turn have children, so that you end up with a multi-branched tree of pages. Each page can have only one parent page.

- **Comments.** Provided they have permission, Confluence users can add comments to any page. This is useful for receiving feedback on documentation during the draft and review process and for receiving feedback from customers after publication. On our documentation wiki, customers often use the comments to help each other discuss and resolve problems.

- **Attachments.** An attachment is any file that you upload to the wiki and associate with a specific page. If you want to display a screenshot or other image on a page, you will upload it as an attachment and then embed it onto the page. You may also want to attach Office documents, multimedia files, and so on.

- **Restrictions.** If you have the relevant permission, you can apply a viewing or editing restriction to specific pages. For example, you can hide a page from general viewing while it is still in draft form. Then when you are ready to publish the page, you just remove the viewing restrictions.

- **Page history.** Confluence keeps track of all updates made to a page. The page history shows previous versions of the page, along with information about past authors, the dates when changes were made, and comments the authors made when updating the page. You can also revert to a previous version or compare any two versions to see what changed.

Spaces

A space is a collection of pages. It offers a way of grouping logically related information. The original designers of Confluence chose the name "space" as an allusion to the term "namespace," used by other wikis to indicate a group of pages where each has a unique name. MediaWiki, for example, uses namespaces as a mechanism for grouping spaces.

The way you define your spaces depends on how you are using the wiki. If your wiki is an intranet, you may give each department its own space. If your wiki is used by just one department, each team could have its own space. If your wiki is a knowledge base, you might define a space per product.

In a technical documentation wiki, there are several ways to use spaces. You might choose to have a space per product or per version. You might even have separate spaces for your user's guide, administrator's guide, and installation guide.

The Confluence space directory lists all the spaces on your wiki site that you have permission to view.

To see the Confluence space directory: Choose "**Browse**" > "**Space Directory**" (see Figure 3.3).

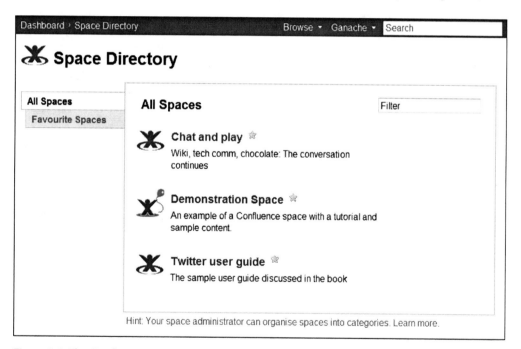

Figure 3.3. The Confluence space directory with just a few spaces on this wiki

Spaces are useful things, in that each space has a certain autonomy. Each space is almost like a wiki in its own right, offering the following functionality:

- **Separate administration.** Each space has one or more space administrators. These are users who have been granted the "space administration" permission. They can assign space permissions to other users, apply themes and styles to the space, and perform other administrative and maintenance tasks on the space.

- **Space permissions.** You can define a separate permission scheme for each space. For example, you can make part of the wiki visible to the general public, while another part is visible to staff members only. Or perhaps you might allow everyone to edit the content of some spaces, but make other spaces read-only, except for privileged users. The permission levels in Confluence are particularly useful for technical communication purposes.

- **Themes and style sheets.** A theme defines the color schemes, fonts and layout of the pages in a space. Some themes offer additional functionality, such as a smart search. Within a Confluence site, each space can have its own theme. When you first install Confluence there is a choice of themes available, and you can add more themes as plugins. You can also edit the CSS style sheets to customize your space.

- **Export to other formats.** Each space offers the option to export its pages to HTML, PDF, or XML format. The HTML and PDF formats are useful for giving your customers an offline version of your documentation. The XML export creates a Confluence-specific XML format and is a way of backing up your content and of transferring it from one Confluence site to another. Plugins provide additional export formats, including DocBook and EPUB, as described in Chapter 11.

Child pages and page trees

Within the space, each Confluence page can have any number of child pages. This means that you can organize your content into a hierarchical structure, or tree, like the sections and chapters of a book. Combined with the division of content into spaces, this gives a flexible framework for structuring your documentation.

Figure 3.4 illustrates a tree of pages in a space. On the right is part of a page from the Confluence demonstration space. The page is titled "Adding a comment to a page." The demonstration space is shipped with Confluence to provide sample content in your wiki. The space uses the Confluence documentation theme, which supplies a neat table of contents in the left-hand panel. In the screenshot, on the left you can see the tree of pages within the demonstration space. The title of the currently-displayed page is in bold, matching the page title on the right.

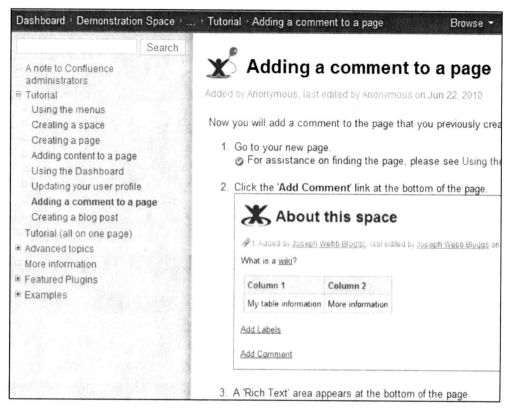

Figure 3.4. A tree of pages shown in the left-hand panel

Blog posts

A blog post is very similar to a page. It exists in a space. It can contain words, images, attachments, comments, and more. The main difference is that blog posts have the date of posting as part of their URL. They are intended to represent a state at a certain point in time. They are a news flash! That is the way people think of them, and that is the way people expect them to behave.

RSS feeds

On request, Confluence will send you an XML file containing information about the most recent updates that people have made to Confluence pages or other content. This is called an RSS feed. There are a number of freely available feed readers that will convert the XML to something you can read.

To build an RSS feed in Confluence 4: Choose "**Browse**" > "**Feed Builder**."

In Confluence 3.5, the feed builder option is on the dashboard.

Chapter 14 has more details about RSS feeds and some examples of feed readers.

Watches and notifications

Another way to keep track of updates is to watch a page or a space. Once you have set a watch on a page, Confluence will send you an email message each time someone changes the page or adds a comment to it. The email message is called a notification. If you watch a space, you will receive a notification whenever anyone adds, updates, or removes content from that space.

To watch a space: Go to a page in the space, then choose "**Browse**" > "**Advanced**" > "**Start watching this space**."

So, if our delightfully helpful Ganache changes the spelling of a word from American to Australian, you can go and change it back again. Chapter 14 describes ways to keep up with wiki activity.

The Confluence site

How do all the bits and pieces of the wiki hang together from a user's perspective? When you install Confluence, you get a website that people can access via their web browsers. In a default installation, the dashboard is the first thing people see when they go to the Confluence URL.

The dashboard is a hive of activity. No two dashboards look the same. People can change the welcome message at the top, add pictures, decide which spaces they want to see displayed, and generally make the dashboard their own. Also, the layout of the dashboard may change at each release of a new Confluence version.

To get to the dashboard: Choose "**Dashboard**" in the breadcrumbs at the top of the screen.

Figure 3.5 shows a dashboard with a fairly standard layout in Confluence 3.5:

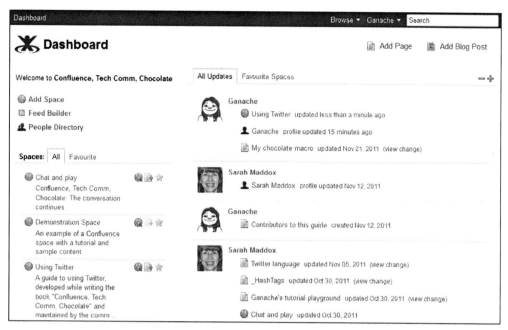

Figure 3.5. A Confluence 3.5 dashboard

On the left side of the dashboard you can see:

- **The welcome message at the top.** In this case it says,"Welcome to Confluence, Tech Comm, Chocolate." The administrator can change the message, add images or videos, and even animate the dashboard with Confluence macros.

- **Action buttons.** These are site-wide options for adding a new space to the wiki, building an RSS feed, and viewing the people directory. In Confluence 4, the "Add Space" option has moved lower down on the dashboard. The feed builder and people directory options have moved into the "Browse" menu.

- **A list of spaces.** The dashboard shows the spaces that you have permission to see. Confluence 4 also lists your favorite pages and the people you are following. Earlier I showed you the space directory, which is an alternative way of seeing the spaces in the wiki.

The most significant element on the right side of the dashboard is an **activity stream**. This shows you the latest news about what people are doing on the wiki. I often drop in on the dashboard just to see who is around at the same time as I am, who is working on the weekend, and what people are talking about!

The Confluence site also has an administration console, accessible only to administrators.

To get to the administration console: Choose "**Browse**" > "**Confluence Admin.**"

You will see that option only if you have the "Confluence administrator" or "system administrator" permission. The administration console offers all the usual configuration settings that you need to run a website. From our point of view as technical communicators, the most significant part is the management of permissions. We ourselves will have control over the permissions at space level, but the system administrator has final say over who can access the wiki as a whole, and who has permission to add spaces on the wiki.

Wiki as subversive technology

Wiki banned for five weeks!

Waxing whimsical about wikis, Mike Cannon-Brookes posted a picture of the front page from the *Daily Telegraph* on January 24, 2005. The headline was, "Wiki banned for five weeks."[8]

A subversive technology indeed! Mike went on to explain that Ruben Wiki was a New Zealand rugby player.

Charles Miller is the architect on the Confluence project. He has been around almost as long as the wiki itself. When Confluence was just a toddler, Charles wrote a thought-provoking post[9] about the wiki as subversive technology. Once you have a wiki in the house, anyone can have their say. The lowliest employee's voice is just as likely to be heard as the CEO's. Speaking from my own experience, that is certainly the case in a company where the CEOs read the wiki and comment regularly on posts that catch their eye.

In his post Charles went on to say:

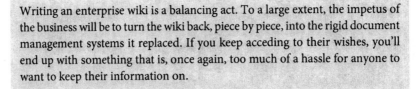

Writing an enterprise wiki is a balancing act. To a large extent, the impetus of the business will be to turn the wiki back, piece by piece, into the rigid document management systems it replaced. If you keep acceding to their wishes, you'll end up with something that is, once again, too much of a hassle for anyone to want to keep their information on.

That is a great paragraph, with two points that are key to technical communication on a wiki. The first thing that struck me is that, even in those early days, Charles was thinking of the wiki in terms of document management. Today, technical documentation is one of the primary uses of Confluence promoted by Atlassian. The other aspect of Charles's comment is the balancing act between simplicity and functional requirements. As a technical communicator, I value the simplicity of content development and collaboration that Confluence provides. I also occasionally chafe at the bit when it costs me time to work through the more complex parts of the technical documentation workflow.

Is a wiki subversive, from our point of view as technical communicators? It does change the way we do things and the way our readers interact with our work. But does wiki documentation overthrow or undermine the more traditional forms of documentation? My answer is no. Instead, a wiki enhances and builds upon existing media and technologies. Our skills as technical communicators are still very much needed. That much has not changed. It is just that we can now apply them to a different medium, a new way of organizing information, and a new way of communicating with our readers.

Why do they call Confluence an enterprise wiki?

Atlassian develops applications for technical teams. The company's mission is to help other developers develop brilliant products. At the same time, Confluence is often called an "enterprise" wiki. What does that mean, and why is it interesting in the technical communication sphere?

Here is a list of the Confluence attributes that qualify it for the description "enterprise wiki."

- **Rich text editor.** To many people, the thought of editing pages in wiki markup is off-putting. Confluence offers a rich text editor with many features similar to those in Microsoft Word. This is a familiar environment for many people, especially those in the business areas, rather than the technical areas, of an organization. In the Confluence 3 series, people can choose whether to use rich text or wiki markup. Confluence 4 has just one editor, combining rich text and smart wiki markup conversion. A comfortable editor makes it easy for our subject matter experts to contribute to a document during the review phase and for support engineers and others to update the document after publication.

- **Spaces.** Not all wikis offer a way to group content into libraries. I have already discussed how spaces are useful in technical documentation. The next chapter describes the points to consider when dividing your content into spaces. Then you will jump in and create your own technical documentation space.

- **Sophisticated user and group management.** Confluence user management includes support for LDAP directories. Users, groups, and permissions can mirror those already set up in the corporate user base.

- **Database support.** Confluence comes with a built-in HSQL database that you can use while evaluating the wiki. For production use Confluence supports a number of mainstream databases, including: PostgreSQL, MySQL, Oracle, DB2 and Microsoft SQL Server. Please check the *Supported Platforms* page[10] in the Confluence documentation for an up-to-date list.

- **Integration with Microsoft SharePoint.** With the Confluence SharePoint Connector you can put SharePoint web parts on a Confluence page and display Confluence pages on SharePoint screens. I imagine that this could be particularly useful if your subject matter experts live in SharePoint! Get them to collaborate on your documentation page from within their comfort zone while you stay in yours. This is not something I have used in earnest myself, although I have played with the connector a few times.

- **Integration with Office documents.** The Office Connector is shipped with Confluence. Let's say that you have an existing document in Microsoft Word. You can import it into Confluence, converting its content to wiki format. You can also attach an Office document to a Confluence page and display its content in Confluence, without converting the content. This is great for spreadsheets in particular. Use the power of an Excel spreadsheet to crunch your numbers and then display the sheet on a wiki page. Readers can see the information without having Office installed on their computers.

- **Integration with JIRA and other Atlassian applications.** In your release notes you might embed a list of the issues fixed in this release, drawn directly from the issue tracker. If the developers change the list in JIRA, the updates are reflected immediately in your document.

- **Atlassian support and documentation.** Atlassian is committed to providing world class support and documentation for Confluence. Please refer to the appendixes in this book, for a list of resources and websites that you can consult if you have questions about or problems with Confluence.

- **Support for plugin developers.** Atlassian recognizes the value of the work done by the plugin developers and is committed to giving them as much support and encouragement as possible. The Atlassian documentation covers the APIs that developers need and provides extensive information about building, deploying, and marketing a plugin. Atlassian provides an online market place, called the *Atlassian Plugin Exchange*, where people can make their plugins available for download and sale.[11] At the time of writing this chapter, there are about 800 plugins available on the plugin exchange, of which around 400 are for Confluence. Annual events keep the developers actively engaged with the company and with the internal engineering teams. AtlasCamp is a codefest (code festival) devoted to plugin developers. CodeGeist is a plugin competition offering prizes targeted at code geeks. Even the grand annual conference, the Atlassian Summit, has a strong focus on the development side of things. This development community is an asset for technical communicators, since we can work with the community developers to add documentation-specific features to the wiki.

- **Atlassian Experts.** A network of accredited experts provides extended support and development services for Atlassian products. They also help Atlassian to provide services in countries where the company does not have its own offices.

Confluence 4

Confluence 4.0, released in September 2011, marks the biggest change to Confluence since its birth way back in 2003.

What is different in Confluence 4? In brief: the editor and the storage format. There are a number of other new features in Confluence 4.0,[12] but for us as technical communicators, the change to the editor and storage format are very interesting indeed:

- **The editor.** In the very early days there was only the wiki markup editor. In late 2005, the rich text editor made its first appearance as part of the Confluence 2.0 release. (For some reason best known to the Confluence team, the Confluence 2.0 release was also known as Yarra.[13] I suspect it had something to do with the Socceroos winning a World Cup qualification match in a town called Yarra.) From that release up to Confluence 4, Confluence has offered two editors: a wiki markup editor and a rich text editor. Users can choose the one they prefer. Now with Confluence 4 there is again just one editor, but it is a vastly different beast to the pre-Yarra days. It combines rich text functionality, similar to that of Microsoft Word, with smart wiki macros, autocomplete and auto-formatting capabilities. A quick foretaste: Try typing "**h2. Chocolate is yummy**" and see it converted automatically to a level-2 heading. More later in this book, as you will have guessed.

- **The storage format.** In all releases before Confluence 4, the storage format is wiki markup. This means that the Confluence database contains wiki markup code to represent the content of pages, comments and blog posts. If you export a Confluence space to XML, you will see the wiki markup code wrapped in XML tags. From Confluence 4.0 onwards, the storage format is XHTML. As always, there are special markup elements to represent the wiki macros that add integration components and dynamic, real-time content to a page.

The Confluence documentation includes a change management guide.[14] This is a special set of documents intended to help customers manage the change to Confluence 4 when their organization decides to upgrade. The change management guide provides material for people who will need to train their staff or update their own internal procedures for the new Confluence.

Controversy around the Confluence 4 editor

The new editor is not perfect. It has bugs and some of its functionality is clunky to use. Moreover, many people just prefer the simple and direct editing experience that a plain text editor provides. Atlassian has received a huge amount of feedback, both good and bad, about the change in Confluence 4. A blog post, titled *Why We Removed the Wiki Markup Editor in Confluence 4.0*,[15] explains the reasons behind Atlassian's decision to remove the wiki markup editor and go with just a single editor.

The primary reason was that it became impossible to provide a reliable editing experience when two different editors could update the content of a page. Confluence stored its content in wiki markup format. Each time someone edited a page in rich text mode, the content had to go through a complex process that converted it to wiki markup for storage, then back to rich text for display in the editor. This resulted in the dreaded "roundtrip errors." You would edit the content and make it look just as you wanted, then save it, but when you opened it again the formatting might be different. This was even more the case if one person edited a page in wiki markup mode, then someone else edited the same page in rich text mode.

Atlassian is committed to improving the editor in Confluence 4. People are talking, both inside the company and out, about whether it is possible to provide some sort of XHTML editor, or perhaps something more similar to Microsoft Word's "show field codes" feature. At the time of writing this book, the discussions are ongoing. Chapter 23 includes a list of problems that a technical writer may run into and some ideas for fixing them.

The Confluence versions covered by this book

The content of this book applies to versions 3.5, 4.0, and 4.1 of Confluence. As the book is going to press, Confluence 4.1 has just been released. I have tested the guidelines against both versions 3.5 and 4.0. There are no changes in version 4.1 that significantly affect the content of the book.

Confluence is now your friend

This chapter introduced Atlassian and a few Atlassians in the team that develops Confluence. It also introduced the question of whether a wiki is a desirable acquaintance or a subversive element in the sphere of technical communication.

You got to know the basic content elements of a wiki including spaces, pages, and blog posts, plus peripherals like comments, attachments and labels. You gained a nodding acquaintance with the dashboard and the administration console.

Confluence introduced itself as an enterprise wiki. On the feature side of things, it offers support for enterprise databases and user management, spaces and page trees for structuring content, a rich text editor and integration with Office and Microsoft SharePoint. Atlassian is committed to world class support of its products. The company has a network of community developers and partners offering additional development and support services.

Do wikis save trees?

Using a wiki, we do all our reviewing and collaborating online. No paper. But there are studies that show printed documents to have a lower carbon footprint than online documents. The paper industry puts trees back in the ground! It is an interesting debate.[16]

Along comes Confluence 4. The new editor combines rich text editing with autocomplete and autoformatting for smart wiki markup conversion. The storage format is now XHMTL with additional Confluence-specific elements to define the macros.

Now you have shaken hands with Confluence and its makers, it is time to start planning your own documentation wiki.

References

These are the tools, websites, blog posts, and other references mentioned in this chapter.

Endnotes

[1] *We finally made it! Confluence 1.0b1!* a blog post by Mike Cannon-Brookes on *rebelutionary*, December 2003: `http://blogs.atlassian.com/rebelutionary/archives/2003/12/-we_finally_made.html`

[2] Confluence product page on the Atlassian website: `http://www.atlassian.com/software/-confluence`

[3] Atlassian documentation wiki: `http://confluence.atlassian.com`

[4] Pricing and license plans for Confluence wiki: `http://www.atlassian.com/software/confluence/pricing.jsp`

[5] *Atlassian Support – The Legend Continues*, a blog post by Donna McGahan about the early days in the Atlassian support team, March 2009: `http://blogs.atlassian.com/2009/03/atlassian-support-the-legend-continues/`

[6] In earlier versions of the *Confluence Notation Guide*, the entry for the `userlister` macro used the names Tyler Durden, Marla Singer and Robert Paulsen (sic) as examples. A PDF copy of the old guide still exists: `http://confluence.atlassian.com/download/attachments/-12877996/Notation+Guide+-+Confluence.pdf`

[7] *Confluence User Community* space on the wiki: `http://confluence.atlassian.com/display/DISC`

[8] *Wiki banned for 5 weeks?* a blog post by Mike Cannon-Brookes on *Wikizen*, January 2005: `http://wikizen.typepad.com/wikizen/2005/01/wiki_banned_for.html`

[9] *Wiki, a Subversive Technology*, a blog post by Charles Miller on *The Fishbowl*, January 2006: `http://fishbowl.pastiche.org/2006/01/27/wiki_a_subversive_technology/`

[10] *Supported Platforms* in the Confluence documentation: `http://confluence.atlassian.com/-display/DOC/Supported+Platforms`

[11] The *Atlassian Plugin Exchange*: `http://plugins.atlassian.com`

[12] Release notes for Confluence 4.0: `http://confluence.atlassian.com/display/DOC/Confluence+4.0+Release+Notes`

[13] Release notes for Confluence 2.0: `http://confluence.atlassian.com/display/DOC/Release+Notes+2.0`

[14] Change management guide for Confluence 4: `http://confluence.atlassian.com/display/-DOC/Planning+for+Confluence+4.0`

[15] *Why We Removed the Wiki Markup Editor in Confluence 4.0*, a blog post by Matt Hodges on *Atlassian Blogs*, November 2011: `http://blogs.atlassian.com/2011/11/why-we-removed-wiki-markup-editor-in-confluence-4/`

[16] *Do wikis save trees?* a blog post by Sarah Maddox on *ffeathers*, December 2008: `https://ffeathers.wordpress.com/2008/01/12/do-wikis-save-trees/`

Developing documentation on a wiki

So, now you want a wiki. How do you get hold of one, get to know it, and then bend it to your will? This part of the book provides an in-depth guide to writing and publishing documentation on Confluence wiki.

4

Planning your wiki documentation

"Everything should be made as simple as possible, but no simpler." That is a quotation cited on *Wikiquote* and popularly attributed to Einstein. It is possible that Einstein actually said this instead: "It can scarcely be denied that the supreme goal of all theory is to make the irreducible basic elements as simple and as few as possible without having to surrender the adequate representation of a single datum of experience."[1]

Planning is an essential part of documentation design and development, but it is not as much fun as getting in the zone, creating a wiki space and writing the words. I will keep this chapter as short as possible, but no shorter. If you want to try out Confluence and create a basic documentation wiki right now, I understand completely! You can safely skip this chapter and come back to it later, without surrendering too many of Einstein's "data of experience."

Let's talk about planning your wiki

This chapter is about analyzing the requirements for the documentation, then choosing and designing the wiki accordingly. The chapter has the following sections:

- **Analyzing your documentation requirements.** Considering how you plan to use the wiki: An all-in-one solution, document development only, or publication only.

- **Choosing a wiki.** A list of questions to ask when choosing a wiki platform.

- **Understanding the server and machine.** Hardware, application server, database and more.

- **Planning your user management and global permissions.** Where your users will come from and what they will be able to do.

- **Dividing your content into spaces.** Factors to consider when splitting your content into wiki spaces: a space per product, per version, for different audiences, or per team.

- **Deciding if you need to add plugins.** Matching your requirements against the core wiki functionality and deciding if you need the extra functionality that plugins can provide.

- **Getting to know your wiki.** Understanding how your wiki works, so that you can bend it to your will.

- **Wikitoria.** A building in Auckland, and the Maori name "Wiki."

- **What you know about planning your wiki.** A summary of this chapter.

- **References.** The tools, websites, blog posts, and other references mentioned in this chapter.

Analyzing your documentation requirements

When choosing any platform, technical communicators look at their audience, environment, and other requirements and then choose the platform that meets their needs. This is no different when deciding on a wiki. The first step is to determine what you are trying to accomplish (the high level goals that are driving you to make any change at all) and how you will know when you have reached that goal. Then comes the choice between a wiki or some other platform. Having decided on a wiki, the next step is to choose the specific wiki software.

I will not describe the details of audience and requirements analysis that we all know so well. And this entire book is the explanation of why a wiki is the platform extraordinaire for technical communication! So now, in this chapter, let's take a look at the considerations that affect your choice of and design of a wiki platform.

Perhaps the essential consideration is how you plan to use the wiki. Will it be your all-in-one solution, just the document development platform, or just the publication platform?

- **Wiki as all-in-one solution.** In this scenario, the wiki is where everything happens. Authors develop the documentation on the wiki and publish it on the same site, often simply by clicking the save button. Readers come to the site to read, comment on, and interact with the content. If necessary, the publishing team also provides downloadable versions of the content in other formats, such as PDF and HTML, for customers who cannot access the site online. This is the way we do it at Atlassian, both for our internal wiki (intranet) and our external documentation wiki. I recommend it heartily and unreservedly as the most efficient way from a document development point of view, and the most rewarding way from a reader's point of view. Provided, of course, that the readers are online.

- **Wiki for documentation development, but publication elsewhere.** In some environments the readers cannot access the wiki, so it does not make much sense to publish the wiki content to the general public. Nevertheless, content developers want the ease of collaboration and the quick turnaround that a wiki promises. If you choose this scenario, the wiki will need to export its content to a transportable format such as Microsoft Word, PDF, XML, HTML, or an epublishing format.

- **Wiki as publication platform, but content development elsewhere.** Perhaps your organization has a large set of existing documents or a number of people highly skilled in a particular tool set. It may make sense to develop the content on other platforms, such as XML or Adobe FrameMaker, and publish it to various formats, one of which is the wiki. Customers and staff can then use the wiki for collaboration and feedback. With clever use of spaces and permissions, the wiki can contain

What part does chocolate play in your life as technical writer?

From Craig Haiss on Technical Writing World:[2] "My favorite breakfast is fresh cantaloupe, strong coffee, and a small bar of dark chocolate. Not only do they taste great together (perfect combination of bitter and sweet), but they seem to provide the spark of energy I need to be productive in the morning."

At work, we have monthly food events where a team member is responsible for providing a sugar fix for everyone. Chocolate shows up often at these events: chocolate-covered strawberries, eclairs, cakes, brownies, etc. We have one team member who has professional-level baking skills, and she makes some serious works of art that taste incredible.

"Our manager understands the connection between chocolate and productivity. She keeps a small dish in the aisles by our cubicles and regularly fills it with Hershey's products and similar fare.

"So, yes, chocolate plays an essential role in my day-to-day life and I'd be far less satisfied without it."

a mix of content added by people and content generated from the document development platform. If this scenario suits your requirements, the wiki will need to import content from your document development tool on a regular basis.

Choosing a wiki

The online *WikiMatrix* at `http://www.wikimatrix.org/` is a useful tool for comparing wikis and their features. It offers two methods of comparison. The first method is to choose two or more wikis by name and run the comparison. The second method is the Wiki Choice Wizard. The wizard asks you a set of simple questions to define your wiki needs, then presents you with a list of wikis that match your criteria.

Following on from *WikiMatrix*, you are probably ready to ask some more in-depth questions. Here is a list that I hope is useful when choosing a wiki for technical communication. Whether a particular capability is a must have or a nice to have depends on your specific requirements.

Questions about environment, support and cost:

1. **Price.** Some wikis are free, others are commercial. Some commercial wikis offer free licenses to open-source and not-for-profit organizations. If the wiki vendor offers commercial licenses, it is worth examining the pricing structure. Is it based on number of users or some other measurement, and how does this fit in with your environment?

2. **Ease of installation and evaluation.** Will you be able to download and install an evaluation wiki without help from an IT department? Some wikis are available in a hosted environment, also called "software as a service" (SaaS). This means that the wiki runs on a server that is managed by the hosting company. It makes things very easy, because you do not need to download and install the wiki software. You can evaluate the wiki in the hosted environment, then choose to buy a hosted license or get the wiki installed on your own servers.

3. **Supported platforms.** Will the wiki run on the operating system and database that is available in your organization? Does it support the web browsers that your readers will use? Can you connect it to your organization's user directory?

4. **Support and documentation.** Examine the documentation to get an idea of the quality of the product. See if the wiki developers or other organizations make any effort to help you out if problems arise in the wiki, either during installation or at some time in the future.

5. **Active development.** Can you see signs of ongoing development of the core wiki platform, and are the developers open to discussion about possible new features and requirements? Check whether the code base is open source or proprietary, and decide whether this is important to you.

6. **Plugins and extensions.** How vibrant are the developer and user communities? Is there any chance of finding someone to help you customize the wiki for your requirements, or are there plugin developers who understand the needs of technical communicators?

7. **Known documentation wikis.** Do you know anyone who is already using this particular wiki for technical documentation, and can you get some tips from them? Alternatively, take a look at some of the documentation already published on that wiki. Chapter 2 has some examples of wiki documentation sites.

Questions about content-related features:

1. **Content structure.** Examine the structural framework that the wiki provides for your content. Does it offer only a flat layer of pages, or is there some way of organizing your manuals, chapters and pages? Can you put the pages into any sequence you like or does the wiki impose an alphanumeric order by title?

2. **Page structure.** How easy is it to structure the information within a page using headings, frames, tables, columns, and styles?

3. **Images, videos and other multimedia.** How easy is it to upload and insert images? Consider whether you want to display videos or play audio files on the pages and whether the wiki supports this functionality.

4. **Integration with other applications.** Can the wiki pages display live content from other websites and applications, and are there other ways of integrating the wiki content with your existing tools?

5. **Content import.** Do you need to convert content from some other format into the wiki format? For example, you may need to do a one-time conversion of your existing documentation and import it into the wiki. Or you may need to configure a regular conversion from a documentation suite developed on another platform. Check whether the wiki is able to import the content you need.

6. **Content export.** Can you convert the wiki content into another medium, such as PDF, XML, or HTML and then download it onto a local drive?

7. **Content reuse.** Does the wiki offer any mechanisms for defining a chunk of content on one page and then reusing it on another page, without having to duplicate the content?

8. **Templates.** Can you define templates for different types of content?

Questions about other functionality:

1. **Permissions and access control.** Does your wiki allow you to restrict access to the pages, and what permission levels does it allow? For example, you might want to prevent anonymous users from seeing anything at all. Or you might want to configure the permissions so that everyone can view your pages, but only logged-in users can comment on them and only users belonging to a certain group can change the content. Does the wiki have a permission framework that supports your requirements?

2. **Social features.** Are social features important? Examples are the ability to share a page, subscribe to a page, follow people, publish, and subscribe to blog posts, post status updates, and view an activity stream.

3. **Version control.** Does the wiki keep a history of changes to a page and allow you to revert to a previous version?

4. **Release management.** Is there some way of taking a snapshot of your documentation for each release of your product, so that you can give your customers a set of guides matching the product version they own?

5. **Workflow.** Is it easy to put a page through the draft, review, and publish workflow?

6. **Search and find.** What tools are there to help your readers find information, such as index, metadata, table of contents, and search?

7. **Customized appearance and presentation.** Does the wiki allow customization of styles and presentation? Some wikis offer themes or skins and adaptable style sheets. Perhaps you can add plugins, change the default page layouts, or even change the source code. You may want to change the appearance of the dashboard (wiki home page), too.

8. **Monitoring, tracking and analytics.** What tools does the wiki provide for keeping track of who's doing what and when? Examples might be RSS feeds, email notifications, and site access statistics. Find out the capabilities for analyzing user behavior, such as support for web analytics tools.

9. **Feedback.** Is there a way of inviting and gathering feedback on the wiki pages?

From this point on, let's assume you have chosen Confluence as your wiki. I think it is an excellent choice. But then, you could say I am biased.

Understanding the server and machine

Confluence is a web application that runs on a server. It stores content and serves it up to users as web pages. People will use a web browser to read those pages. You can choose to install Confluence on your own server or sign up for a Confluence site that is hosted on Atlassian's or another company's servers. If you decide that you want to install Confluence on your own machine, you can run Confluence quite happily on your own laptop or desktop while you are

What part does chocolate play in your life as technical writer?

From Matthew Helmke on Technical Writing World:[3] "Try dark chocolate with a wee dram of peaty scotch whisky…amazing."

evaluating it. When you are ready, you can move it to a production-ready server. Chapter 5 tells you how to install Confluence or sign up for a hosted site using Atlassian's Confluence OnDemand offering.

If you are already at the stage when you want a production installation on your own servers, have a chat with the system administrators about the following points:

- **Hardware.** Confluence will run quite happily on your laptop or desktop computer when you are playing with it for evaluation or demonstration purposes. In a production environment, and depending on the number of users and the amount of content, your Confluence site may need a sizable box. The *Server Hardware Requirements Guide*[4] can help here.

- **Application server.** Confluence is a Java application that runs within an application server. The Confluence installation includes an application server, Apache Tomcat. A system administrator may choose to deploy Confluence into an existing application server rather than adding a new server to the environment. In that case, the administrator will use the WAR (Web ARchive) distribution of Confluence.

- **Database.** Confluence stores its content and administrative data in a database. When you install an evaluation version of Confluence, you can use the built-in HSQL database. When you move your wiki to a production environment, you will need to use one of the supported databases: PostgreSQL, MySQL, Oracle, DB2, or Microsoft SQL Server. Please check the *Supported Platforms* page[5] for an up-to-date list.

- **Backups, server security and other administrative matters.** The *Confluence Administrator's Guide*[6] has all the information that your system administrators will need.

Planning your user management and global permissions

Think about where the users will come from. Are they staff members only, or community and customers too? Do you want to allow people to sign up for a username themselves?

- **Staff and/or community users.** Perhaps the wiki will be visible only to people within the organization. Maybe customers and community members will have access too, or even the general public.

- **Corporate and/or wiki-specific user base.** Confluence can manage its own users and groups, storing them on the Confluence database. In that case, the users and groups are specific to the wiki. People will need to remember their Confluence username and password as well as those for other systems in the organization. Alternatively, the system administrator can hook Confluence up to the organization's LDAP user repository, so that it shares the common pool of users and groups.

  If you are using JIRA, then Confluence can piggy back off JIRA's user management system. You can even hook Confluence up to two ore more user management systems at once. For example, if you want to allow both staff members and customers access, then Confluence could get the staff user data from an LDAP repository and it could store the customer user data on its own database.

- **Anonymous and/or logged-in users.** The wiki is a website. It can be visible to anyone who enters the web address into their browser. People who have not logged in are called anonymous users. Alternatively, you can make the site's content visible only to people who have a username and password and have logged in.

- **Public signup or controlled user base.** The administrator can configure Confluence to allow public signup. That means that anyone can choose the "sign up" option, request a username and password, be granted one automatically on the spot, and log in immediately. Alternatively, if public signup is not allowed, then the administrators must assign a username and password before a person can log in.

Once you have the answers to these questions, discuss the requirements with your system administrators. They have control over the global (site-wide) permissions and user management for the wiki.

Dividing your content into spaces

Start by considering the overall structure of your documentation suite, much as you would plan a set of documents written in any other tool. Then plan how that structure will fit into the wiki's framework. Spaces are the primary way of organizing content in Confluence. When readers come to the site, the first breakdown they see is the spaces. Chapter 3 gives an overview of spaces and how content fits into them.

Let's consider our undaunted hero, Ganache. She plans to write a user's guide and an administrator's guide for *ChocTactics*, a chocolate assessment tool. Each guide consists of sections, chapters, and pages. Will Ganache put each guide into a separate space? Or perhaps she should use a single space to collect all the guides for *ChocTactics*. When the next version of *ChocTactics* is released, Ganache may need to put the manuals for the new version into a new space.

On the Atlassian documentation wiki, we have chosen to put all the guides for a single product into one space. For example, the Crowd space[7] contains the user's guide, administrator's guide, and installation guide for the Crowd application. The JIRA space[8] has similar guides for JIRA, as does the Confluence space[9] that documents Confluence itself. We also create a new space for each major version of each product.

On our intranet wiki, each team has its own space to manage its procedures and guidelines. Some teams, such as human resources and finance, publish procedures for the company as a whole. Everyone in the company can see all spaces, and most of the spaces are open to updates by everyone, too. Needless to say, the technical writers have the most organized space in the company. There are also spaces for more esoteric concepts such as games, running, and social events.

Here are some considerations to help you decide how to divide your content:

- **Permissions.** In Confluence, you can manage users' permissions at the space level. Site administrators can also manage global (site-wide) permissions, and authors can restrict viewing and editing rights at the page level, too. When considering how to divide up your content, the space permissions are the most relevant. Chapter 5 describes space permissions in detail. For example, if you want to set different permissions on your developer-focused documentation and your user guides, you will need to put the guides in separate spaces. Or

Can you guess why the Confluence space has a key of DOC?

A space key is a unique identifier for the space and is part of the URL for accessing pages in that space. The Crowd space has a key of CROWD. The JIRA space key is JIRA. Why is the Confluence space key DOC and not CONF?

Answer: This was the first documentation space on the wiki. No-one thought that we might add documentation for other products in later years!

you may want to create a space for community-contributed content, separate from the official product guides. On an intranet, one team may want to protect its content from being updated by the rest of the organization, and another team may be happy for everyone to contribute.

- **Themes, color scheme and stylesheets.** Each space can have its own theme, sometimes known as a skin, which defines the look and feel of the space: colors, layout, font, and more. In addition, you can define a separate color scheme for each space or customize the CSS style sheets for the space. You may decide to use different styles for different products or for different audiences, and this would affect the division of content into spaces.

- **Number of products.** If you only need to document one product, then it may be useful to have a space for each manual. But if there are many products, each with its own set of manuals, it will probably be simpler for the readers if all of the documentation for a product is contained in a single space.

- **Product versions.** Will your readers need a separate set of documentation for each version of the product? In some cases this is not necessary. For example, if the customers are using an online hosted service then they will all be using the same version of the product, and it will always be the latest version. But if there is a chance that some customers will retain an older version of the product, then they will need the guides for their version. Spaces are useful to manage different versions of the documentation.

- **Developer-focused documentation and product documentation.** Developers are very different animals from regular customers! The information they need bears no resemblance to the user's guide and administrator's guide that everyone else hankers after. It is often useful to put the APIs and developer guides in a separate space.

- **Duplicate page names.** Confluence does not permit duplicate page names within a single space. This is one very good reason why you would put the documentation for different versions in different spaces. Otherwise you would need to add the product version number to the title of every page.

Deciding if you need to add plugins

A wiki is an extensible platform. As wiki users, we tend to view the wiki as the user interface that lets you add content and retrieve it and as the software that provides the smarts to respond to our requests. From the wiki developers' point of view, that is only part of it. They are also building a platform with "plugin points" and APIs (application programming interfaces) that make it possible for other developers to extend the wiki.

When you are planning your wiki, you will need to match the core wiki functionality against your requirements and install any plugins you may need for extra functionality.

This book helps you with that decision. Chapter 2 introduced the concept of plugins, add-ons and extensions. As you go through this book, setting up your documentation space and developing your content, you will learn about the plugins that are essential for doing technical documentation on a wiki. At the end of each chapter is a list of useful plugins relevant to the content of that chapter.

Getting to know your wiki

Play with the wiki. Get intimate with its capabilities and tools. This will help you quickly find a solution to any questions that arise when other authors and readers hop on board.

Like other platforms, wikis offer many advantages and a few quirks. Treat your wiki right, and it will come through for you. Often it is a matter of recognizing the wiki's strengths and using them in a creative way to perform or even replace a more traditional function. For example, labels present interesting ways of organizing content. People can add labels, or tags, to wiki pages. You can then collect and display the labeled pages to form a list of related topics or an index. (See Chapter 9.) I hope this book will give you many more ideas too.

At this stage, the important thing is to rush in and hug your wiki.

Wikitoria

Here is a photograph that I took in Auckland in September 2010. This was my first ever trip to New Zealand. I went to Wellington to attend TCANZ 2010, a great conference run by the Technical Communicators Association of New Zealand. After the conference, my husband and I spent a week driving up the North Island from Wellington to Auckland. Imagine my delight when I spotted a building boldly labeled "**Wikitoria!**"

Figure 4.1. Wikitoria in Auckland, New Zealand

Since seeing the building, I have done some research. It turns out that Wikitoria is a girl's name, the Maori equivalent of Victoria. What's more, it is often shortened to "Wiki" and means victorious. That has a nice ring to it, especially now when you are about to create your first wiki documentation space!

What you know about planning your wiki

Finding out what your wiki can do, and plugging any gaps to match your requirements, form the most important part of planning a documentation wiki. The first thing is to put your requirements down on paper, so that you know what questions to ask. Then examine the core functionality of the wiki and extra features that plugins offer. This chapter listed some useful questions to ask about support, cost, environment, features and functionality. We considered using the wiki for all your writing and publication requirements or just for either the writing or the publication side of things. You learned about the backbone of the wiki server, including the application server, database and user management, so that you now have some guidelines about what to discuss with your system administrators.

The chief take away of this chapter is this: Become the wiki whisperer in your organization.

References

Here are some tools, websites, blog posts, and other references relevant to the content of this chapter.

Further reading

Janet Swisher's presentation, *(Things to Think About) Before You Pick a Wiki*, is well worth a read. She has presented it at a few conferences in 2011, and has made the slides available at `http://www.slide-share.net/janetswisher/things-to-think-about-before-you-pick-a-wiki`. Her references and resources are at `http://www.janetswisher.com/index.php?itemid=283`.

Endnotes

[1] Page about Albert Einstein on *Wikiquote*: `http://en.wikiquote.org/wiki/Albert_Einstein`

[2] *What part does chocolate play in your life as technical writer*, a discussion on *Technical Writing World*, started in May 2011: `http://technicalwritingworld.com/forum/topics/what-part-does-chocolate-play`

[3] *What part does chocolate play in your life as technical writer*, a discussion on *Technical Writing World*, started in May 2011: `http://technicalwritingworld.com/forum/topics/what-part-does-chocolate-play`

[4] *Server Hardware Requirements Guide* in the Confluence documentation: `http://confluence.at-lassian.com/display/DOC/Server+Hardware+Requirements+Guide`

[5] *Supported Platforms* in the Confluence documentation: `http://confluence.atlassian.com/-display/DOC/Supported+Platforms`

[6] *Confluence Administrator's Guide*: `http://confluence.atlassian.com/display/DOC/Confluence+Administrator%27s+Guide`

[7] Documentation for *Atlassian Crowd*: `http://confluence.atlassian.com/display/CROWD`

[8] Documentation for *Atlassian JIRA*: `http://confluence.atlassian.com/display/JIRA`

[9] Documentation for *Atlassian Confluence*: `http://confluence.atlassian.com/display/DOC`

5

Getting started on the wiki

Choc Ink, employer of our hero Ganache, wants to encourage staff members to explore the potential of the social media and Web 2.0. Ganache has noticed that many people are keen on the idea but do not know where to start. She decides to write a Twitter user's guide. What's more, she will write the guide on a wiki. Two social media at one blow!

Follow in Ganache's footsteps as she installs Confluence and then creates a documentation space for her manual.

Let's get a documentation wiki up and running

By the end of this chapter you will have a documentation space on Confluence, ready and waiting for content and collaboration. This chapter has the following sections:

- **Installing your own Confluence or signing up for Confluence OnDemand.** Choose whether you want to run Confluence on your own machine or use a hosted Confluence site.

- **Adding users and groups.** When you are ready to invite other people onto the wiki, you will need to give them usernames. It is also useful to assign people to groups, for easier management.

- **Adding a documentation space.** Create a space on the wiki specifically designed for technical documentation.

- **Setting permissions.** Understand global and space permissions and set the permissions for your documentation space.

- **Adding the space description.** Explain the purpose of your space to other wiki users.

- **Designing the space home page.** Replace the default content of the space home page with something that describes your content.

- **Adding pages and creating the table of contents.** Create the skeleton table of contents for your manual, useful for iterative design and brainstorming.

- **Kay's chocolate cake: part 1.** The first part of the recipe for a delicious treat.

- **What you have done so far.** A summary of this chapter.

- **References.** The tools, websites, blog posts, and other references mentioned in this chapter.

Installing your own Confluence or signing up for Confluence OnDemand

Confluence is a web application that you can download and install on your own computer. Alternatively, you can use Atlassian's "software as a service" (SaaS) offering of Confluence, called Confluence On-Demand. This means that you get a Confluence site hosted on Atlassian's servers.

There is a thirty-day free trial available for both the downloadable and the SaaS versions of Confluence. This is for people who wish to evaluate Confluence before buying it.

Both the SaaS and the installation options are described below. I will show you where to sign up for a SaaS site. Ganache, meanwhile, has decided to install Confluence on her Windows computer. We will follow in her footsteps too.

Signing up for a SaaS Confluence site

Atlassian offers a thirty-day free trial of Confluence OnDemand. Details and signup options are at `http://www.atlassian.com/software/confluence/try/`.

A note about plugins: If you are thinking of using a Confluence OnDemand site, it is important to make a list of the plugins that you need and then check to see whether you will be able to install them on your Confluence OnDemand site. At the end of each chapter in this book, there is a list of useful plugins relevant to the content of that chapter. Please check the current plugin policy for Atlassian OnDemand.[1]

Installing Confluence on your own computer

Atlassian offers a thirty-day free trial of the downloadable Confluence installation too. The recommended way to install Confluence is via the automatic installers, supplied for Windows and Linux. There are alternative installation packages available for specific configurations. If in doubt, use the automatic installer.

The installation and home directories

There are two directories that make up Confluence on your computer: The "installation" directory, also called the "destination" directory, and the "home" directory. Do not delete either of them. Confluence runs from the installation directory and keeps its data in the home directory.

- The automatic installer sets up a standard installation on Windows or Linux, including an Apache Tomcat web server.

- As an alternative to the automatic installer, you can download an archive file (a zip archive for Windows, or a tar.gz file for Linux or Mac OS X) and unzip it into the location of your choice.

- For people who wish to install Confluence inside an existing web server, Atlassian provides a WAR (Web ARchive) distribution of the Confluence installation files.

- For very large organizations, Atlassian provides a Confluence installation that runs on a server cluster.

The *Confluence Installation Guide*[2] has instructions for the various installation options.

Let's take a look at the automatic installers for Windows and Linux, since these are the recommended installation options. We will also see how to install Confluence from an archive file on Mac OS X.

Installing Confluence on Windows

Our intrepid Ganache decides that she wants to run Confluence on her own Windows computer. She makes sure that she has a good stock of chocolate nearby, then gets started. The first step is to download

Confluence from the Atlassian website and install it onto her computer. Ganache decides to use the automatic installer, which makes the installation as easy as one, two, wiki.

 The command window

> When you start Confluence on Windows a command window will appear, probably called "Tomcat." Tomcat is the server that is now running Confluence. Do not close the command window until you are ready to shut down your Confluence server for the day. When you want to shut down Confluence, go to the Windows Start menu and choose the option to stop the Confluence server (or go to the command window and press Ctrl+C on your keyboard). Next time you want to use Confluence, first start the Confluence server from your Windows Start menu.

To download and install Confluence on Windows:

1. Go to the Atlassian Confluence download center at `http://www.atlassian.com/software/-confluence/download`.

2. Download the **Confluence x.x.x - Windows Installer** .exe file, where "x.x.x" is the Confluence version on offer. Choose the 64 bit installer or the 32 bit installer to match your version of Windows. Save the .exe file onto your computer.

3. Double-click the .exe file to run the installer.

4. If Windows asks if you want to allow the installer to make changes to your computer, click **"Yes."**

5. Choose **"Create a new Confluence installation."**

6. The installer will ask you to set some options. It is safe to leave all the options at their default values, but you can change them if you wish. Here are some guidelines on what the options mean:

   - The **destination directory**, also called the **installation directory**, is where Confluence will store its program files.

   - The **home directory** is where Confluence will store its configuration files and the wiki data. Do not put your Confluence home directory inside the Confluence installation directory, because this may cause problems when you need to upgrade to a new version of Confluence.

   - The **start menu folder** determines where you will find the Confluence startup options in your Windows Start menu.

- The **TCP ports** are the HTTP connector port and a control port that Confluence will operate on. It is safest to leave these settings at their default values unless you have good reason for changing them.

- The **Run as service** setting means that Confluence will start automatically each time you start Windows. Choose this option if you will need Confluence every day. If not, it may be better to start the Confluence service yourself whenever you need it, because Confluence can use up quite a bit of machine resources. You will be able to start Confluence from the Windows Start menu.

7. When the installation is complete, the installer will start the Confluence server on your computer and open the Confluence setup wizard in your default web browser. The steps in the setup wizard are described in the section titled "Running the setup wizard."

Installing Confluence on Linux

Atlassian provides an automatic installer for Linux environments. Alternatively, you can use the tar.gz archive file that is also supplied.

To download and install Confluence on Linux:

1. Go to the Atlassian Confluence download center at `http://www.atlassian.com/software/-confluence/download`.

2. Click the "**Linux**" tab.

3. Download the **Confluence x.x.x - Linux Installer** .bin file, where "x.x.x" is the Confluence version on offer. Choose the 64 bit installer or the 32 bit installer to match your version of Linux. Save the .bin file onto your computer.

4. Open a Linux console and go to the directory where you saved the .bin file. (Use the `cd` command to go to the directory.)

5. Run the .bin file.

6. Choose "**Create a new Confluence installation**."

7. The installer will ask you to set some options. It is safe to leave all the options at their default values, but you can change them if you wish. Here are some guidelines on what the options mean:

- The **destination directory**, also called the **installation directory**, is where Confluence will store its program files.

- The **home directory** is where Confluence will store its configuration files and the wiki data. Do not put your Confluence home directory inside the Confluence installation directory, because this may cause problems when you need to upgrade to a new version of Confluence.

- The **TCP ports** are the HTTP connector port and a control port that Confluence will operate on. It is safest to leave these settings at their default values unless you have good reason for changing them.

- The **Run Confluence as a service** setting means that Confluence will start automatically each time you start Linux.

8. When the installation is complete, the installer will start the Confluence server on your computer and open the Confluence setup wizard in your default web browser. The steps in the setup wizard are described in the section titled "Running the setup wizard."

Installing Confluence on Mac OS X

Mac OS X is not a supported environment for a Confluence installation. This means that Atlassian support does not cover the running of a Confluence server on a Mac, although of course it does cover the use of Confluence via web browser on a Mac.

Nevertheless, at time of writing this section, the Atlassian website offers an installation for Mac OS X. This is available for people who want to try out Confluence on their own computers before installing it in a Windows or Linux environment. Here are some guidelines on how to install Confluence on a Mac. The first thing you will need to do is install Java.

To check and install your Java Development Kit (JDK) on Mac OS X:

1. Check the Atlassian documentation to see which version of the JDK is required. This information is on the Confluence *Supported Platforms* page.[3] At time of writing, the required version is the Oracle JDK 1.6.

2. Check whether you have a JDK installed and what version it is: Run Terminal (go to Applications > Utilities > Terminal). In Terminal, run `java -version`.

3. If you do not have Java or your version is not the correct release, install the JDK:

   - Go to the Oracle Java download site at `http://www.oracle.com/technetwork/java/-javase/downloads/index.html`.

   - Download the Java Platform (JDK) installer that matches your operating system. The name of the download starts with "jdk" and a version number, such as "jdk-6u26...."

- Run the Java installer that you have downloaded. When it asks you to choose a directory for your Java installation, make a note of the directory you choose. You will need it later, to set up your JAVA_HOME environment variable.

Oracle provides detailed Java installation instructions on the *Java Platform Installation Page.*[4]

To download and install Confluence on Mac OS X:

1. Go to the Atlassian Confluence download center at http://www.atlassian.com/software/-confluence/download.

2. Click the "**Mac OS X**" tab.

3. Download the "**Confluence x.x.x (TAR.GZ Archive)**" file, where "x.x.x" is the version number on offer, and save the file onto your computer.

4. Unpack the tar.gz archive into a directory of your choice. That directory is now your Confluence installation directory. Let's call that directory {CONFLUENCE_INSTALL}.

5. Follow these steps to set up an environment variable which tells Confluence where to find Java:

   - Inside the {CONFLUENCE_INSTALL} directory, go to the bin directory and find the setenv.sh file.

   - Open the setenv.sh file in TextEdit.

   - Add the following line as the first line of the file:

     ```
     export JAVA_HOME=/Library/Java/Home
     ```

   - Save the file.

6. Now you will tell Confluence where to put its "home" directory. This is where Confluence will store its configuration data.

   - Create a new folder in your user's home folder. A good name for the new folder is data. This new folder will be the Confluence home directory. Do not put your Confluence home directory inside the Confluence installation directory, because this will cause problems when you need to upgrade to a new version of Confluence.

   - Use TextEdit to edit the properties file at {CONFLUENCE_INSTALL}/confluence/WEB-INF/classes/confluence-init.properties.

- Add a line at the end of the file, specifying the name and path of your Confluence home directory. For example: `confluence.home=/Users/myname/data`

- Save the file.

7. Start your Confluence server:

   - Run Terminal. (Go to Applications > Utilities > Terminal).

   - In Terminal, run `{CONFLUENCE_INSTALL}/bin/startup.sh`.

8. Go to your web browser and start the Confluence setup wizard, as described below.

Running the setup wizard

Our dauntless Ganache now has Confluence up and running on her computer. Next she needs to set up an administrative account (by defining an administrator's username and password) and specify other default settings on the Confluence site. She will do this using a web browser such as Microsoft Internet Explorer or Mozilla Firefox. The Confluence installation includes a web server, now running on Ganache's computer ("localhost"). By default, Confluence runs on port 8090. This means that Ganache can access Confluence via a web browser at the address `localhost:8090`.

To run the Confluence setup wizard:

1. Start up your web browser, such as Microsoft Internet Explorer or Mozilla Firefox.

2. Type this address into the browser address bar: `http://localhost:8090`.

3. The Confluence setup wizard will start.

4. Enter your Confluence license key into the "**License Key**" field. If you do not already have a Confluence license, follow the prompts on the screen to generate an evaluation license. The license will last for thirty days and gives you full access to all Confluence functionality for that period.

5. Click "**Evaluation Installation**" under "Choose Installation Type." This option gives you a built-in database, useful for a quick start to Confluence. When you decide to move your site to production, you can easily transfer your content to an external, enterprise-ready database. The evaluation installation also includes a demonstration space containing sample pages and a tutorial.

6. If you decided to go for the production installation instead of the evaluation installation, the setup wizard will offer you options to configure your database, load content into the database, set up user management and connect to JIRA. We will skip all those steps, going for a simple default installation.

7. The "Setup System Administrator" screen will appear. Enter the following information:

- Username: An identifier for the user that will have administrative permissions on your Confluence site.

- Password: A password that the administrator will need to use when accessing Confluence.

- Confirm: The same password again, to confirm it.

- Name: The name of the administrator.

- Email: The administrator's email address. It is a good idea to give your own email address here, so that you will receive any messages that Confluence sends when something needs your attention.

8. Click the "**Next**" button.

9. The "Confluence Setup Successful" screen will appear. Click "**Start using Confluence now**."

10. If prompted, log in to Confluence using the administrator's username and password that you supplied earlier.

11. If the "What's new in Confluence" popup window appears, select "**Don't show again**" and click "**Close**."

12. The home page of the Confluence *Demonstration Space* will appear.

The Confluence installation and setup procedures may change over time. If you run into problems following the above steps, please check the latest instructions in the *Confluence Installation Guide*.

At this stage, Ganache pops out for a quick hot chocolate with her *Choc Ink* colleagues. If you like, you can take the opportunity to have a look around the *Demonstration Space* and the rest of your new Confluence site.

Adding users and groups

When you are ready to invite other people onto the wiki, you will need to give them usernames. Alternatively, you can allow people to sign up and choose their own usernames. Confluence can store its users on its own database or draw them from a corporate user repository, such as Microsoft Active Directory.

It is also useful to assign people to groups for easier management.

The *Confluence Administrator's Guide* has the details on user and group management.[5] Here is a quick guide to adding users and groups.

To add a group: Choose "**Browse**" > "**Confluence Admin**" > "**Manage Groups**" > "**Add Group**."

To add a user: Choose "**Browse**" > "**Confluence Admin**" > "**Manage Users**" > "**Add User**."

For example, you may want to add a group called `techcomm` and then add all the technical communicators to that group. This will make it easy to give all the technical communicators administrative rights over all the documentation spaces.

If you have followed the steps in the previous section and installed Confluence, then you are currently logged in as a system administrator. You have total control over the Confluence site. When you give other people access to the site they will not have as much power as you, unless you make them system administrators too.

Adding a documentation space

Chapter 3 described how a space fits into a Confluence wiki. Armed with that conceptual overview, Ganache is now ready to create the space that will house her Twitter user's guide. She chooses the Documentation theme because it gives the look and feel she wants for her technical guide: a smidgen of sobriety in the colors, a touch of sophistication in the fonts, a dose of sweetness in the left-hand navigation panel, plus plenty of power in the customizable header and footer.

Ganache has the "create space" permission in Confluence. If you do not have that permission, you will not see the option for adding a space. Have a word with your Confluence administrators. There is more about permissions later in this chapter.

To add a documentation space:

1. Go to the Confluence dashboard and click "**Add Space**."

2. Enter the "**space name**." Ganache enters "Using Twitter." This name will appear on the dashboard, in the space directory, and in the breadcrumbs at the top of each page. You can change the space name at any time. This will not affect the links that point to your space or pages.

3. Enter the "**space key**." Ganache enters "DOC." This is a short identifier for your space. It is used in the URL for accessing pages in the space and for actions performed on the space. No two spaces can have the same key within the Confluence site. You cannot change this key after adding the space, except by copying all the content to a new space.

4. Under "**Who can use this space?**," set the preliminary permissions for your space. Ganache decides to let all registered users see her space, but for the moment only she can add content. You can refine these settings immediately after creating the space, and you can also change them at any time.

5. Choose the "**theme**" for your space. Ganache chooses the Documentation theme. Another name for a theme is a skin. It determines the look and feel of the space.

6. Click "**OK**."

Confluence creates the space and opens the home page of that space in your browser.

These screenshots show the values that Ganache entered when creating her space.

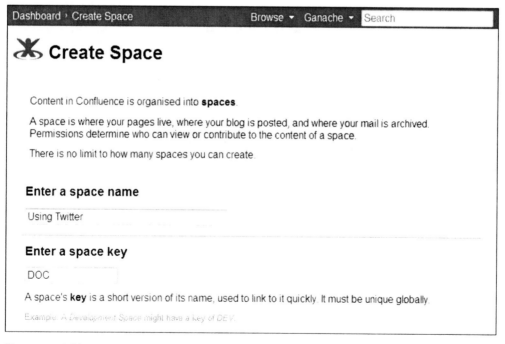

Figure 5.1. Adding a space - part 1

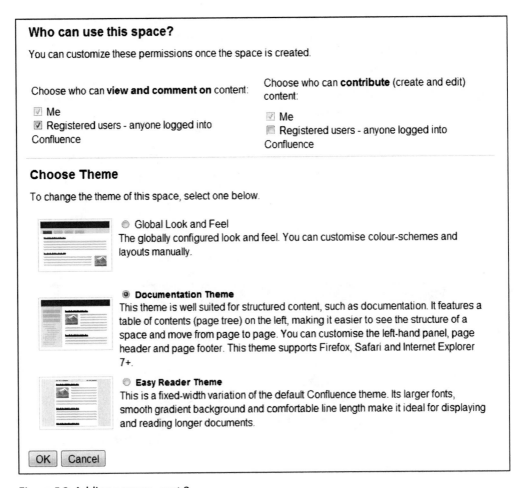

Figure 5.2. Adding a space - part 2

Setting permissions

When Ganache added the space, she set some preliminary permissions. Now she wants to refine those settings, telling Confluence who can add pages or blog posts, view and update existing content, comment on the pages, and perform more advanced functions. First, Ganache takes a step back and reads up on the overall Confluence permission framework.

Confluence has three levels of permissions:

- Global permissions, described below.

- Space permissions, also described below.

- Page restrictions. Authors can restrict the viewing and editing rights for specific pages, provided they themselves have the permission to restrict pages. This is useful when you are drafting a page and you do not yet want feedback on it. I will cover page restrictions later in the book.

Global permissions and groups

Only the Confluence system administrator can view and set the global permissions. If you installed your own version of Confluence, then you are a system administrator. The global permissions determine site-wide access, such as the ability to add spaces and administer the site itself (see Figure 5.3).

To set global permissions: Choose "**Browse**" > "**Confluence Admin**" > "**Global Permissions**."

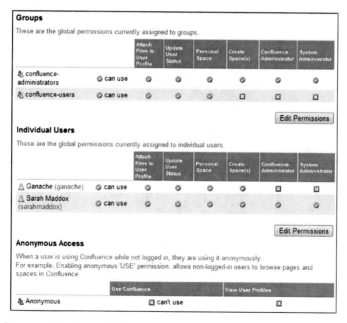

Figure 5.3. Confluence global permissions

You can assign global permissions to groups of users, individual users, or anonymous users. An anonymous user is someone who is viewing the wiki pages but has not logged in.

The permissions at global level determine who can perform the following functions:

- Use Confluence. This permission has the odd but functional label, "can use."

- Attach files to their user profile.

- Update their user status message, similar to a Twitter or Facebook status.

- Create their own personal spaces.

- Create spaces. These are normal Confluence spaces, as opposed to personal spaces.

The *Confluence Administrator's Guide* describes the permissions in detail.[6] For our purposes, the most important two to know about are the permission to use Confluence and the permission to create spaces.

In addition Confluence has two preset roles, which it also calls permissions:

- Confluence administrator.

- System administrator.

The *Confluence Administrator's Guide* describes the powers of each. For now, it is enough to know that a system administrator is more powerful than a Confluence administrator. It is not possible to add any other roles or to change the powers of the preset roles.

To make it easier to manage the permissions, you can define groups of users. It is much quicker to assign permissions to a group of people rather than to a number of people individually. When you install Confluence, there are already two groups defined:

- The `confluence-administrators` group. This group has nothing to do with the Confluence administrator role. People in the `confluence-administrators` group are super users. They can do anything on the Confluence site. The global permission settings do not affect these people. When you install Confluence, you define the initial administrator who is by default a member of this group. You can add other people to this group too.

- The `confluence-users` group. When you add users to Confluence, they are automatically in this group. People must be in this group in order to use Confluence. The number of people in this group is the number that counts for your Confluence license. To stop someone from using Confluence, just remove them from the group. This is useful if you want to remove someone's access but cannot delete their username because they have added content to the wiki.

You can add any other groups you like.

What if a user is in a group with a certain level of permissions and there are also specific permissions assigned to the user individually? The highest permission wins.

In Figure 5.3 people in the group called `confluence-administrators` can do everything. People in the group called `confluence-users` can do everything except create spaces or perform administrative actions. The user "Ganache" can create spaces, but is not a Confluence administrator or a system administrator. (In case you are wondering, Ganache did not install this Confluence site!) The user "Sarah Maddox" can do everything. Anonymous users cannot access the site at all.

Space permissions

Space administrators can set the permissions at space level. If you created a space, you are automatically its administrator. You can also make other people administrators of the space. The space permissions determine who can update content and perform other actions in the space (see Figure 5.4).

To set space permissions: Go to any page in the space, then choose **"Browse"** > **"Space Admin"** > **"Permissions."**

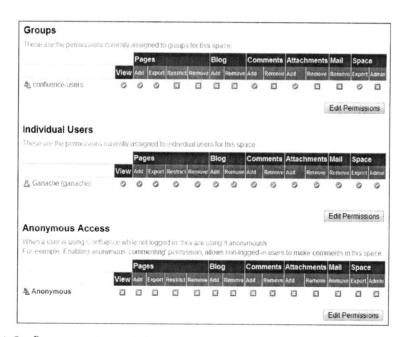

Figure 5.4. Confluence space permissions

Just as with the global permissions, you can assign space permissions to groups of users, individual users, or anonymous users.

The space permissions determine who can perform the following functions:

- View the content of the space.

- Add, edit, or delete pages; export the content of pages to other formats; and restrict the viewing and editing rights on individual pages.

- Add, edit, or delete blog posts.

- Add, edit, or delete comments on pages.

- Add or delete attachments such as screenshots, other images, and other files.

- Remove mail archives.

- Export the content of the space to other formats, including PDF, HTML, and XML.

- Administer the space.

The "Add" permission covers editing rights as well as the right to add a page, blog post, or comment.

You can change the permissions on the space at any time. Think about how much power you want to give people and who should be able to do what.

As you can see in Figure 5.4, Ganache has set her space permissions so that all logged-in Confluence users can add and update the content. Only Ganache herself can delete content. Anonymous users cannot see the space.

Hint: Can't find a specific function in Confluence?

If you cannot find one of the menu options or other functions mentioned in this book, the reason is most likely one of the following:

- You do not have permission to use the option. First check to see if you have logged in to Confluence. If you are not logged in, you will have only the rights assigned to an anonymous user. If you still cannot see the option after logging in, ask your Confluence administrators to give you the required permission.

- You are using a different version of Confluence. This book is based on Confluence 3.5, 4.0 and 4.1. Consult the Atlassian Confluence documentation directory to find the guides for your version of Confluence.[7]

- Your Confluence site has a custom theme, which has put the option in a different spot. Contact your Confluence administrators or consult the documentation for the Confluence theme.

- There is a bug in Confluence. It happens, but this is the least likely cause. Search the Confluence issue tracker at `http://jira.atlassian.com` to see if someone has already reported this problem. If not, you can log the bug yourself.

Who do you allow to update the wiki?

Much of the power of a wiki lies in its simplicity and flexibility. People value the easy way they can collaborate by editing a page and building on someone else's content. Try not to lay down too many rules. Allow as many people as possible to add and edit the pages in your wiki.

On the Atlassian documentation wiki, all staff members have full editing rights. Other people, including anonymous users, can add comments and view pages. In addition, community authors can update the content too. On the Atlassian intranet wiki, most spaces are open to editing by anyone in the company.

Adding the space description

As well as a name and key, you can add a description to each space.

To update the space description: Go to any page in the space, then choose **"Browse"** > **"Space Admin"** > **"Space Details."**

The description should contain a short sentence or two about the content and purpose of the space. It appears on the Confluence dashboard and in the space directory.

Designing the space home page

When Ganache added her *Using Twitter* space, Confluence created a home page for the space with some default content (see Figure 5.5). It looks pretty ordinary. Ganache wants a good-looking home page, tailored to her content and audience.

Figure 5.5. Default space home page

To edit your space home page: Go to the page, then choose "**Edit.**"

The page will open in edit mode. Select all the content and delete it, then replace it with your own. The next chapter guides you through the process of using the editor to develop content on a page. If you do not know what you want at this stage, just add a short description of your space.

Ganache starts with a short welcome message, followed by a table of contents showing what is in the guide (see Figure 5.6). The table of contents will be dynamically generated from the titles of the pages in the space. At the moment it is empty because Ganache has not yet added any pages. Lastly, since her documentation is all about Twitter, Ganache decides to display a live stream of Tweets on the page. That will get her readers' attention!

The table of contents, as shown under the heading "What's in the guide" on Ganache's page, is produced by the children macro. Chapter 7 describes this and other macros that are useful in adding structure to your wiki documentation.

Figure 5.6. Updated home page

Ganache used Confluence's widget macro to add the Twitter stream. It gets its results from a Twitter search URL that picks up all Tweets containing the keyword "techcomm2." Details are in Chapter 21, all about Twitter and the wiki. As a foretaste, here is the wiki markup for the widget macro on Ganache's page:

```
{widget:url=http://search.twitter.com/search?q=techcomm2|width=300|height=300}
```

Adding pages and creating the table of contents

When designing a new manual, a technical writer often starts with a skeleton structure showing the chapters or top-level headings in the document. This is useful for brainstorming and iterative review with product managers and the development team. In Confluence, it is easy to create the skeleton table of contents. Just add the pages to the space, with very little or no content.

Keyboard shortcuts

Confluence offers a number of useful keyboard shortcuts. To add a page, press C on your keyboard. To see a list of all keyboard shortcuts, press ? on your keyboard. The latter works when you are viewing a page, on the dashboard or on the administration screens. If you are in the rich text editor, click the question mark icon ⦿ on the toolbar.

You can add a page as a child of another page, such as the space home page. In fact, you can add any number of child pages. The space home page is then the "parent" of the new pages. You can add a child to any other page too. This means that you can build a hierarchy, or tree, of pages to reflect the sections and chapters of a manual.

Let's assume that you are designing a complete documentation suite for a product. You will add the top-level pages for all the usual sections, including the installation guide, administrator's guide, user's guide, developer's guide, FAQ, and troubleshooting. These top-level pages will all be children of the space home page. Tackling the installation guide first, you might then add a child page for the Windows installation steps, another page for UNIX, and so on. Moving on to the administrator's guide, you would add child pages for instructions on configuring permissions, user management, backup and restore procedures, and so on.

To add a page: Go to the parent page, then choose "**Add**" > "**Page.**"

Ganache starts outlining the content of her Twitter guide by adding empty pages to the space. Figure 5.7 shows her first draft.

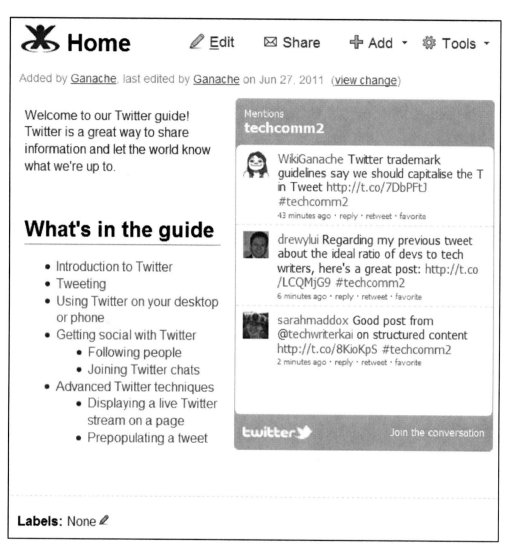

Figure 5.7. Twitter guide home page including draft table of contents

Because Ganache has chosen the Documentation theme for her space, she also has a left-hand navigation panel that shows the table of contents (see Figure 5.8).

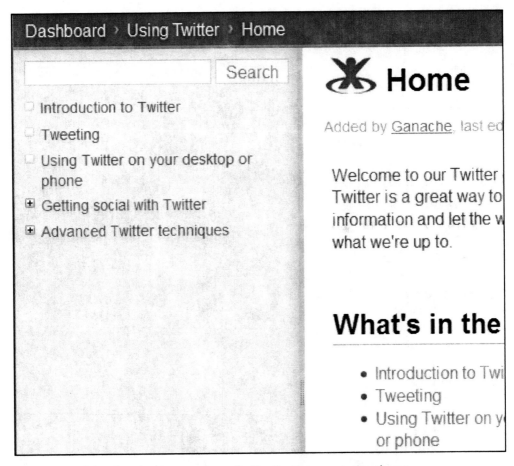

Figure 5.8. Left-hand navigation panel supplied by the Documentation theme

Ganache will refine her table of contents as she goes along, consulting her subject matter experts and encouraging input from the community too. It is an iterative process. In fact, you will possibly see the table of contents change in later screenshots throughout the book.

Have you added some pages and discovered that they are in alphabetical order rather than the logical sequence that you want? You can put them into the right order. Chapter 7 covers page ordering as part of the discussion on structure and style. Quick hint: Choose "**Browse**" > "**Pages**" > "**Tree**," then open the branches of the tree and drag and drop the pages into position.

Kay's chocolate cake: part 1

My mother, Kay, taught me how to bake this cake when I was a little girl. Up to now, the recipe has lived in our heads. I suspect it has evolved over the years. This is the first time I have written it down.

Oven temperature: 350° F, 180° C

Implements:

- 2 round, oven-proof cake pans, diameter 8.5 inches (21 cm), height 1.5 inches (3.5 cm).
- 2 cooling racks, to put the hot cakes on.
- A small, clean plastic bag and some scissors, to make a piping tool.
- A mixing bowl and spoon or electric mixer.
- A sieve.
- A deep bowl, a knife and the usual kitchen implements.

Ingredients for cake:

- Small amount margarine and flour to spread on cake pans.
- 6 ounces margarine or butter (170 g)
- 6 ounces light brown sugar (170 g)
- 3 eggs
- 5 ounces self-raising flour (140 g)
- 1 ounce cocoa (30 g)
- 1 teaspoon baking powder
- Pinch of salt
- Approximately 4 tablespoons milk
- 1 teaspoon vanilla essence

Ingredients for icing and filling:

- 3 heaped teaspoons cocoa
- Boiling water
- 1 packet of icing sugar (finely powdered white sugar, also called confectioner's sugar – approximately a pound or 500 g)
- A small jar of raspberry or blackberry jam (approximately 5 ounces or 150 g)

This recipe is in three parts. Read fast to get to the next part of the recipe in Chapter 16!

What you have done so far

You have a Confluence site up and running! Perhaps you chose to set up an evaluation installation on your own computer, or you signed up for a hosted Confluence site running in the cloud. Either way, you can use this Confluence site free of charge for thirty days and then decide how to continue.

Setting up a documentation space on the wiki is similar to setting up a space for any other purpose. Select the Confluence "Add Space" option, then choose the Documentation theme to take advantage of its documentation-specific features. One of the first things to do is to refine the default permission scheme for the space. Decide who should be able to see the content, who should be able to update it, and who should be able to add comments. You may decide to limit the permissions to a specific group of people at first, then open it up later when the content is ready for general consumption.

The space has a home page with some default content, which you will probably want to fix up immediately. In Chapter 6 we will look at the editor in more detail, so that you can make your home page compelling and informative.

To define the structure of your manual, just add empty pages with titles for the administrator's guide, user's guide and so on. Whatever your manual needs. The table of contents will appear in the left-hand navigation panel as you add the pages. This is a great way of creating a framework for discussion and iterative development.

Ganache's manual is already taking shape. Would you like to see the Twitter guide in action? It is on display on the *Confluence, Tech Comm, Chocolate* site. Come, take a look and join the conversation at `http://wikitechcomm.onconfluence.com`.

In the meantime, onward to the editor.

References

These are the tools, websites, blog posts, and other references mentioned in this chapter.

Endnotes

[1]   Plugin policy for Atlassian OnDemand: `http://confluence.atlassian.com/display/AOD/-Atlassian+OnDemand+Plugin+Policy`

[2] *Confluence Installation Guide*: `http://confluence.atlassian.com/display/DOC/Confluence+Installation+Guide`

[3] *Supported Platforms* in the Confluence documentation: `http://confluence.atlassian.com/display/DOC/Supported+Platforms`

[4] *Java Platform Installation Page* by *Oracle*: `http://www.oracle.com/technetwork/java/javase/index-137561.html`

[5] Page about user management in the *Confluence Administrator's Guide*: `http://confluence.atlassian.com/display/DOC/User+Management`

[6] Overview of global permissions in the *Confluence Administrator's Guide*: `http://confluence.atlassian.com/display/DOC/Global+Permissions+Overview`

[7] Guides for all versions of Confluence, in the Confluence documentation directory: `http://confluence.atlassian.com/display/ALLDOC/Confluence+Documentation+Directory`

6

Developing content

Now that you have a Confluence space to hold your content, it is time to get acquainted with the heart and soul of the wiki: the editor. Whenever you edit a page, add a new page, or post an entry in the Confluence blog, the editor does all the work. In fact, the editor springs into action when you add a comment, too.

As I write these words, I am soon to head off on a treasure hunt through the city, à la Indiana Jones and courtesy of the Atlassian end-of-financial-year party. At least, that is what I think we will be doing. It is all a big secret. Our only clue to date is a 30-second video! I hope that this chapter is a great kickoff point for you to start your adventures in Confluence. It is not intended to be an exhaustive guide to the editor. Instead, my aim is to give you a sound understanding of what the editor looks like and what it can do for you. I have also given some background to the editors and wiki markup, because it is interesting to me and I hope to you, too. If at any point you feel that I have led you astray, take refuge in the excellent *Confluence User's Guide*.[1] In tribute to the immortal Professor Henry Jones, "You call *this* wikiology?"

Getting acquainted with the Confluence editor

There is a fairly significant change in the editor between Confluence 3.5 and Confluence 4. This chapter covers both versions of the editor. You will uncover the following treasures in this chapter:

- **The editor in Confluence 3.5.** Two editing modes are available. People can choose to work in the rich text editor or the wiki markup editor.

- **The editor in Confluence 4.** Just one editor provides the rich text editing experience while responding to wiki markup cues, too.

- **Wiki markup explained.** This section goes into detail about the benefits and weaknesses of wiki markup, provides a quick reference guide to the syntax, and discusses the demise of wiki markup in Confluence 4. Wait, does wiki markup die? Surely, as in the case of the great Indy himself, the fans will not let that happen!

- **Editor quick reference guide.** For power users, this is a quick start guide to the shortcut keys, autoformatting, and autocomplete functionality in the Confluence editors.

- **Editing a page.** Follow Ganache as she adds content to her first page.

- **Adding screenshots and other images to your page.** See how to add an image using the Confluence image browser and then manage the images as attachments to the page.

- **Drawing diagrams.** Let the fun begin! Install the Gliffy plugin and add a diagram to your page.

- **What you have learned about developing content.** A summary of this chapter.

- **References.** The tools, websites, blog posts, and other references mentioned in this chapter.

The editor in Confluence 3.5

Confluence 3.5 offers two editors to choose from. Each person can decide to work in either the rich text editor or the wiki markup editor. Note that the Confluence system administrators can disable the rich text editor. If they do that, only the wiki markup editor will be available.

The rich text editor

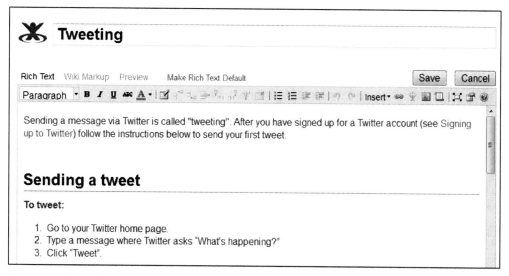

Figure 6.1. The rich text editor in Confluence 3.5

Figure 6.1 shows a page in edit mode, using the Confluence 3.5 rich text editor. The page title is *Tweeting*. The tabs under the title offer the following options:

- **Rich Text**. This tab is currently selected. The editor is open in rich text mode.

- **Wiki Markup**. This tab opens the wiki markup editor.

- **Preview**. The preview option is useful if you want to see what the page will look like in view mode before you save the page.

- **Make Rich Text Default**. Click this option if you want the editor to open in rich text mode by default, instead of wiki markup mode.

The rich text editor has some WYSIWYG (what you see is what you get) characteristics, in that it renders much of the text formatting in the editing pane. For example, bold text, italic text, headings, and lists will all be displayed as if the page were in view mode. However, the more complex markup provided by macros is not rendered in the editing pane.

The toolbar across the top of the editing pane offers a number of options. Their use is similar to most such editors. For example:

- To apply a paragraph or heading style to some text, select the text and then select the style from the dropdown menu on the left of the toolbar.

- To apply bold formatting to some text, select the text and then click "**B**."

- Use the table section of the toolbar to add and edit tables.

- Use the other options to apply list formatting, add hyperlinks, insert images, and more.

A really useful feature is the ability to open the editor in full screen mode. Click the ⬚ icon on the toolbar.

Later in this chapter is a quick reference guide to the editor functionality and shortcuts.

The wiki markup editor

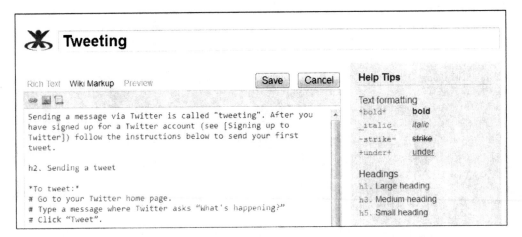

Figure 6.2. The wiki markup editor in Confluence 3.5

The wiki markup editor in Confluence 3.5 is a plain text editor (see Figure 6.2). It does not render any text formatting. The minimalist toolbar offers the options to add hyperlinks, images and macros.

The editor in Confluence 4

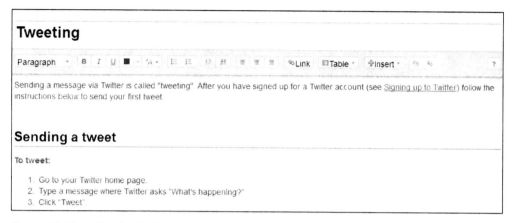

Figure 6.3. The editor in Confluence 4

Confluence 4, starting with version 4.0, has just one editor (see Figure 6.3). I have tested the guidelines in this book in the Confluence 4.0 editor. As this book going to press, Confluence 4.1 has just been released. There are no changes in version 4.1 that significantly affect the content of the book.

At first glance, Confluence 4 presents a souped-up rich text editor with better display of macros and many more options in the toolbar. As you start using the editor, you realize that it responds to wiki markup, too. For example, if you type an asterisk at the beginning and end of some text, the text format will automatically change to bold. This is the autoformatting feature in action. In that way, the Confluence 4 editor is a combination of the two editors from earlier versions of Confluence. The difference is that the wiki markup is for input only. As soon as you enter the wiki markup, Confluence converts it to rich text format. From that point onwards, you must edit the text in rich text format. You cannot edit the wiki markup that you used originally.

Let's take a look at wiki markup in more detail, then fly through a couple of quick reference guides. After that, you will be ready to pilot the Confluence editor yourself!

Wiki markup explained

What part does chocolate play in your life as technical writer?

From Eddie VanArsdall on Technical Writing World:[2] "I've always used the old bumper sticker analogy: I brake for chocolate. Or is it I BREAK for chocolate?"

It doesn't matter. I just love the stuff, and the darker the better. For me, milk chocolate simply isn't chocolate. I share your passion for chocolate that has been infused with red wine. Sometimes I even infuse myself with red wine when eating dark chocolate!

"Other great combinations: Dark chocolate with spices and a twist of orange; dark chocolate with chilis."

Wiki markup. People either love it or hate it. I love it, and I suspect that many technical writers are in the same boat. As you have probably guessed, wiki markup is a markup language. That means that you apply styles and other smarts to your document by adding recognized codes to the text. HTML is a markup language, as is its stricter relative, XML. Another name for wiki markup is wikitext language.

Not all wikis use the same wiki markup syntax. Confluence's wiki markup is based on Textile.[3] MediaWiki uses its own syntax.[4] In 2006, a group of people got together at the International Symposium on Wikis to kick off the design of a common wiki markup language called Creole.[5] A few wikis support Creole, either natively or via plugins.

The benefits of wiki markup

The designers of wiki markup aim for a simple, lightweight language tailored to allow people to edit a web page in their browsers. Here are some of the benefits of wiki markup, as opposed to HTML or XML:

- Wiki markup is simple to learn and to use. It provides a small set of tags and codes chosen to be those most useful for people who want to create content on a web page.

- Wiki markup is simple to read. There is as little clutter as possible. Wiki markup often does away with closing tags. The markup may even look a little like the end product, or at least call it to mind. For example, an asterisk, followed by a space and some text, represents a bullet point. A series of pound signs (hash signs) denotes a numbered list.

- Wiki markup provides a secure environment for editing pages in the web browser. It does not expose the HTML elements that are vulnerable to XSS, XSRF, and other attacks.

Wiki markup weaknesses

Wiki markup may be simpler to learn, use, and read than HTML, but it is more complex than a rich text editor. Many people find wiki markup daunting, especially people in the business areas of an organization and people who do not work on the wiki regularly. It is a lot of effort to learn a new syntax if you do not use it every day. This can be a barrier to collaboration. Also, the wiki markup editor tends to present you with a wall of text.

However, there are problems allowing people to edit content in rich text mode and then converting the content back to wiki markup for storing in the wiki database. Sometimes when you see the page in view mode, the layout is not exactly the same as it appeared in the rich text editor. This problem is a result of those dreaded "roundtrip errors," which you may hear wiki developers discussing in heated tones. It is one of the primary reasons for Atlassian's decision to offer a new format (XHTML instead of wiki markup) and a single editor in Confluence 4.

A wiki markup quick reference guide

This is my quick reference guide (Table 6.1). It is not exhaustive, but it covers the markup a technical writer needs most often. The *Confluence User's Guide*[6] offers a complete reference.

Table 6.1 – Wiki markup summary

What you want	Wiki markup	Rendering
Bold text	*my text*	**my text**
Italic text	_my text_	*my text*
Underlined text	+my text+	<u>my text</u>
Monospaced text	{{my text}}	`my text`
Superscript	^my text^ **Hint:** If you want to put the superscripted text right up against other text, enclose the circumflex in curly brackets: normal text{^}my text^	normal text$^{my\ text}$
Subscript	~my text~	normal text$_{my\ text}$
Line break	Usually just an empty line is enough. If you need to force a line break, use a double forward slash: //	(Not illustrated.)
En dash	--	–
Heading level 1	h1. My heading	My heading

What you want	Wiki markup	Rendering
Heading level 2	h2 . My heading *And so on, up to h6 .*	My heading

Bulleted list	* List level 1
	** Level 2
	*** Level 3
	And so on.
		■ List level 1
		  ■ Level 2
		    ■ Level 3

Numbered list	# List level 1
	## Level 2
	### Level 3
	And so on.
		1. List level 1
		  a. Level 2
		    i. Level 3

Mixing bullets and numbers	* Bullet item 1
	*# Numbered item 1
	*# Numbered item 2
	* Bullet item 2
	*# Numbered item 1
	*# Numbered item 2
	# Bullet
	
	# Numbered list point 1
	#* Bullet
	#* Another bullet
	#** Second-level bullet
	# Numbered list point 2
	#* Bullet
	#* Another bullet
	#** Second-level bullet
	And so on.
		■ Bullet item 1
		  1. Numbered item 1
		  2. Numbered item 2
		■ Bullet item 2
		  1. Numbered item 1
		  2. Numbered item 2
		    ■ Bullet

Hyperlink	[http://google.com]	http://google.com
Hyperlink with alias	[Google|http://google.com]	Google

What you want	Wiki markup	Rendering
Table	|| Heading 1 || Heading 2 || Heading 3 || | My cell 1 | My cell 2 | My cell 3 |	(Not illustrated. This quick reference guide is in a table.)
Image	!myImage.png!	(Not illustrated.)
Escaping a special character	\[Dear Confluence, please display the square brackets. This is not a link.\]	[Dear Confluence, please display the square brackets. This is not a link.]
Emoticon	Try the commonly-recognized emoticon symbols, including a smiley, a wink and more. For example, :)	(Not illustrated.)
Advanced formatting	Use Confluence macros to draw panels around blocks of text, color text and add dynamic content to your page. The next chapter discusses macros in more depth. For example, {my-macro}my text{my-macro}	(Not illustrated.)

Introducing macros

Macros are not really part of wiki markup, but people often think of them that way, so I will include them here. In Confluence wiki markup, a macro is a word or words in curly brackets. You can also insert a macro via the editor menu. A macro represents a small piece of code or an entire application. It can do something as simple as display a piece of text or draw a panel around a block of text, or it can give you the full power of a drawing tool on your web page. We will look at macros in more detail in Chapter 7. For now, here are a couple of examples to whet your appetite:

- {info} – Puts your text inside a blue box with an information icon.
- {toc} – Displays a list of the headings that appear on the page, thus forming a table of contents.
- {children} – Displays the child pages of the current page.

There's extra wiki markup on this site!

Here is an interesting thing about the wiki markup available on your Confluence site: some Confluence sites offer more wiki markup than others! When you first install Confluence, you get the core set of markup. If you add plugins or user macros to your site, you will gain additional wiki markup, usually in the form of macros. Because this markup is provided by plugins and add-ons, it is not documented in the *Confluence User's Guide*. However, each Confluence site provides an online quick reference guide to the markup available on that site.

To see the wiki markup guide for your site:

1. Go to this URL in your web browser: `http://MY.CONFLUENCE.SITE/renderer/notation-help.action`, where MY.CONFLUENCE.SITE is the address of your Confluence site. If you are running Confluence on your own computer, the address is `http://localhost:8090/renderer/notationhelp.action`. (For Confluence 3.5, use port `8080` instead of `8090`.)

2. The notation guide will appear. Choose "**ALL**" or a specific section in the left-hand panel.

Where does the information in the notation guide come from? The guide to the core wiki markup is shipped with Confluence. Plugin developers provide the additional information, by adding the required XML elements to their plugin code.

Special characters

Both Confluence 3.5 and Confluence 4 provide a menu option for inserting special characters or symbols. This option presents you with a set of characters to choose from.

In Confluence 3.5, you can add the code for HTML entities (character entity references) onto your Confluence page. Confluence will render the relevant character. For example, type © or © to display a copyright sign ©. Use ♪ to display a musical eighth note . *W3C* has a comprehensive list of the HTML entities[7] and *mistywindow.com* has a good list of special characters.[8]

The death of wiki markup?

Confluence 4 does not have a wiki markup editor. Even more significantly, the storage format is XHTML instead of wiki markup. The content of the pages is stored in XHTML on the Confluence database, whereas in Confluence 3.5 it is stored in wiki markup.

But wiki markup never dies! Many people love the simplicity and speed of typing an asterisk and getting a bulleted list, or bracketing text between two asterisks to make it bold, rather than having to select the text and apply the formatting as separate steps. Such actions still work in the Confluence 4 editor. The autoformatting feature recognizes wiki markup and applies it as you type.

Some people, and I am one of them, like to make notes for themselves while they are on the bus. There is nothing more satisfying than to pull up a note pad on your phone and type in that blindingly brilliant piece of wording before it slips your mind, complete with formatting. When you get to work, just copy and paste it into the wiki. That's it, done. The Confluence 4 editor has an "insert wiki markup" option for just this occasion. As with autoformatting, it is a one-way conversion. You cannot edit the wiki markup later. But it does the job.

Wiki markup and technical communication

How does the demise, or at least the fading into the background, of wiki markup affect us as technical communicators?

If wiki markup had disappeared entirely, it would have made many of us sad. We, like other power users of Confluence, need that rapid response and all-in-one-action editing experience that a plain text editor provides. Luckily we can breathe easy on that front.

We will see the greatest benefit of the Confluence 4 editor when we are collaborating with our subject matter experts to develop and review our documentation. The business users in our organizations will find it much easier to work with the rich text features of the editor than with wiki markup. In earlier versions of Confluence it was possible for some people to work in wiki markup and others to work in rich text mode. This could lead to problems, however, when editing a page with a sophisticated layout. If you carefully crafted a page in wiki markup, and then someone edited that page using the rich text editor, those infamous roundtrip errors could mess up the layout of the page. In Confluence 4, the day of the roundtrip error is over.

The same applies if we open up our documentation wiki to contributions from community authors. They will not need to learn wiki markup, and we will all be using the same editor to update a page.

Some organizations will stay on earlier versions of Confluence, postponing the upgrade to Confluence 4 until they are ready. For that reason, we still need to understand wiki markup.

When Confluence 4.0 was first released, it had some problems that made it not yet ready for use as a platform for technical documentation. See Chapter 23 for a description of these issues and how technical writers are working with the developers to plug the gaps.

What part does chocolate play in your life as technical writer?

From Anindita Basu on Technical Writing World, complete with markup:[9]

[deep sigh]
None, whatsoever. I am the exception to this rule. :)
[/deep sigh]

Editor quick reference guide

Technical writers are power users. We just need to know how to get started, then we can figure out the rest. Table 6.2 is a quick start guide to the shortcut keys, autoformatting and autocomplete functionality in the Confluence editors. Anything that you can do via a shortcut, you can also do via the menus.

Table 6.2 – Editor quick reference

What you want to do	How to do it	More about it
Get quick help	Press ? or click on the ? icon on the editor toolbar.	If you are viewing a page, press ? In the editor, press the question mark icon. (This option is not available in the Confluence 3.5 wiki markup editor.) A popup window appears with a list of the keyboard shortcuts available. In Confluence 4 it also shows the autoformatting and autocomplete trigger characters.
Add a page	Press C	Go to the parent page, press C, type the text, press Ctrl+S to save the page.
Edit a page	Press E	Go to the page, press E, type the text, press Ctrl+S to save the changes.
Format the text as bold	Press Ctrl+B	
Make a bulleted list	Press Ctrl+Shift+B	
Apply other text formatting	Use standard keyboard shortcuts.	Try the standard keyboard shortcuts that you know from other rich text editors such as Microsoft Word. The options are more extensive in Confluence 4 than in 3.5.

What you want to do	How to do it	More about it

Let Confluence auto-format the text	Type the wiki markup formatting characters. Confluence will format the text as you type. For example:
	
	■ Type an asterisk followed by a space to trigger a bulleted list:
	
	`* my list item`
	
	■ Type an asterisk followed by text, then another asterisk, to get bold formatting:
	
	`*my bold text*`
		Autoformatting is available only in Confluence 4. You can disable the autoformatting feature. Go to the "**Editor Settings**" section of your user profile.

Use autocomplete	Just start typing the wiki markup then follow the prompts. For example: ■ Start typing `{children` and then follow the prompts to insert the children macro. ■ Start typing `[my text` and follow the prompts to insert a hyperlink.	Confluence autocomplete options are more extensive in Confluence 4 than in 3.5. This feature is useful for more advanced formatting and for adding macros. As you type, a dropdown list will appear with options matching what you type. Use the arrow keys or the mouse the choose the appropriate option.

Figure 6.4 shows the autocomplete feature in action in the Confluence 4 editor. I want to add a hyperlink, so I type a square bracket and some text. Confluence offers me some pages and images that I may like to link to. If none of those is what I want, I can bring up the link browser by choosing one of the options from the bottom section of the dropdown list.

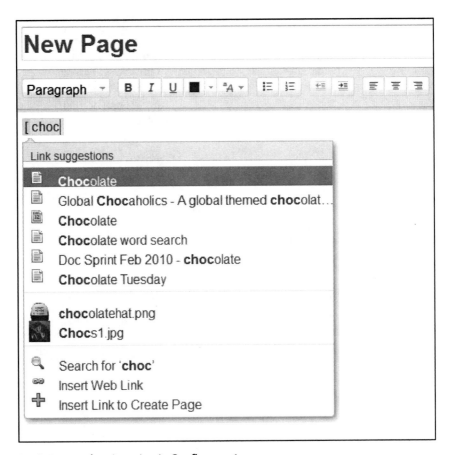

Figure 6.4. Autocomplete in action in Confluence 4

Editing a page

Ganache has gone right ahead and written her first page, called *Tweeting*. Let's see how she did it. First, here is a screenshot of the page in view mode (Figure 6.5).

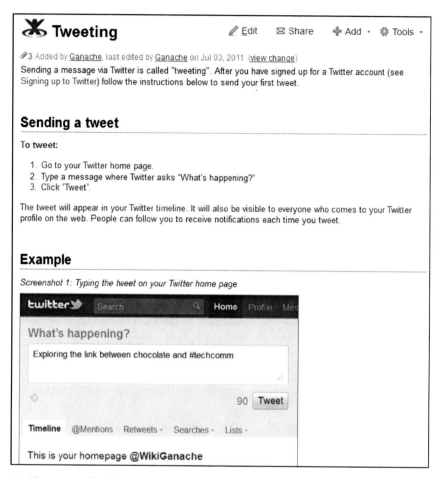

Figure 6.5. The page called "Tweeting" in Ganache's Twitter guide

It has all the usual things you would expect:

- An introductory paragraph.
- Headings.
- The step by step guide.
- Illustrations.
- Hyperlinks to related web pages and topics.

Example 6.1 shows the wiki markup for Ganache's page.

```
Sending a message via Twitter is called "tweeting". After you have signed
up for a Twitter account (see [Signing up to Twitter]) follow the
instructions below to send your first tweet.

h2. Sending a tweet

*To tweet:*
# Go to your Twitter home page.
# Type a message where Twitter asks "What's happening?"
# Click "Tweet".

The tweet will appear in your Twitter timeline. It will also be visible to
everyone who comes to your Twitter profile on the web. People can follow
you to receive notifications each time you tweet.

h2. Example

_Screenshot 1: Typing the tweet on your Twitter home page_

!FirstTweet.png|border=2,bordercolor=gray!
\\
\\
\\
_Screenshot 2: The tweet in your Twitter timeline_

!TwitterTimeline.png|border=2,bordercolor=gray!
\\
\\
\\
_Screenshot 3: The tweet seen at its own URL_

!FirstTweetSent.png|border=1!

You can see Ganache's tweet at this URL:
[http://twitter.com/#!/WikiGanache/status/87432838726696961]

h2. Want to know more?

[Using Twitter on your desktop or phone]
```

Example 6.1 – Wiki markup for Ganache's Twitter page

That is the original markup, as Ganache first wrote it. You can see the page in its current state on the *Confluence, Tech Comm, Chocolate* wiki.[10] It has probably changed a bit by now!

Let's assume that you want to recreate Ganache's page on your own Confluence site. You can do that either by adding a page or by editing an existing page.

First, a quick refresher on how to add a page: Go to the parent page, then choose "**Add**" > "**Page**."

To edit a page: Go to the page, then choose "**Edit**."

If you cannot see the add or edit options, the most likely reason is that you do not have the "Add page" permission in the space. The "Add page" permission covers editing as well as adding pages. **Hint:** Check that you are logged in to Confluence.

To add the wiki markup to your page:

- If you are using Confluence 3.5: Edit the page, choose the "**Wiki Markup**" editor, and type or paste the text into the editing pane. Click "**Save**."

- If you are using Confluence 4: Edit the page, choose "**Insert**" > "**Wiki Markup**," and type or paste the text into the wiki markup pane. Click "**Insert**" then "**Save**."

Now you have the text, but what about the screenshots? On to the next section.

Getting the wiki markup or source of a page

In Confluence 3.5, you can grab the wiki markup of any page that you have permission to view. Go to the page, then choose "**Tools**" > "**View Wiki Markup**." Copy the text and paste it into your own page.

In Confluence 4 you can similarly grab the source of a page. Go to the page, then choose "**Tools**" > "**View Source**."

Adding screenshots and other images to your page

One way to display an image is via the "Insert Image" option in the editor's toolbar. The image browser will pop up and guide you through the steps.

To add an image to your page:

- In Confluence 3.5: Edit the page and place the cursor where you want the image. Choose the image icon on the toolbar, browse to the image on your file system and upload it, then choose "**Insert**."

- In Confluence 4: Edit the page and place the cursor where you want the image. Choose "**Insert**" > "**Image**" on the toolbar, browse to the image on your file system and upload it, then choose "**Insert.**"

At this point it is useful to know about attachments. You can attach a file to a Confluence page and then use the attached file, or "attachment," in various ways. If the attachment is an image, then the most common use is to display the image on the page. If you insert your images via the image browser, it will attach the image to the page for you. Once the file is attached to the page, you can see and manipulate it on the page's "Attached Files" view.

To see and manage the attachments on a page: Choose "**Tools**" > "**Attachments.**"

Here are some more hints about working with images in Confluence:

- Instead of browsing to an image file and uploading it, you can simply drag and drop it onto the Confluence screen. Try it in edit mode, in view mode, and on the page's attached files view. You can even drag and drop an image file onto the image insertion dialog.

- In Confluence 4, you can copy and paste an image directly into the editor.

- The image does not have to be attached to the page. You can display an image that is attached to another Confluence page or an image from somewhere on the web via a URL.

Drawing diagrams

This is where things start to get really exciting! Up to now, you have added some text and a static image to your page. That is standard stuff. Any content management system can do that.

How about a tool that lets you draw a diagram directly on the web page using just your browser? Would you like to build a wireframe as part of your user interface design? The drawing tool is part of the web page. You do not need to install any software on your computer. You can edit the diagram from any computer, anywhere, any time. Anyone who has permission to edit the page can also edit the diagram. Collaboration has just taken a big step forward.

This is where Confluence plugins come into their own. Let's take a look at a plugin called *Gliffy*, which adds the ability to design and develop diagrams on a Confluence page. Then I will list other useful plugins that perform similar wonders.

Gliffy is developed and sold by a company called Gliffy. It is not an Atlassian product, and it is one of the many excellent plugins developed to extend the functionality of the wiki. There are a few ways to use *Gliffy*. You can go to the *Gliffy* website at `http://www.gliffy.com` and draw diagrams online. This offers a way to collaborate with your colleagues and publish the diagram online or export it as an image for incorporation into a document. You can install the *Gliffy plugin* into Atlassian's JIRA issue tracker, and draw diagrams on your JIRA screen. You can use *Gliffy* in Google Apps. And, of most interest to us in this book, you can install the *Gliffy plugin* into Confluence.

Installing the Gliffy plugin

In order to use a plugin, you first need to install it into your Confluence site. Skip this section if you already have *Gliffy* installed.

You can install and use *Gliffy* for free for an evaluation period of thirty days. (This is correct at time of writing. Please refer to the *Gliffy* website for up to date information.) Installing *Gliffy* is the same as installing any other Confluence plugin. You need Confluence system administrator permission to do this.

To install a plugin:

1. Log in as a Confluence system administrator.

2. Choose "**Browse**" > "**Confluence Admin**" > "**Plugins**" > "**Install.**"

3. Type the name of the plugin (in this case, "**Gliffy**") into the search box and click "**Search.**"

4. Click the name of the plugin in the list of search results, to open the panel showing the plugin details.

5. Click "**Install Now.**"

The *Confluence Administrator's Guide* has the details on installing plugins.[11]

Adding a Gliffy diagram to your page

Ganache decides to add a diagram to her Twitter guide, illustrating her vision for *Choc Ink*'s documentation wiki and community. Figure 6.6 shows what her diagram looks like to someone viewing the Confluence page.

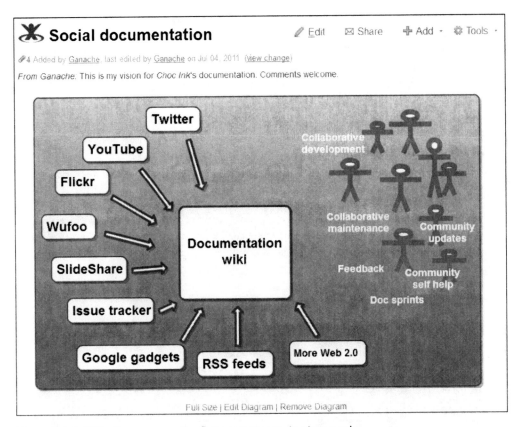

Figure 6.6. A Gliffy diagram on a Confluence 3.5 page, in view mode

To add a diagram or edit an existing diagram, you need permission to edit the Confluence page.

To add a Gliffy diagram to a page: Go to the page and choose "**Add**" > "**Gliffy Diagram.**"

The way you edit a *Gliffy* diagram is slightly different in Confluence 3.5 and Confluence 4. In Confluence 3.5 you can open the *Gliffy* editor while viewing the Confluence page. In Confluence 4 the *Gliffy* editor is more closely integrated with the Confluence editor so that you can add or edit a diagram both when the page is in view mode and when it is in edit mode.

To edit a Gliffy diagram in Confluence 3.5: Go to the page containing the diagram, and click the "**Edit Diagram**" link under the image.

To edit a Gliffy diagram in Confluence 4: Go to the page containing the diagram, move your mouse pointer over the diagram to invoke the edit options, then click "**Edit**." If you are editing the page, click somewhere on the diagram to invoke the edit option.

Once you are in the diagram editor, you will feel right at home. So at home, in fact, that you may feel a moment of disorientation. It is as if you have been transferred magically to the world of desk-top drag-and-drop diagram development, complete with smart connectors, variable arrow heads, and libraries of shapes to fit every need, from the the elliptical to the esoteric.

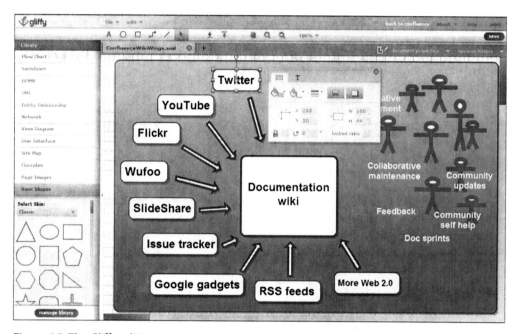

Figure 6.7. The Gliffy editing environment

Figure 6.7 shows the *Gliffy* editing environment using *Gliffy 3.5.2* in Confluence 3.5. Here is a short tour of what you can see:

- The right-hand pane is the canvas, where you draw your diagram. You can drag the vertical separators between the panes so that the canvas area suits your needs.

- On the left are the shape libraries. The "Basic Shapes" library is currently open. Other libraries provide shapes for flow charts, swimlanes, UML, and more. You can drag and drop shapes from the library onto your canvas.

- The "Twitter" shape is selected. The tool box next to it displays the properties and options available for the shape currently selected.

- The toolbars at the top offer more options, including text insertion, standard shapes, connectors, and so on.

Note that the *Gliffy* editor is constantly being enhanced. Drop in on the *Gliffy* website to see the latest news.

What you have learned about developing content

Now you know the answer to the big question, "Who killed wiki markup?" Nobody did. Confluence 4 has just one editor, whereas earlier versions offered two editors to choose from. Confluence 4 stores all content in XHTML, whereas earlier versions stored it in wiki markup. But the Confluence 4 editor recognizes wiki markup on data entry and converts it on the fly to rich text. You can also insert a block of wiki markup code using the **"Insert Wiki Markup"** option.

Why has wiki markup taken a back seat in Confluence 4? The simple, easy-to-read, fast-to-use syntax is popular with power users, but it forces collaborators to learn a new syntax that they may use only infrequently. In this chapter you examined the strengths and weaknesses of wiki markup and learned how it affects technical communicators.

Armed with quick reference guides to wiki markup and the Confluence editor, you followed Ganache as she added text and images to her page. You saw how to manage images when they are attachments on a page. To add extra shine to your editing experience and to your end product, you took a look at the plugins available to extend the Confluence editor. One example is *Gliffy* for drawing diagrams right there on the web page, as discussed in depth in this chapter. Other plugins are available on the *Atlassian Plugin Exchange*.

As a bonus, you saw how to install a plugin into Confluence, something that only Confluence system administrators can do. Now, to misquote Professor Henry Jones again, let's leave this chapter just as it is becoming interesting. Our next quest is for structure. The search for structure is the search for the divine in all of us!

References

These are the tools, websites, blog posts, and other references mentioned in this chapter.

Useful plugins

These plugins extend the Confluence editor so that you can draw diagrams and user interface (UI) mockups from within your browser.

- *Balsamiq Mockups For Confluence*, a Confluence plugin for mocking up a user interface design: `https://plugins.atlassian.com/plugin/details/256`

- *Creately for Confluence*, a Confluence plugin for creating diagrams: `https://plugins.atlassian.com/plugin/details/326582`

- *Gliffy Confluence Plugin*, an online tool and Confluence plugin for creating diagrams: `http://www.gliffy.com/`

- *LucidChart for Confluence*, a plugin for creating diagrams in Confluence and inserting diagrams from an existing LucidChart account: `https://plugins.atlassian.com/plugin/details/350288`

- To find the full list of diagram tools at the *Atlassian Plugin Exchange*: Go to `http://plugins.atlassian.com`, type "**diagram OR mockup**" in the search box and select "**Confluence,**" then click "**Search.**"

A new plugin for Confluence 4 is currently under development by Arsenale that will allow authors to edit the XHTML source of the page: *Arsenale Invisible Ink* (`https://plugins.atlassian.com/plugin/details/918877`). This plugin will offer an alternative to the default Confluence 4 rich text editor. At the time that this book is going to press, the plugin is in alpha testing. The initial release is planned for January 2012. The aim is to provide power users with a rapid text editor, including a quick way of finding and editing macro parameters.

Endnotes

[1] *Confluence User's Guide*: `http://confluence.atlassian.com/display/DOC/Confluence+User%27s+Guide`

[2] *What part does chocolate play in your life as technical writer*, a discussion on *Technical Writing World*, started in May 2011: `http://technicalwritingworld.com/forum/topics/what-part-does-chocolate-play`

[3] *Textile* markup language and tool on *Textism*: `http://textism.com/tools/textile/`

[4] *MediaWiki*'s markup language: `http://en.wikipedia.org/wiki/Help:Wiki_markup`

[5] *Creole*, a common wiki markup language: `http://wiki.wikicreole.org/`

[6] *Confluence User's Guide*: `http://confluence.atlassian.com/display/DOC/Conflu-ence+User%27s+Guide`

[7] HTML entities, on the *W3C* website: `http://www.w3.org/TR/html4/sgml/entities.html`

[8] Special HTML characters, on *mistywindow.com*: `http://www.mistywindow.com/reference/-html-characters.php`

[9] *What part does chocolate play in your life as technical writer*, a discussion on *Technical Writing World*, started in May 2011: `http://technicalwritingworld.com/forum/topics/what-part-does-chocolate-play`

[10] Ganache's Twitter guide, on the *Confluence, Tech Comm, Chocolate* wiki: `https://wikitech-comm.onconfluence.com/display/DOC/Tweeting`

[11] Page about installing plugins, in the *Confluence Administrator's Guide*: `http://confluence.at-lassian.com/display/DOC/Installing+a+Plugin`

7

Structure and style

Isn't a wiki just a puddle of chaos? Doesn't it always look like an unimaginative scrabble of words, with no form to enhance the meaning? Not necessarily! If you have not yet had a chance, take a look at the wiki documentation sites mentioned in Chapter 2, for some great examples of stylish, structured wikis.

In search of structure, style and branding

Let's see how to add structure, style, and branding to your wiki documentation. We start with pages, then move up to spaces, then go to the wiki as a whole. This chapter has the following sections:

- **Adding structure to a page.** A wiki page is a web page, so it supports the standard HTML layout features. Confluence macros provide additional layout capabilities.

- **All about macros.** Become a macro magician, with the help of a quick reference guide and some specific examples of macros that are handy when designing a page layout. Bring a smile to a Confluence developer's lips by mentioning the cheese macro in context.

- **User macros.** A user macro is a macro that you can define yourself via the Confluence administration screens, without having to add a plugin. Let's take a look at the expand macro for Confluence 3.5, and try our hand at writing a chocolate macro.

- **Structuring a space with the Documentation theme.** The simplest and most effective way to customize the look and feel of a space is to apply a theme. The Documentation theme is supplied with Confluence. Learn how to add a customized header, footer, and navigation panel to all the pages in your space.

- **Structuring your chapters and sections.** Build your manual by defining a tree of pages, each branch representing a manual or a section in the manual. Learn how to move pages around in the tree and how to put pages into a logical order.

- **Structuring your wiki.** Spaces are the primary tool for structuring a Confluence wiki. This section is a quick refresher on dividing your content into spaces.

- **Categorizing your spaces.** See how to put your spaces into categories so that readers can choose to see just the spaces relevant to them.

- **Customizing the appearance of the wiki.** It may be useful to apply your own branding, styles, and colors to the documentation site. These are the customizable areas of Confluence: Themes, Velocity layouts, style sheets (CSS), color schemes, dashboard, site home page, and logos.

- **What you now know about structuring and styling your wiki.** A summary of this chapter.

- **References.** The tools, websites, blog posts, and other references mentioned in this chapter.

Adding structure to a page

Because a wiki page is basically a web page, it provides everything that HTML provides, including heading levels, bulleted and numbered lists, and tables. The previous chapter described how to add such formatting in the wiki markup editor and the rich text editor.

Confluence provides macros that bundle together some formatting attributes for use as warnings, tips, notes, and information callouts. Other macros add content that will be generated when the page is rendered, such as a table of contents or a list of child pages. A number of macros are part of the core Confluence installation. You can add more macros by installing plugins, and you can even define your own macro on your Confluence site by adding a user macro. For example, you can use a user macro to define a standard way of formatting a company logo.

All about macros

A Confluence macro is a small piece of code – or an entire application – that does something. Its purpose may be as simple as displaying a piece of text or drawing a panel around a block of text. Or it may open up a new application, such as a drawing tool, within your web browser.

Wiki markup is even less dead than we first thought

It is worthwhile learning wiki markup. There are a few places in Confluence 4 that still accept and retain wiki markup in input fields, so that you can both enter wiki markup and edit it again later as wiki markup. The configuration screen for the Documentation theme is one of those places. Other examples are the user profile, the site welcome message, the default space content screen, and page templates (discussed in the next chapter). At some stage a future release of Confluence may provide an alternative format, such as rich text, for these fields.

Remember also that the editor's auto-formatting and autocomplete features will respond to wiki markup cues. And you can insert wiki markup into a page at any time, using the "Insert Wiki Markup" option on the editor toolbar. In this case, the wiki markup is converted to the rich text format used by the editor. The wiki markup code is no longer available for editing. See Chapter 6 for more about these options.

In wiki markup and in the Confluence 3.5 editor, a macro is represented by a word or words in curly brackets. The Confluence 4 editor provides a more graphical representation of a macro, using "placeholders" to draw a border around the macro and its contents.

A macro quick reference guide

Table 7.1 shows some useful macros for adding structural elements to your page and for representing your wiki's structure on the page.

Table 7.1 – Macro quick reference guide

What you want	Macro	Example in wiki markup
A panel around a block of text	panel	{panel:title=My title}My text.{panel}
An information panel	info	{info:title=My title}My info.{info}
A tip panel	tip	{tip:title=My title}My hint.{tip}
A note panel	note	{note:title=My title}My note.{info}
A warning panel	warning	{warning:title=My title}My caution.{warning}
Text displayed exactly as entered	noformat	{noformat}My unformatted text.{noformat}
Sections and columns	section column	{section} {column}My column 1{column} {column}My column 2{column} {section}
A list of the headings that appear on the page, thus forming a table of contents for that page	toc	{toc}
A list of the child pages of the current page, or of a given page	children	{children}
A list of pages tagged with a label	content-by-label	{contentbylabel:label=techcomm}
A list of all spaces on the wiki	space-list	{spaces}

As you can see, wiki markup does not differentiate between the opening and closing tags of an element. In many cases, you do not need a closing tag. Where required, the opening and closing tags are the same, with the exception that any parameters occur in the opening tag only. The panel macro, shown in the above table, is an example of a macro that has an optional parameter ("title") and a closing tag.

The *Confluence User's Guide* has the full list of macros shipped with Confluence.[1] As noted in Chapter 5, your Confluence site may acquire new macros when you install plugins. These additional macros are not documented in the *Confluence User's Guide*, since they are not shipped with the wiki. The plugin developer probably provides documentation on their own website. You can also pull up the online quick reference guide to the markup for your own Confluence site.

To see the quick markup guide for your site:

1. Go to this URL in your web browser: `http://MY.CONFLUENCE.SITE/renderer/notation-help.action`, where MY.CONFLUENCE.SITE is the address of your Confluence site. If you are running Confluence on your own computer, the address is `http://localhost:8090/renderer/notationhelp.action`. (For Confluence 3.5, the default port is `8080` instead of `8090`.)

2. The notation guide will appear. Choose "**ALL**" or a specific section in the left-hand panel.

The information will be in the guide if the plugin developers have coded it into their plugins.

How to add a macro to a page

If you are a wiki markup magician using Confluence 3.5, just enter the macro code and its parameters between curly brackets: {`mymacro:parameter1|parameter2`}. Alternatively, the macro browser is available in both Confluence 3.5 and Confluence 4.

To add a macro using the macro browser in Confluence 3.5:

1. Edit the page and place your cursor in the right spot.
2. Click the "**Insert/Edit Macro**" icon in the editor toolbar.
3. A list of macros will appear. Scroll down or enter text to find the macro you need.
4. Click the macro block and supply values for the parameters as prompted.
5. Click "**Insert**" to add the macro to the page.
6. Save the page.

To add a macro using the macro browser in Confluence 4:

1. Edit the page and place your cursor in the right spot.
2. Click "**Insert**" > "**Other Macros**" in the editor toolbar.
3. A list of macros will appear. Scroll down or enter text to find the macro you need.
4. Click the macro block and supply values for the parameters as prompted.
5. Click "**Insert**" to add the macro to the page.
6. Save the page.

The autocomplete functionality, discussed in the section titled "Editor quick reference guide," works for macros too. Instead of opening the macro browser via the toolbar, just start typing a curly bracket and the macro name. A list of options will appear, offering a shortcut to the macro browser.

Using a macro to add a table of contents to your page

In her Twitter guide, Ganache has written a page about Twitter linguistic conventions.[2] There are a number of sections on the page, so Ganache decides to add a list of sections at the top of the page. This will give her readers an idea of the structure and content of the page. Ganache uses the table of contents macro, or toc macro, instead of creating the list manually. In future, when other people add more sections or rename existing sections, there will be no need to change the list at the top of the page.

Figure 7.1 shows what Ganache's page looks like in view mode.

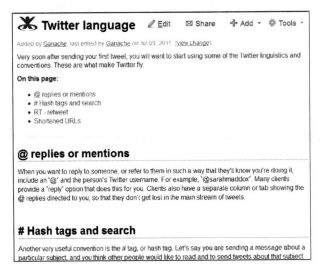

Figure 7.1. The table of contents at the top of the "Twitter language" page

In Confluence 3.5, in both the wiki markup editor and the rich text editor, you will see the {toc} macro markup. Figure 7.2 shows the wiki markup editor.

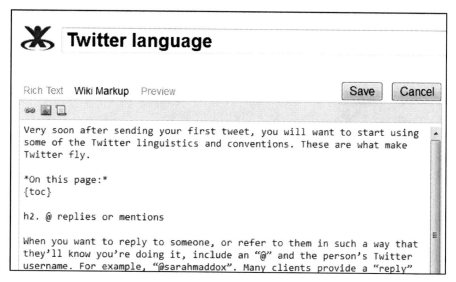

Figure 7.2. The toc macro markup in the Confluence 3.5 wiki markup editor

In the Confluence 4 editor, the macro is displayed as a visual element on the page (see Figure 7.3).

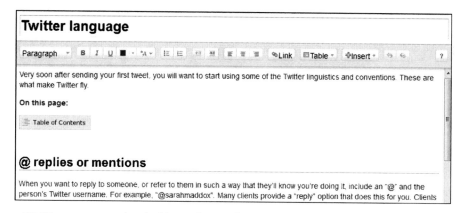

Figure 7.3. The toc macro placeholder in the Confluence 4 editor

Using macros to add columns to your page

To divide your content into columns on a page, you need to define a section and then define the columns within the section. You can have as many sections as you like on the page and as many columns as you like within each section.

Ganache has used the section and column macros to divide her *Introduction to Twitter* page[3] into two columns.

Figure 7.4 shows the page in view mode. The left-hand column contains the text, while the right-hand column displays two images of Tweets. Each Tweet is inside a panel.

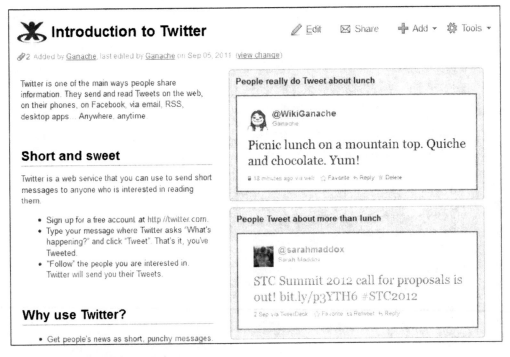

Figure 7.4. A page with two columns

In Confluence 3.5, in both the wiki markup editor and the rich text editor, you will see the markup for the section, column and panel macros. Figure 7.5 is a screenshot of the wiki markup editor.

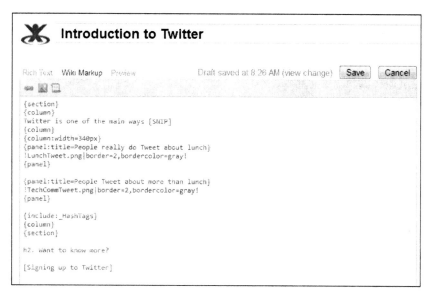

Figure 7.5. The Confluence 3.5 wiki markup editor showing the section, column and panel macros

In Confluence 4, macros are wrapped in a placeholder to set them apart visually (Figure 7.6).

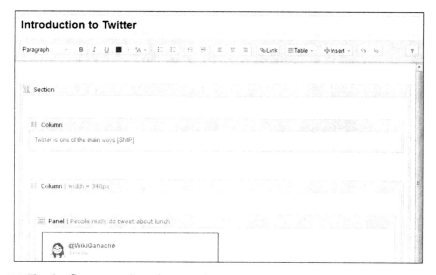

Figure 7.6. The Confluence 4 editor showing the section, column and panel macros (partial)

There are plans afoot to add "page layouts" to the Confluence 4 editor, as a better way of defining columns and sections. At the time of writing this chapter, development is underway but we do not know yet when the feature will be released in Confluence.

A bit more about Gliffy

In Chapter 6, Ganache added a Gliffy diagram to her page. What we did not mention is that Gliffy works by adding a macro and two attachments to the page. Once you know that, you have the power to move the Gliffy diagram to a different place on the page or even copy the macro and attachment to another page.

If you edit a page that contains a Gliffy diagram in Confluence 3.5, you will see a macro that looks something like this:

```
{gliffy:name=ConfluenceWikiWings.xml|align=left|size=L|version=2}
```

In Confluence 4 you will see the diagram itself right there in the editor.

In both versions of Confluence, you will see the Gliffy data attachment (XML) and image file (PNG) attached to the page.

The cheese macro

Ganache had a bit of a chuckle when she discovered the cheese macro in Confluence. All it does is display the words, "I like cheese!" If you ask Atlassian about the purpose of the macro, the official line is that it provides a simple, quick example for people who want to try out a macro on their wiki page.

As an Atlassian myself, I know that the macro has become an "in" joke. If you know about it, you are one of the inner circle. If you can go up to a Confluence developer and say "I like cheese" with a straight face, you have made a friend for life.

 Origin of the cheese macro

Charles Miller is a long-time Atlassian, one of the very first developers in the Confluence team, and currently the Confluence architect. When I asked, he admitted with characteristic charm and candor that he developed the cheese macro:

"That would be me.

"The cheese macro is a piece of temporary testing code that I forgot to delete. I was writing the feature that allows you to define user macros in regular plugin XML files instead of having to enter them through the web interface, and to test that it worked I stuck a simple 'print out some random text' macro at the end of one of Confluence's internal macro definition files.

"Then I committed that file to source control. I realized my mistake pretty quickly, but some things are too funny to remove.

"Developers have to come up with a /lot/ of random text for testing and examples. It's easier to have a couple of standby names and phrases so you don't have to think about it. 'I like cheese' is the Charles equivalent of 'lorem ipsum.'"

User macros

A "user macro" is a macro that you (the "user") can define by typing the code into a form on the Confluence administration screen. Confluence will store the code in the database and execute it whenever someone views a page that uses the macro. It is called a user macro to distinguish it from macros that are shipped with Confluence or added via plugins.

In most cases, user macros perform simple functions such as formatting text or displaying a piece of standard text. For example, you may want to define a standard way of showing your company's name wherever it appears on a wiki page. Or, you may want to add a simple feature that is not in the Confluence user interface, such as opening a link in a new window or providing a tool tip when users hover their mouse over a section of text. Many people write user macros and make them available to other Confluence users. The Atlassian documentation wiki has a section for these shared user macros.[4]

You need Confluence system administrator rights to add a user macro. Once you have added the user macro, other people can use it on their pages. The *Confluence Administrator's Guide*[5] has the details.

Writing the chocolate macro

Inspired by the cheese macro, Ganache decides to write a user macro that displays "I like chocolate." Ganache's macro does not need any information passed via parameters, so she adds the @noparams instruction to the macro template. The macro is a little more sophisticated than the cheese macro, in that it displays its text inside a panel, with width restricted to 150 pixels.

There are some differences in the method of defining user macros in Confluence 3.5 and Confluence 4. The primary change is that, whereas Confluence 3.5 accepts wiki markup in the macro definition, Confluence 4 expects XHTML. The indomitable Ganache has tried it in both.

To add the chocolate macro in Confluence 3.5:

1. Choose "**Browse**" > "**Confluence Admin**" > "**User Macros**" > "**Create a User Macro.**"

2. Enter the macro attributes:
   - Macro Name: `chocolate`
   - Visibility: `Visible to all users in the Macro Browser`
   - Macro Title: `chocolate`
   - Description: `Displays "I like chocolate!" in a panel`
   - Categories: `Confluence Content`
   - Icon URL: (Not necessary.)
   - Documentation URL: (Not necessary.)
   - Macro Body Processing: `No Macro Body`
   - Output Format: `Wiki Markup`
   - Template:

```
## @noparams
{section}
{column:width=150px}
{panel}*I like chocolate!*{panel}
{column}
{column}
{column}
{section}
```

3. Click "**Save.**"

To add the chocolate macro in Confluence 4:

1. Choose "**Browse**" > "**Confluence Admin**" > "**User Macros**" > "**Create a User Macro.**"

2. Enter the macro attributes:
   - Macro Name: `chocolate`
   - Visibility: `Visible to all users in the Macro Browser`
   - Macro Title: `chocolate`
   - Description: `Displays "I like chocolate!" in a panel`
   - Categories: `Confluence Content`
   - Icon URL: (Not necessary.)
   - Documentation URL: (Not necessary.)
   - Macro Body Processing: `No Macro Body`

- Template:

```
## @noparams
<div class="panel" style="border-width: 1px; width: 150px">
  <div class="panelContent">
    <p><b>I like chocolate!</b></p>
  </div>
</div>
```

3. Click "**Save**."

The screenshot in Figure 7.7 shows the top part of the user macro form in Confluence.

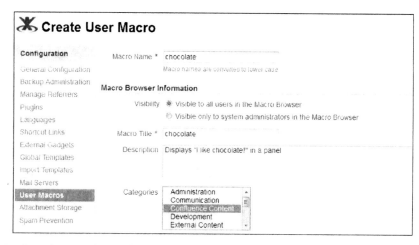

Figure 7.7. Partial screenshot - adding a user macro

Next Ganache adds the chocolate macro to a page (Figure 7.8). You can find this very page in the *Chat and play* space on the *Confluence, Tech Comm, Chocolate* wiki.[6]

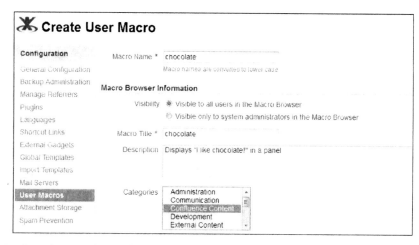

Figure 7.8. The chocolate macro in action

Adding the expand macro for use on your Confluence 3.5 site

The expand macro (see Figure 7.9) is a very useful macro, particularly for technical documentation. You can use it to define a section of text that the reader can show and hide by clicking a link. The expand macro is developed and supported by Atlassian.

In Confluence 4 the expand macro is bundled with Confluence. In Confluence 3.5, you can add it yourself as a user macro.

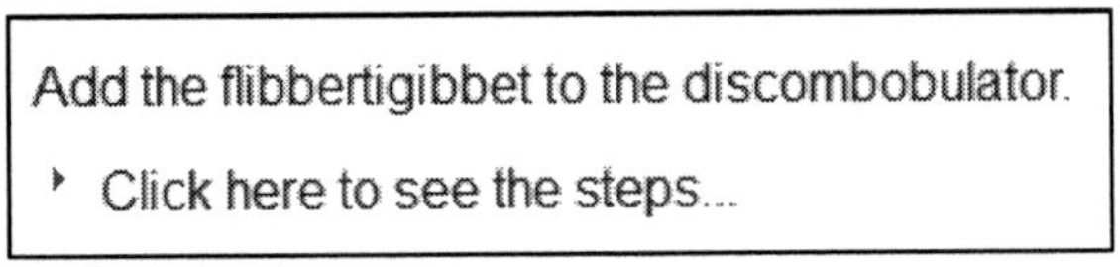

Add the flibbertigibbet to the discombobulator.

▸ Click here to see the steps...

Figure 7.9. The expand macro in action

To add the expand macro to your Confluence 3.5 site, follow the instructions in the *Confluence Administrator's Guide* at `http://confluence.atlassian.com/display/CONF35/Writing+the+Expand+User+Macro`.

Structuring a space with the Documentation theme

The structure of chocolate

Did you know that the structure of the chocolate particles affects the taste and texture of the chocolate and that a good part of the chocolate-making process is designed to change its structure? Refining, conching, and tempering affect the crystallization of the cocoa butter particles. Cadbury[7] and *wikiHow*[8] tell us how it is done. Gummy or snappy, dry-as-dust crumbly or melt-on-the-tongue smooth, it is all in the structure.

Now that Ganache has the structure of her pages under control, she turns her attention to her documentation space.

The simplest and most effective way to customize the look and feel of your space is to apply a theme. When you install Confluence, you get a set of default themes. You can also add other themes in the form of plugins. See the section titled "Customizing the appearance of the wiki," later in this chapter, for a detailed description.

The Documentation theme is bundled with Confluence. You do not need to install it separately. However, it is not the default theme, so you need to apply it to your site or your space. As with most themes, you can choose to set it up for the whole site or for individual spaces. In this section I will show you how to configure the Documentation theme for a single space. For instructions on configuring a theme at site level, see the section titled "Themes," later in this chapter.

To change the theme of a space: Go to a page in the space, choose **"Browse"** > **"Space Admin"** > **"Themes,"** then choose the theme.

Ganache applied the Documentation theme when she added the space for her Twitter guide (See Chapter 5). Now she adds a copyright statement to the footer of every page by configuring the options in the Documentation theme.

To configure the options in the Documentation theme for a space:

1. Go to a page in the space and choose **"Browse"** > **"Space Admin"** > **"Themes."**
2. If the Documentation theme is not already the current theme, apply it now.
3. Choose **"Configure theme"** to customize the theme options.

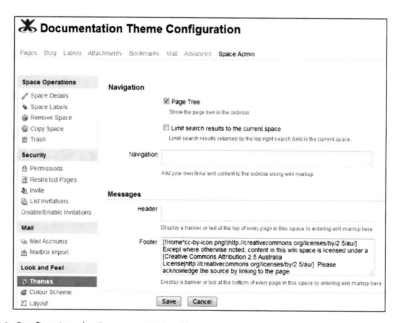

Figure 7.10. Configuring the Documentation theme

There are options (see Figure 7.10) to show or hide the table of contents and to restrict the Confluence search to the current space only. There are also text boxes to specify the content for the left-hand panel and the page header and footer. These settings apply to the entire space.

Ganache types the text for her copyright statement into the footer box. She uses wiki markup to include an image (the cc-by icon), a hyperlink, and the text.

```
[!Home^cc-by-icon.png!|http://creativecommons.org/licenses/by/2.5/au/]
Except where otherwise noted, content in this wiki space is licensed under
a [Creative Commons Attribution 2.5 Australia License|
http://creativecommons.org/licenses/by/2.5/au/]. Please acknowledge
the source by linking to the page.
```

Did you notice that the image (cc-by-icon.png) is attached to a page called Home? Ganache has attached the image to the *Home* page in the space, so Confluence can find it. Now every page in Ganache's Twitter guide has a copyright statement at the bottom (see Figure 7.11).

Figure 7.11. Copyright statement in the page footer

Structuring your chapters and sections

In Chapter 5, you learned that when you add a page, it becomes a child of the page you are currently viewing. You can build up a family, or tree, of pages to form the structure of your manual. For example, one page may introduce the administration guide, another page at the same level may introduce the installation guide, and yet another the user guide. The child pages of the administration guide introduce the sections in the guide and their children hold the detailed instructions.

The screenshot in Figure 7.12 shows part of the table of contents in the documentation space for a product called *Crowd*. The following guides are at the top level of the tree:

- *Crowd 101*
- *Crowd Administration Guide*
- *Crowd Installation and Upgrade Guide*
- *Crowd User Guide*

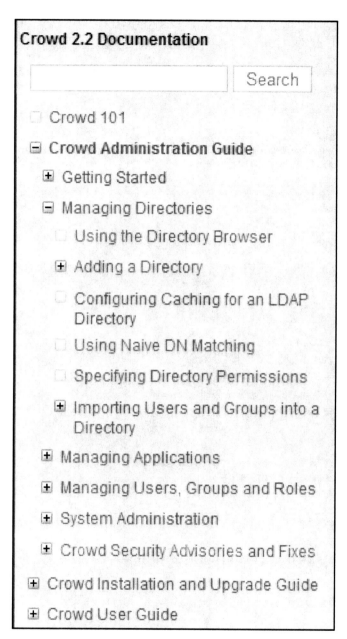

Figure 7.12. A page tree showing the Crowd documentation

The plus signs in the screenshot represent branches that have pages within them. On the user interface, people can click a plus sign to expand the branch and click a minus sign to collapse the branch.

Once you have added your pages, you can move them around within the space or into another space. For example, if you decide that *Crowd 101* is actually part of the user guide, you would move it into the appropriate branch of the tree.

The sequence of the pages is also very important in technical documentation. In some wikis, the only possible page order is alphabetical. In Confluence, the default order is alphabetical, but you can define a logical order by moving the pages into position. Confluence will respect the new order in all the relevant places, including the page tree, the macro that lists the children of a page, and the PDF export.

To move pages around within a space:

1. Go to a page in the space, then choose "**Browse**" > "**Pages**" > "**Tree**."

2. Click the plus signs to expand the branches of the page tree.

3. Drag and drop the pages into position.

Hint: To abandon the drag and drop action, press Escape on your keyboard. The page that you are dragging will return to its previous position.

You can also reorder a page using the "Move" dialog.

To move and reorder a page:

1. Go to the page.

2. Choose "**Tools**" > "**Move**."

3. Choose the new location, or leave the location as it is.

4. Tick the "**Reorder**" check box.

5. Click "**Move**."

6. Move the cursor over the list of pages. A horizontal bar will appear. Put that bar in the place where you want the page to be.

7. Click "**Reorder**."

Let's take a look at Ganache's Twitter guide. When she created her documentation space in Chapter 5, she added some skeleton pages to form her draft table of contents (see Figure 5.7). Since then, she has added more pages. Now she wants to put them in the right order.

Figure 7.13 shows what Ganache's page tree looks like at the moment:

Figure 7.13. The Twitter guide page tree before reorganization

Ganache moves the page called *Getting started on Twitter* up into position just after *Introduction to Twitter*. Next she moves three pages underneath *Getting started on Twitter* so they become children of that page. In the screenshot below, she is in the process of moving the page called *Using Twitter on your desktop or phone*. When you drag a page over another page, a gray highlighted area appears for you to drop the child pages into (see Figure 7.14).

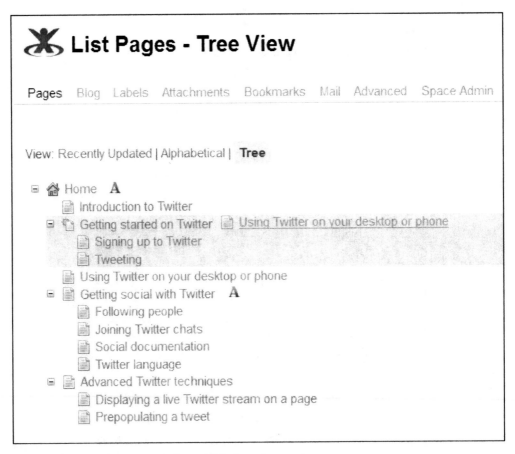

Figure 7.14. Dragging a page to be a child of another page

When Ganache has finished organizing her pages, the Twitter guide's left-hand navigation panel looks like Figure 7.15:

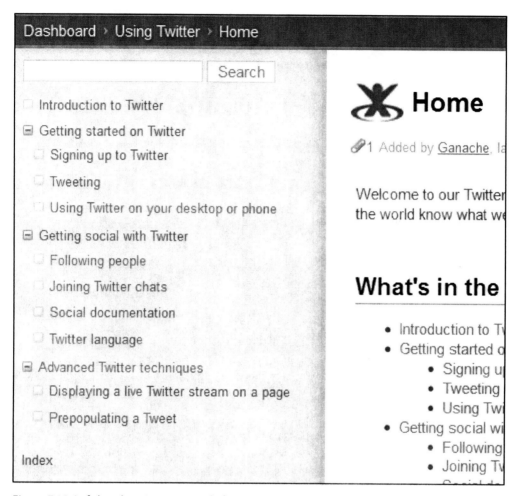

Figure 7.15. Left-hand navigation panel after reorganization

Note: It is not possible to move pages around directly in the left-hand panel. Instead, use the space's page tree view or the page's "Move" dialog as described above.

Structuring your wiki

The sections above give some ideas for structure within each page and then within each space. Now let's broaden our view to the wiki as a whole.

One of the primary tools that Confluence provides for structuring the wiki is the space. Chapter 4 described some techniques for dividing your content into spaces. The specific division of topics into spaces will depend on your own requirements and the scope of your documentation. Chapter 4 mentioned the most important things to consider: permissions, themes, and style, the number of products documented, version control, audience, and the limitation that page names must be unique within a space.

Once you have a number of spaces, how can you group related spaces together? There is no concept of nested spaces in Confluence. A page can be a child of another page, but a space cannot be a child of another space. Instead, you can put spaces into categories.

Categorizing your spaces

As your organization uses its Confluence site over time, the site will acquire more and more spaces. People, teams, and departments will quickly become enthusiastic and add their own spaces to hold their information, communicate with customers, and collaborate on content creation.

If there are a number of unrelated spaces on the same wiki, readers may find it confusing. They need a way to choose just the spaces that are relevant to them. To help them do that, you can categorize the spaces. Readers can then filter the selection of spaces that they see in the Confluence space directory.

The first task is to decide how you want to categorize your spaces. It is a good idea to plan your categories beforehand so they make sense to your readers and so there are not too many spaces in any one category. The example below may help.

First, here are the mechanics of categorizing spaces.

To add a space to a category: Go to a page in the space, then choose **"Space Admin"** > **"Space Labels"** > **"Space Categories."**

Adding a space to a category is the same thing as adding the category to the space. Each space can appear in any number of categories, and each category can have any number of spaces.

An example from the Atlassian documentation wiki

The Atlassian documentation wiki[9] has a number of spaces containing user guides for each product. The wiki also has a number of spaces that do not contain technical documentation.

It therefore makes sense to have at least the following categories:

- "Documentation" – the latest documentation spaces for each product.

- "Confluence," "Crowd," "JIRA," and more – a category for each product.

The screenshot in Figure 7.16 shows the space directory on the Atlassian documentation wiki with the "Documentation" category selected.

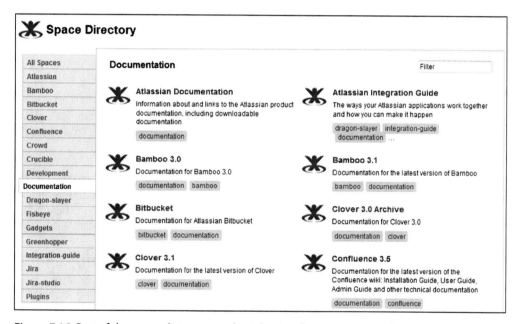

Figure 7.16. Part of the space directory on the Atlassian documentation wiki

To see the space directory on your Confluence site: Choose "**Browse**" > "**Space Directory.**"

The space directory is visible to all readers of the wiki. It shows the spaces that they have permission to see. People can choose to see all the spaces or limit their list to just one category. They can further reduce the list by typing text into the "**Filter**" box.

Customizing the appearance of the wiki

It is often useful to apply corporate branding to your wiki by customizing the site home page, colors and logos. These are the customizable areas of Confluence at space level and over the entire site:

- Themes
- Velocity layouts
- Style sheets (CSS)
- Color schemes
- Dashboard and site home page
- Site and space logos

Themes

A theme defines the color schemes, fonts, and layout of the pages in a space. Another name for a theme is a skin. Some themes offer additional functionality, such as a smart search or a left-hand navigation panel. When you first install Confluence there is a choice of themes available, and you can add more themes as plugins.

Earlier in this chapter we talked about applying the Documentation theme to change the style of a space. You can configure a theme to apply across the entire site instead of just one space. Some themes will change the appearance of the dashboard as well as all spaces. Other themes will affect the spaces only.

If the theme that you want is not bundled with Confluence, you will need to install it before you can apply it to the site or to a space.

To see what themes you already have on your Confluence site: Log in as a Confluence system administrator, then choose "**Browse**" > "**Confluence Admin**" > "**Themes**."

It is useful to search for themes on the *Atlassian Plugin Exchange* and read about them before making your choice of theme to install.

To find Confluence themes on the Atlassian Plugin Exchange: Go to http://plugins.atlassian.com, type "**theme**" in the search box and select "**Confluence**," then click "**Search**."

If you are a code ninja, or have some of those on tap, you may want to develop your own theme. To do that, you will need to develop a Confluence plugin. The *Atlassian Developers* site has a quick-start guide to developing with the Atlassian Plugin SDK[10] and a tutorial on creating a theme.[11]

Installing a theme is the same as installing any other Confluence plugin. You need Confluence system administrator permission to do this.

To install a theme onto your site:

1. Choose "**Browse**" > "**Confluence Admin**" > "**Plugins**" > "**Install**."

2. Type the name of the theme into the search box and click "**Search**."

3. Click the name of the plugin in the list of search results, to open the panel showing the plugin details.

4. Click "**Install Now**."

The *Confluence Administrator's Guide* has the details about installing plugins.[12] After installing the theme, you can apply it to the site or to a given space.

To apply a theme across the entire wiki site: Choose "**Browse**" > "**Confluence Admin**" > "**Themes**," then choose the theme.

To change the theme of a space: Go to a page in the space, choose "**Browse**" > "**Space Admin**" > "**Themes**," then choose the theme.

Velocity layouts

For the fearless code ninjas amongst us, Confluence offers some intriguing ways of customizing the layout of a space. If you know HTML and the programming language called Velocity (http://velocity.apache.org/), you can edit the space layouts or site layouts directly. These layouts, also called "decorators," contain the HTML and Velocity code that Confluence uses to generate the HTML that renders the page. There are a number of layout files, one for each area of a Confluence page or site.

To edit the space layouts: Go to a page in the space, choose "**Browse**" > "**Space Admin**" > "**Layout**," then choose the layout that you want to change.

Note that the layouts are not available for updating if the space uses a custom theme. They are available only with the default Confluence theme. Moreover, you need Confluence system administrator permission to edit the layouts. The *Confluence User's Guide* has the details.[13] Similarly, you can edit the layouts for the entire site.

My recommendation is to steer clear of customizing the Velocity layouts because you will need to re-apply your custom code each time you upgrade to a later version of Confluence. Moreover, the upgrade process will not upgrade any customized code. That means that you will have to upgrade each layout separately.

Style sheets (CSS)

You can adjust the style sheets used by the Confluence theme for each space and the global stylesheets used for the site as a whole. See the *Confluence User's Guide*.[14] I advise against customizing the CSS too heavily because of the potential that the theme's CSS will change with the next release of Confluence or of the theme plugin. Moreover, CSS is quite finicky. I am not sure that the result is worth the effort, unless your customization is very simple.

Color schemes

Confluence comes with a default map that associates a color with each element in the user interface (UI). Examples of such UI elements are the top bar on the screen, the text of the space name, the text of a heading, links, items in the page menu, and so on. This map is called a "color scheme." One color scheme applies across the site as a whole, and there are also separate color schemes for each space.

When using the default Confluence theme, you can choose different colors for each of the UI elements. When using a custom theme, you can use the color scheme belonging to that theme or create your own color scheme.

To change the color scheme across the entire wiki site: Choose "**Browse**" > "**Confluence Admin**" > "**Color Scheme**."

To change the color scheme of a space: Go to a page in the space, choose "**Browse**" > "**Space Admin**" > "**Color Scheme**."

Dashboard and site home page

There are two ways to customize the site home page. The first is to add your own content to the "site welcome message," which appears at the top left of the dashboard. The second is to change the default destination for people coming to the site so they go to a space's home page instead of the dashboard.

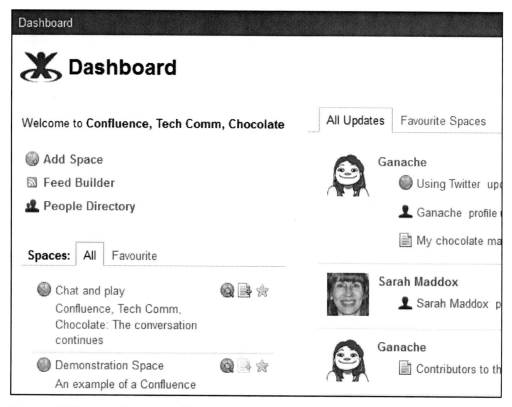

Figure 7.17. The top left section of the dashboard

The site welcome message occupies the top left section of the dashboard. When you first install Confluence, the message has some text that welcomes you to Confluence and explains a bit about the wiki. In Figure 7.17, the welcome message says, "**Welcome to Confluence, Tech Comm, Chocolate.**"

To change the site welcome message:

1. Choose "**Browse**" > "**Confluence Admin**" > "**General Configuration.**"

2. Click "**Edit.**"

3. Remove the existing content from the "**Site Welcome Message**" and add your own.

4. Click "**Save.**"

The "Site Welcome Message" field accepts text and wiki markup.

Hint: Design your site welcome message on a normal Confluence page using rich text. You can add images or videos and even animate the content with Confluence macros. Then use an include macro to insert the content of the page into the site welcome message. From that point onwards, any changes to the page will affect the site welcome message. Use page restrictions or space permissions to control who can update the page.

For example, let's assume your page is in the MYSPACE space, and has a title of "My Dashboard Welcome Message." Then the "Site Welcome Message" field would contain just this wiki markup:

```
{include:MYSPACE:My Dashboard Welcome Message}
```

Instead of customizing the dashboard, you may prefer to direct people to a different home page for your site. By default, the site home page is the dashboard. This means that when people go to the base URL of the site, such as `http://wikitechcomm.onconfluence.com`, they will see the dashboard.

To change the site home page:

1. Choose "**Browse**" > "**Confluence Admin**" > "**General Configuration**."

2. Click "**Edit**."

3. Choose a space in the dropdown list next to "**Site Homepage**."

4. Click "**Save**."

Of course, clever people will discover that they can change their own default destination. And, people being people, they will do that. Confluence users can set their personal preferences in their user profile. One of those preferences is the page they want to see each time they log in. Personal preferences override the site-wide settings.

Site and space logos

By default, the Confluence logo appears at the top left of the dashboard and next to the title of every page. You can change the logo for the site as a whole, as well as for each space. You can also decide not to display a logo at all.

To change the site logo: Choose "**Browse**" > "**Confluence Admin**" > "**Global Logo**."

To change the logo for a space: Go to a page in the space, choose "**Browse**" > "**Space Admin**" > "**Change Space Logo**."

What you now know about structuring and styling your wiki

Far from being a dog's breakfast, a documentation wiki can present a highly organized structure that helps readers find the information they need. This chapter started at the lowest level of the content model in Confluence, the page. Since a wiki page is a web page, it offers all the layout tools that HTML offers. In addition, Confluence macros are useful to bundle sets of layout properties into meaningful clusters. For example, the information macro draws a blue panel around a chunk of text and adds an optional icon and title. The warning macro draws a similar panel, but red instead of blue. The section and column macros divide part of the page into columns. The cheese macro does nothing much useful at all.

User macros are a handy way to add your own standard formatting rules or text to a page. You can define a user macro on the Confluence administration screens, whereas more complex macros need a plugin to define their functionality. The expand macro is an example of a popular user macro for Confluence 3.5.

Moving up a level, spaces are another area where we can add structure and standardize formatting. The easiest way to customize the look and and feel of a space is by applying a theme, such as the Documentation theme. Another aspect of space structure is the division of pages into sections and subsections and the order of the pages within those sections.

Looking at the wiki as a whole, the principle content element for defining structure is the space. A space contains the information relevant to a particular subject or a particular audience. Moreover, you can collect related spaces into groups called space categories.

You can also apply your own branding and customize the look and feel of the wiki. You can apply a theme to the entire site or to individual spaces, choose your own color schemes, or tweak the CSS for a small but vital customization. You can brand the wiki by replacing the Confluence logo with your own, and add your own site welcome message to the dashboard. Alternatively, you can bypass the dashboard and direct all comers to a space home page for even more flexible formatting, style, and content.

Next we will look at setting up a framework for content reuse and show how templates help ensure that other authors follow the structure and format you have designed for your documentation wiki.

References

Here are some tools, websites, blog posts, and other references relevant to the content of this chapter.

Useful plugins

- *Builder Theme* by Adaptavist, a Confluence theme and a number of tools to customize the look and feel of the wiki site: `https://plugins.atlassian.com/plugin/details/274`

- *Content Formatting Macros* by Adaptavist, offering various options for formatting the content of a page: `https://plugins.atlassian.com/plugin/details/247`

- *RefinedWiki Original Theme*, a Confluence theme especially tuned for designing an intranet site: `https://plugins.atlassian.com/plugin/details/15231`

- *Zen Foundation*, a theme by Stepstone Technologies, designed to create an attractive and easy-to-use site for non-technical online communities: `https://plugins.atlassian.com/plugin/-details/32965`

Further reading

- *Expert Wiki Style Guide*, a style guide created collaboratively for authors on the *Expert Documentation Wiki* at Missouri State University. The primary authors are Krista Parker and Britney Shryer: `http://experts.missouristate.edu/display/wug/Expert+Wiki+Style+Guide`. Krista mentioned this style guide in a post on *Technical Writing World*: `http://technicalwriting-world.com/xn/detail/6394117:Comment:20721`. I like this style guide because it tackles just the points that are relevant and important for the wiki-based documentation. The guide is lightweight, and assumes the right level of knowledge. It is an excellent example of a collaborative document that grows to suit its users' needs.

- *Wiki web fonts*, a blog post by David Simpson on *dvdsmpsn*, December 2011: `http://david-simpson.me/2011/12/09/wiki-web-fonts/`. In this post, David explains how to use Google Web Fonts to change the style of your headings in Confluence. Google Web Fonts is a directory of open-source web fonts, available for everyone to share and customize.

Endnotes

[1] List of macros shipped with Confluence, in the *Confluence User's Guide*: `http://confluence.at-lassian.com/display/DOC/Working+with+Macros`

[2]   Ganache's page about Twitter linguistic conventions, on the *Confluence, Tech Comm, Chocolate* wiki: `https://wikitechcomm.onconfluence.com/display/DOC/Twitter+language`

[3]   Ganache's *Introduction to Twitter*, on the *Confluence, Tech Comm, Chocolate* wiki: `https://wikitechcomm.onconfluence.com/display/DOC/Introduction+to+Twitter`

[4]   A page of shared user macros on the Atlassian documentation wiki: `http://confluence.atlassian.com/display/DISC/Shared+User+Macros`

[5]   Guide to writing user macros, in the *Confluence Administrator's Guide*: `http://confluence.atlassian.com/display/DOC/Writing+User+Macros`

[6]   Ganache's page using the chocolate macro, on the *Confluence, Tech Comm, Chocolate* wiki: `https://wikitechcomm.onconfluence.com/display/CHAT/My+chocolate+macro`

[7]   Guide to making chocolate, by Cadbury: `http://www.cadbury.com.au/About-Chocolate/Chocolate-Making.aspx`

[8]   Guide to making chocolate, on *wikiHow*: `http://m.wikihow.com/Make-Chocolate`

[9]   Atlassian documentation wiki: `http://confluence.atlassian.com`

[10] A quick-start guide to developing with the Atlassian Plugin SDK: `https://developer.atlassian.com/display/DOCS/Developing+with+the+Atlassian+Plugin+SDK`

[11] Tutorial on developing a Confluence theme: `https://developer.atlassian.com/display/CONFDEV/Plugin+Tutorial+-+Writing+a+Confluence+Theme`

[12] Guide to installing plugins, in the *Confluence Administrator's Guide*: `http://confluence.atlassian.com/display/DOC/Installing+a+Plugin`

[13] Guide to customizing space layouts, in the *Confluence User's Guide*: `http://confluence.atlassian.com/display/DOC/Customising+Space+Layouts`

[14] Guide to styling Confluence with CSS, in the *Confluence User's Guide*: `http://confluence.atlassian.com/display/DOC/Styling+Confluence+with+CSS`

8

Content reuse and templates

Consistency and efficiency are the bread and butter of technical communication. Content reuse and templates go a long way towards putting the loaf on the table. Anyone for a lightly toasted chunk of banana bread with a thick coating of chocolate spread?

The quest for consistency and efficiency

This chapter describes content reuse and templates in Confluence, two mechanisms for making good use of technical writers' time and ensuring consistency in the information we give to readers. We'll begin with content reuse, in the following sections:

- **An overview of content reuse.** When technical writers talk about reusing content, they mean they will write a piece of content only once, store it in a single location, and use it again and again. We'll examine the pros and cons of this practice.

- **Confluence macros for content reuse.** The include macro dynamically copies a whole page into another page when someone reads the second page. The excerpt-include macro copies only the chunk of content defined by an excerpt macro.

- **Best practices for content reuse.** Build an inclusions library to hold your reusable content and give your pages funny names to let people know they are special.

- **Reusing images and Gliffy diagrams.** Put screenshots, pictures, and diagrams into your inclusions library.

- **Content reuse in Confluence 4.** The initial releases of Confluence 4 have some problems that make content reuse awkward. A plugin will soon be available to fix the problems.

- **A simple example of content reuse.** Ganache adds a list of Twitter hash tags to a few of the pages in her Twitter guide.

- **A case study of content reuse on a wiki.** We'll examine an inclusions library designed for the user management documentation in two products (Confluence and JIRA).

- **Other interesting bits and pieces.** We'll see how to hide an excerpt from readers, look at various problems with search results, see how to use excerpts to provide summaries of a page, and discover what happens if you nest one inclusion inside another.

- **Enhanced content reuse via plugins.** Plugins provide additional macros for sophisticated content reuse.

Then we'll see how to use templates in Confluence, as described in these sections:

- **Templates.** In terms of wiki-based technical documentation, it is useful to talk about page templates and space templates. A page template defines the structure and content of a page. A space template contains the framework for a space.

- **Page templates.** Learn how to import predefined templates and how to define your own. See how a wiki author can use the template to add a new page.

- **Plugins for enhanced templates.** Plugins extend the standard Confluence page templates.

- **Creating a space template.** Confluence does not have the concept of a space template. Instead, you can create your own by copying your documentation space and nominating the copy as a template.

Wrapping up:

- **What you now know about content reuse and templates.** A summary of this chapter.

- **References.** The tools, websites, blog posts, and other references mentioned in this chapter.

An overview of content reuse

Content reuse means using the same piece of content in more than one place, without rewriting the content or permanently copying it to another location. This chapter uses the term "content reuse" to mean the reuse of text or pages within the wiki. This is not the same thing as single source publishing of content across multiple media, which is covered in Chapter 11.

Let's assume that you are documenting the software products developed by your company. The developers build common modules that are used in more than one of the products. These modules may be user interface plugins or backend libraries that execute the business rules. When you document a module for one of the products, you know that much of your work will apply to the other products too.

On an intranet, an example would be when you are developing two or more sets of internal procedures that have a number of steps in common.

One option is to copy your words, paste them into the documentation for the next product or procedure, and then adapt them to suit. Another option is to include just a link pointing from one set of documentation to the other and force readers to hop around the documentation suite. A third option is content reuse. The first option is sometimes best, but not always. The second option is almost always not good. How about the third?

There are arguments for and against content reuse. On the plus side, it can save time and promote efficiency. It can save translation costs. It facilitates consistency and prevents problems that can occur when you need to update duplicated content in many different places. On the other hand, the content can lose so much contextual information that it becomes difficult or impossible for the reader to understand. Readers may prefer to read different words about the same topic because a new way of describing something may make it easier to understand. Sometimes, the very mechanism of content reuse requires so much upkeep that it undoes any good it has done.

In my opinion, content reuse has its place. The most important thing is to get the design right. It must be clean and simple, so that the technical writers have minimal overhead in maintaining the reusable sections. It must also avoid the problems caused to readers due to loss of context. The solution is smart design of the reused chunks, achieved by consultation and iterative design.

Let's see how to do content reuse in a wiki!

Confluence macros for content reuse

Out of the box, Confluence supplies three macros for content reuse:

- Use the **include** macro to include a whole page on another page.
- Use the **excerpt** macro to define a reusable chunk of text (an "excerpt").
- Use the **excerpt-include** macro to include an excerpt on another page.

First, let's focus on these three macros. Later, I will introduce plugins that add more functionality.

The include and excerpt-include macros will dynamically include content from one page into another page. By "dynamically," I mean that the macro will fetch the content from page 1 and include it in page 2 when page 2 is rendered. Rendering usually means displaying the page in a browser, but it can also mean exporting the page to PDF, HTML, XML, or some other format. Figure 8.1 shows how the **include** macro works, and Figure 8.2 shows how the **excerpt-include** macro works.

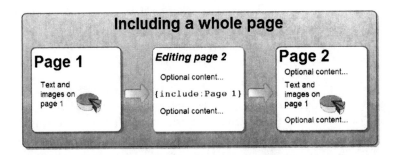

Figure 8.1. The include macro includes an entire page into another page

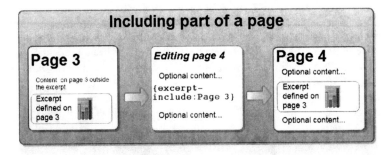

Figure 8.2. The excerpt-include macro includes part of a page into another page

People have logged requests for improvement of these macros, so things may change in future releases of Confluence. Please refer to the *Confluence User's Guide*[1] to see how the macros work in your version of Confluence.

Table 8.1 shows the differences between the include macro and the excerpt-include macro at the time of writing this book.

Table 8.1 – Comparison of the include and excerpt-include macros

Include macro	Excerpt-include macro
Displays the entire content of one page on another page.	Displays part of one page on another page. The part must be defined by an excerpt macro.
Adds one or more block-level elements.	Can add an inline element, as defined by the excerpt macro.
Can include content from a page in a different space.	Works only if both pages are in the same space.
Does not display a panel around the included content.	Can display a panel around the included content.

Note: You can only define one excerpt on each page. In other words, each page can have only one excerpt macro. However, you can include as many excerpt-include macros on a page as you like. The workaround for the limitation of one excerpt per page is to define your reusable chunks in a separate library, as discussed in the next section. Another workaround is to install the *Multi Excerpt plugin*, which allows you to define more than one excerpt on a page and include excerpts from other spaces.

Best practices for content reuse

At Atlassian, we have an extensive set of technical documentation on our wiki at `http://confluence.atlassian.com`. We document a number of products and a number of versions of each product. Sometimes different products share a common feature, so it makes sense to document the functionality once and reuse the words in each product. There are also smaller common elements in the documentation, such as the copyright statement in the footer, the version number of the product, and so on.

Over the years we have built a framework and a set of techniques that work well. In a nutshell:

- Keep reusable chunks separate from other information. We create a special area of each documentation space, which we call the "inclusions library." There is no Confluence feature called an inclusions library. This is just a convention that we have adopted.

- Use special page titles to make it obvious that the reusable chunks are special pages.

- Spend a lot of time designing your reusable chunks and deciding on their size and content.

For an example of an inclusions library, take a look at the User Management space at `http://confluence.atlassian.com/display/USERMAN/_InclusionsLibrary`. We will cover this space in a case study later in the chapter.

Here is a bit more detail about our inclusions libraries:

- **Page location and search.** The pages in the inclusions library are located at the root of the wiki space, not under the space's home page. This means that they will not appear in the table of contents shown on the left of every page, and they will not be picked up by the search box in the left-hand navigation bar. The pages will be picked up by other searches because they are just normal wiki pages. Are you wondering how to add a page at the root of a space? In the next section we will follow Ganache as she does just that.

- **Page titles.** We use an underscore at the start of the page names to remind us that the pages are special. This helps to prevent us and other authors from editing the reusable content by mistake. Most importantly, it reminds people not to change the page name, which would break all the include macros. An irregular page name also helps to hide the pages from the Confluence search. Where possible, remove space characters from the page name. For example, a page about user management could be called *_UserManagement*.

- **Excerpts.** Avoid using the excerpt and excerpt-include macros, because they limit you to a single space. Put each reusable chunk onto its own page in the inclusions library and use an include macro to add it to another page. If you need to include an inline element instead of a block-level element, then the excerpt and excerpt-include macros are good.

- **Labels.** Do not add labels to the pages in the inclusions library because labels will cause the page to appear higher up in the Confluence search results.

- **Design of the reusable chunks.** Keep them as small as possible for greater flexibility. The case study later in this chapter may help.

Reusing images and Gliffy diagrams

What about screenshots and other images – does Confluence make it possible to reuse them? Yes. Depending on your requirements, it can be useful to keep all your images in the inclusions library. Just attach them to a page in the library.

There are two ways to reuse an image:

- Attach each image to its own page in the inclusions library so you have one page per image. Also display the image on the page. When you want to use the image on another page, use the include macro to include the entire page along with the image. Using this technique, you can include a bit of text, such as a caption, too.

- Alternatively, attach all your images to a single page in the inclusions library. When you want to display an image on the page, use the Confluence image display functionality to display the image from the page where it is attached. In wiki markup: `!MyPageName^MyImageName.png!`. You can use the image browser in Confluence 4 to do the same thing.

What about Gliffy diagrams? If you want to use a diagram in more than one place, put the page containing the diagram into your inclusions library. Then use the include macro to include that page, with its diagram, wherever you need it.

Content reuse in Confluence 4

As I am writing this section, the initial releases of the Confluence 4 series have some problems that make content reuse awkward. In particular, it is difficult to keep track of reused images and pages. Chapter 23 has the details of these problems and some suggestions on how to work with the Confluence team to fix them. In addition, Chapter 23 describes the *Scroll Versions* plugin, currently under development by K15t Software, that adds cross-referencing to Confluence's content reuse functionality.

A simple example of content reuse

Sweet Ganache has identified a need for content reuse in her Twitter guide. She wants to add a list of hash tags that people can use in their Tweets. This list will be useful on more than one page.

First, Ganache adds the top-level page in her inclusions library. As recommended, she puts the page at the root level of the space, so that it is not a child of the space's home page and will not appear in the left-hand table of contents.

To add a page at the root of the space:

1. Go to a page in the space and choose "**Browse**" > "**Pages.**"
2. The "List Pages" screen will appear. Choose "**Add**" > "**Page.**"
3. Add content and save the page as usual.

Ganache names her page _InclusionsLibrary. She then adds a child page named _HashTags. In the tree view, she can now see her two new pages at the root level along with the space's home page.

To see the tree view of a space:

1. Go to a page in the space and choose "**Browse**" > "**Pages.**"
2. Choose the "**Tree**" tab.

Figure 8.3 shows the tree view of Ganache's space with the new inclusions library.

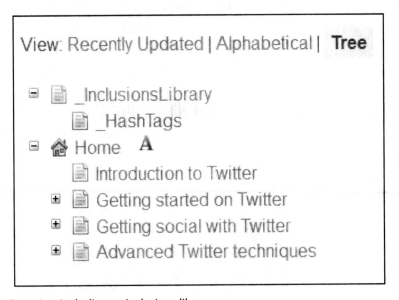

Figure 8.3. Tree view including an inclusions library

The _*HashTags* page contains Ganache's list of hash tags (Figure 8.4).

Figure 8.4. A page in the inclusions library

Ganache adds an include macro to a few pages in the Twitter guide so the pages display the list of hash tags. In wiki markup the include macro looks like this: {include:_HashTags}.

Figure 8.5 shows the result on one of Ganache's pages:

Figure 8.5. A page that includes content from the inclusions library

To see the pages in action, visit Ganache's guide to *Using Twitter* on the *Confluence, Tech Comm, Chocolate* wiki: https://wikitechcomm.onconfluence.com/display/DOC.

A case study of content reuse on a wiki

The subject of the case study is the documentation for a user management framework that was recently incorporated into two of our products. When designing the documentation, one of the primary goals was efficient and effective content reuse. The reusable chunks of documentation are in the *User management* (USERMAN) space[2] on the Atlassian documentation wiki. The chunks are used in the *Confluence Administrator's Guide*[3] and the *JIRA Administrator's Guide*.[4]

In this case, the reusable chunks are held in a space of their own rather than as part of another space. This is not always necessary. Often the reusable chunks are held in the same space as the pages that use them. Either way, the use of the inclusions library is the same.

The content in this section is derived from a post on my *ffeathers* blog written in April 2011, called *Content reuse on a wiki – a case study*.[5] I have summarized the post and added new bits that are relevant to the book.

What the documentation is about

The documentation describes a software framework for managing user directories. A user directory is a place where you store information about users and groups. Using the software, administrators can connect various types of user directories to their application and manage the directory connections. A directory can be an internal directory with data stored on the product database, an external LDAP directory such as Active Directory, or even another application that manages the user information and directory connections.

Why content reuse makes sense for this documentation

The user management framework that we are documenting consists of a set of backend libraries and a user interface (UI) plugin. The libraries and UI plugin are currently used in two separate products, Confluence and JIRA. The business rules and the UI experience are therefore the same in both Confluence and JIRA. In the future, other products may join the party. It makes sense to share the documentation across the products, too. This will ensure a consistent experience for our readers and improve our efficiency when maintaining the documentation.

More about the USERMAN space

The *User management* (USERMAN) space contains the reusable chunks of documentation. Its home page looks a little different from our standard documentation home pages. A standard home page introduces readers to the documentation and gives them links to the user's guide, administrator's guide, and so on.

Instead, the USERMAN home page (Figure 8.6):

- Lets readers know that "this documentation is not intended to be read independently of the JIRA and Confluence documentation" and gives them links to those documentation spaces. This is necessary because many people come to the documentation via Google searches. If they land on this home page, we want them to find their way to the relevant documentation quickly.

- Tells authors where they can find the reusable chunks within the space.

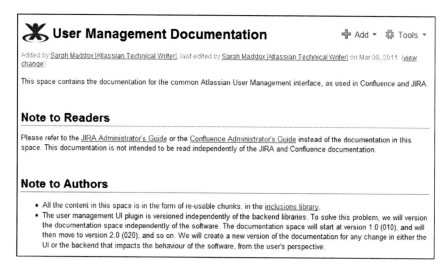

Figure 8.6. Home page of the USERMAN space

What about the contextual information and procedures that are specific to the product?

In the documentation pages for Confluence and JIRA, there are sections that are specific to the product and sections that are common to both products.

These bits are specific to the product, and therefore, not reusable:

- **Step-by-step procedures:** The primary user interface is different in each product, and therefore, the path the user takes (click here, select this, click that) to get to the directory connection screens is specific to each. Once the user has reached the directory connection screens, the user interface is the same for both products.

- **Context:** It is a good idea to start each page with a mention of the product (Confluence or JIRA), and other orienting information, so the reader knows what the page is about. After that, it is safe to dive into the reused content which, of course, is free of contextual information.

- **More context:** The path within the documentation itself differs. For example, we need to give people links to related topics and parent pages.

Therefore, each documentation page has one or more sections of reused content, each wrapped with product-specific content.

A simple example from the case study

Let's start with a simple topic, configuring the LDAP connection pool. The reusable chunk is a page in the USERMAN space called _LDAP Connection Pool Settings._[6] Figure 8.7 is a screenshot of the content of that page:

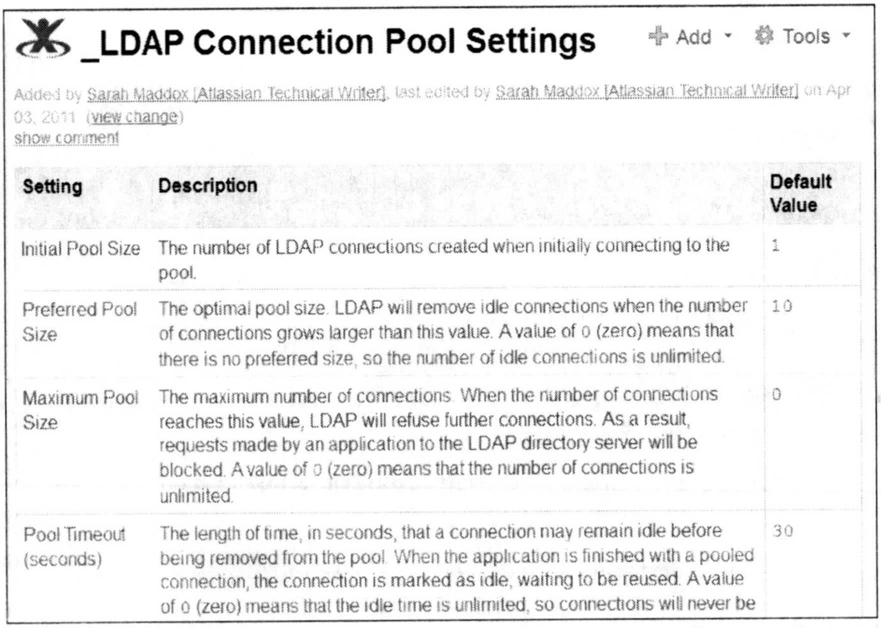

Setting	Description	Default Value
Initial Pool Size	The number of LDAP connections created when initially connecting to the pool.	1
Preferred Pool Size	The optimal pool size. LDAP will remove idle connections when the number of connections grows larger than this value. A value of 0 (zero) means that there is no preferred size, so the number of idle connections is unlimited.	10
Maximum Pool Size	The maximum number of connections. When the number of connections reaches this value, LDAP will refuse further connections. As a result, requests made by an application to the LDAP directory server will be blocked. A value of 0 (zero) means that the number of connections is unlimited.	0
Pool Timeout (seconds)	The length of time, in seconds, that a connection may remain idle before being removed from the pool. When the application is finished with a pooled connection, the connection is marked as idle, waiting to be reused. A value of 0 (zero) means that the idle time is unlimited, so connections will never be	30

Figure 8.7. A chunk of reusable content

The page shown in Figure 8.7 consists entirely of reference information. There is no orientation at the top or any other contextual information.

The content of the page is reused in both the JIRA documentation (*Configuring the LDAP Connection Pool*[7]) and a similar page in the Confluence documentation.[8]

Figure 8.8 shows the JIRA page.

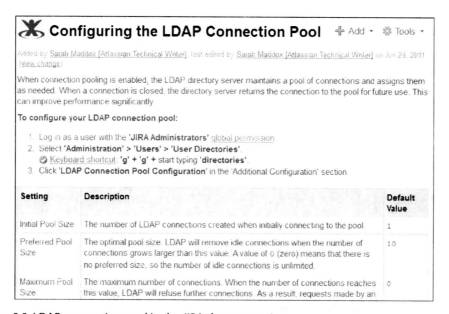

Figure 8.8. LDAP connection pool in the JIRA documentation

The JIRA page in Figure 8.8 contains:

- Contextual information at the top of the page.

- Step by step instructions that are specific to the JIRA product.

- Reused content: The chunk of reference information, dynamically copied in from the USERMAN space when the page loads into the browser.

- More contextual information, in the form of "related topics," at the bottom of the page. (Not shown on the screenshot.)

The tree view of the pages in the USERMAN space shows that the space home page (marked with the house icon at the bottom of the screenshot in Figure 8.9) has no child pages. Instead, all the content of the space is in the inclusions library.

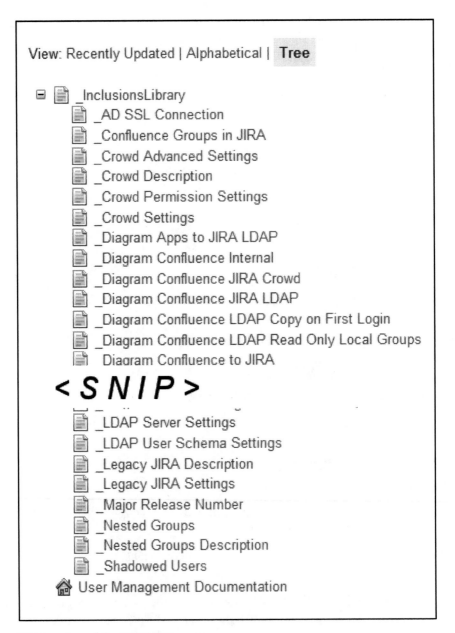

Figure 8.9. Tree view of the USERMAN space

A more complex example from the case study

In the simple example above, there was just one section of reused content and a small amount of contextual information. The user management documentation contains much more complex examples, too. For example, the topic about connecting to an LDAP directory is built up of a number of sections, some of which are from reusable chunks and others which are specific to the product.

The topic has a number of sections:

- A short introduction to the page.
- An overview of the LDAP connections provided and why you would use them.
- A step-by-step procedure.
- Seven sections containing details of specific settings.
- Diagrams of possible configurations.
- Related topics.

Three of the sections contain content specific to the product (JIRA or Confluence):

- The short introduction.
- The step-by-step procedure.
- The related topics.

All the other sections are drawn from reusable chunks:

- An overview of the LDAP connections provided and why you would use them.
- Seven sections containing details of specific settings.
- Diagrams of possible configurations.

The magic is in the way the reusable content is split into chunks and in the use of include macros to pull content into the relevant part of each page. If you are interested in the details, please take a look at the *ffeathers* blog post mentioned above.

Deciding the size of the reusable chunks

On the one hand, the smaller the chunks the better. Small pieces of content give us much more flexibility. We can use them in different sequential orders and in different contexts. We can add product-specific and contextual information between the chunks. We can leave out bits in one product and include them in another. We are less affected by changes in the user interface, such as when the order of the input fields changes or entire sections move to different screens.

On the other hand, smaller chunks are more difficult to manage. Every time we need to update the page inclusions, such as when we need to move to a new version, there are a large number of include macros to update. This reduces the benefit of efficient maintenance. Too much flexibility may lead to confusion in the final output that the reader sees, reducing the benefit of cross-product consistency.

You could go to one extreme and have a reusable chunk per input field or concept (in the above case study, there might be a chunk for "setting the optimal size for your LDAP connection pool"). Or you could go to the other extreme and define one large reusable chunk per user task (again from the case study: "connecting an LDAP directory").

After discussion with the product manager and development team, I decided to go for a middle-of-the-road approach. Most of the reusable content is reference material and conceptual material rather than "how to" steps. We decided to divide the content into blocks, each of which corresponds to a logically-related set of input fields or a concept.

The result is:

- For the simpler task-based topics, there are two page inclusions: one for the concept and one for the reference material.

- For the more complex topics, the number of inclusions ranges from five to ten: one or two conceptual chunks, a number of reference chunks, and one or two illustrative chunks.

Developing the design

The exercise of designing the reusable content was interesting in a couple of ways.

- The end result would be a set of reusable chunks only loosely resembling the documentation that most subject matter experts are used to seeing. How could I prototype the design and the content with them?

- The content would be used in two different sets of documentation, for two different products: Confluence and JIRA. I needed to work with all the technical writers managing those documentation suites to make sure they had input into the design and were happy with the end result.

When designing a new documentation suite, the usual process of consulting and collaborating with subject matter experts goes something like this: Draft the table of contents, review it with product managers and developers, apply changes, add detail, and refine it until we are all happy.

The difference now was that there were **two** tables of contents:

- The list of topics and their hierarchical structure as they would appear to readers. Let's call this the *compiled* documentation.

- The list of reusable chunks that we would plug together to form the above readable material. Let's call these the *reusable chunks*.

My strategy was to discuss and review the planned table of contents for the compiled documentation in great detail, making sure that I covered all topics and was consistent in the amount of attention given to each subject. Then I designed the reusable chunks to best satisfy the requirements so discovered. When designing the reusable chunks, I relied more on my technical writing and information design skills than on consultation with the subject matter experts.

As soon as we had a basic design, I created the draft content in reusable chunks and hooked them together to create the compiled documentation for one of the products.

Then I walked the technical writing team, product managers, and developers through that first draft. I showed them the reusable chunks and discussed the design strategy (size of the chunks, use of wiki spaces, setting of permissions, consistency, removal of context, and so on). But most of our attention focused on the human-readable documentation. The review session, as always, gave me much valuable feedback that I applied to the draft.

After that initial review, I focused on each individual topic and went through the usual iterative process of drafting, reviewing, and updating the content.

Other interesting bits and pieces

Inclusions, excerpts, and excerpt inclusions are weird and wonderful beasties. Here is a collection of notes from my close encounters with inclusions.

Hiding the excerpt from readers

The thing about a wiki is that people may stumble across the pages in the inclusions library. They may wander in via Google search and not realize that this page is just part of a set of instructions, not the entire sequence. Sure, the page has a weird name. But hey, this is a wiki, so what do you expect?

It may even be a bit traumatic for people who start following the instructions without having performed the preceding steps. One way around that is to use the `hidden=true` parameter in the excerpt macro.

```
{excerpt:hidden=true}This would be traumatic if not seen in context.{excerpt}
```

When the excerpt macro contains `hidden=true`, the content between the {excerpt} tags will not be displayed on the page where the excerpt is defined. It will, however, be displayed on the page where the excerpt is included by the excerpt-include macro. I am not sure if that is what this parameter was designed for, but it makes a good story.

Search results and content reuse

One of the problems with content reuse in Confluence is that the Confluence search results pick up the content where it is originally written, but not where it is displayed.

For example, Ganache's handy Twitter hash tags are contained in a reusable page that is included into a number of other pages. When people search for "Twitter hash tags," they will find the page called _HashTags, but not *Getting started on Twitter* or any of the other pages that include the content.

There are some workarounds:

- To force the "real" pages to appear in the Confluence search, add labels to the pages. If practical, add headings within the pages to introduce the reused content.

- To make the reusable page less significant in any search, refrain from adding labels and give the page a funny name.

There is a more complex workaround for making sure that the Confluence search results include all the reused content on the pages where it is used. Install the *Cache plugin* for Confluence. Then wrap your include macros with the cache macro, using the `index=true` parameter to make the content of the include macros searchable.

```
{cache:index=true}
{include:mySPACEKEY:My page name}
{cache}
```

Bob Swift develops and maintains the *Cache plugin* for Confluence. He also provides the support for the plugin. A big thank you to Emily Johnson and David Peterson for finding and sharing this solution. Details are in a post on my *ffeathers* blog written in May 2011, called *Making the include macro's content appear in the Confluence search results.*[9]

Using excerpts as summaries of a page

Certain Confluence macros display a list of pages. And some of them recognize an excerpts parameter that prompts them to display any excerpts that are defined on the pages in their list. (In wiki markup, `excerpt=true`.) For example:

- The children macro, which displays a list of the child pages of a given page. If you specify that excerpts should be included, the list will also show any content defined by an excerpt macro on the child page(s).

- The content-by-label macro, which displays a list of pages tagged with a given label or labels. If you specify that excerpts should be included, the list will also show any content defined by an excerpt macro on the tagged page(s).

What part does chocolate play in your life as technical writer?

From Sarah Maddox and Arnold Burian on Technical Writing World:[10]

Sarah: "Here's a combination sensation: Lightly toasted banana bread thickly coated with Nutella. Yummmm! Has anyone else tried that?"

Arnold: "I believe Nutella is a hazelnut spread. And much like bacon, you can put it on anything to make it better :)"

This is useful in a glossary or FAQ list. The Atlassian Crowd documentation has an example (Figure 8.10) of a glossary constructed using the children macro with excerpts included.[11] The wiki markup looks like this: `{children:excerpt=true|style=h3}`.

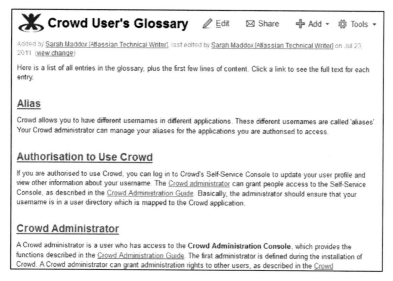

Figure 8.10. Glossary built with the children macro and excerpts

Nesting one inclusion inside another

It is possible to nest one piece of reused text inside another. For example, you can do this:

- Add an excerpt macro on page A, to define a reusable chunk.

- Add an excerpt-include macro on page B, which copies the information from page A.

- Add an include macro on page C, which copies the contents of page B into page C. Page C will then contain the contents of the excerpt on page A as well as the full contents of page B.

It works equally well when you include page A inside page B, then page B inside page C.

I tried to confuse Confluence by having pages C and D include each other. I was hoping for an infinite loop. Instead I got a polite error message: "Unable to render {include} Already included page D, stopping."

Enhanced content reuse via plugins

Two plugins provide extra macros for enhanced content reuse. The *Multi Excerpt plugin* adds a multi-excerpt macro, which you can use to define more than one excerpt on a single page. The *Builder Theme plugin* adds the builder-show macro, which includes conditional logic. You can define the page or pages on which the reused content should appear. Jodie Miners has written an excellent blog post[12] describing her use of the macros provided by the *Multi Excerpt plugin* and the *Builder Theme plugin*.

Scroll Versions is a new plugin, currently under development by K15t Software, that extends Confluence's content reuse functionality to include version control and cross referencing. See the section titled "K15t Software's focus on wiki documentation tools" (p. 436) for more details about the functionality of this plugin.

Templates

Templates are what technical writers dream about. Good templates ensure efficiency, consistency, and simplicity. Simplicity is what our readers dream about. Well, maybe not. But if they thought about it as much as we do, they would surely dream about it too. (People certainly have nightmares about complexity! You know those dreams where you feel you are swimming through treacle, getting nothing done, unable to find the way to your goal?)

The job of a template is to ensure that the language, structure, and style of a page do not distract the reader from the important stuff: the content. That is one of our main goals as technical communicators, too. In most cases, we want to fade into the background. The document itself is transparent, and the information leaps out at the reader. We aim to turn "uh?" into "aha!" in the quickest time possible.

Consistency is also aesthetically pleasing.

In terms of wiki-based documentation, a template is useful in two areas:

- **Page templates.** These give guidelines on the content and layout of each page. We need a number of templates, one for each type of documentation. For example, we may define separate page templates for the "how to" topics, FAQ entries, knowledge base articles, installation guides, financial procedures, human resources procedures, and so on. You can define page templates in Confluence. More information below.

- **Space templates.** These provide the skeleton structure of a space. For a standard set of product manuals, the space template would have a framework of pages for the user's guide, administrator's guide, installation guide, FAQ, and other parts that occur in all of the organization's manuals. The template would also include the mechanics needed for content reuse (described earlier in this chapter), PDF export, and anything else that is standard across your documentation. Confluence does not provide a mechanism for space templates, but you can easily make your own. See below.

Note: The terminology above differs from that in the Confluence documentation.[13] The documentation deals with page templates only, as defined above. But it refers to them as either "global templates" or "space templates," to indicate their scope. A "global template" is a page template that is available across the entire Confluence site. A "space template" is a page template that is available in a given space only. To avoid confusion when I need to differentiate these two types of page template, I will call them "global page templates" and "space-specific page templates."

Page templates

In Confluence, you can define templates that give authors guidelines on the content and style of a page. Confluence also comes with a set of pre-defined page templates. They are not visible to the general user until the administrator makes them available by using a procedure called "importing" the templates.

Importing templates and downloading additional templates

The term "importing" is slightly confusing. In this sense it means simply making the templates available for use, either across the entire Confluence site or for a specific space. The templates are already part of the Confluence installation. It is also possible to download additional templates from the *Atlassian Plugin Exchange*, then import them to make them available for use.

You need Confluence administrator permissions to import templates.

To make a pre-defined page template available for use:

1. Choose "**Browse**" > "**Confluence Admin**" > "**Import Templates.**"

2. Select the templates you want to make available to wiki users.

3. Choose "**Global Templates**" to make the template available across the entire Confluence site, or select a space to make a template available for that space.

People have developed additional page templates and made them available on the *Atlassian Plugin Exchange*. Some are free, others are commercial. You can download template bundles and install them onto your own site. The process of installing the template bundle is the same as manually installing a plugin. You need Confluence system administrator permissions to do that.

To download and install additional templates:

1. Go to the *Atlassian Plugin Exchange* and search for Confluence templates: `https://plugins.at-lassian.com/search/category/20177`.

2. Download the template bundle that you want, and save it on your computer. It will be a .jar file.

3. In Confluence, choose "**Browse**" > "**Confluence Admin**" > "**Plugins**" > "**Install.**"

4. Click "**Upload Plugin,**" browse to the template bundle file that you downloaded and upload it into Confluence.

5. Now the bundle of templates is on the site, and you can make the templates available by "importing" them as described above.

Adding your own templates

You can define your own template either at site level (global page templates) or for a specific space (space-specific page templates). You need Confluence administrator permissions to define a global page template and space administrator permissions to define a space-specific page template.

To add a global page template: Choose **"Browse"** > **"Confluence Admin"** > **"Global Templates"** > **"Add New Global Template."**

To add a space-specific page template: Go to a page in the space, then choose **"Browse"** > **"Advanced"** > **"Templates"** > **"Add New Space Template."**

The contents of a template

The method of defining page templates is the same in Confluence 3.5 and Confluence 4. The page template is one of the remaining places in Confluence 4 that still accept and retain wiki markup. In other words, when defining a template you can enter wiki markup and also edit it again later as wiki markup – but only when editing the template, not the page. At some stage a future release of Confluence may provide an alternative format, such as a rich text form.

At its simplest, a page template contains some standard headings and sections and hints to authors about the type of content they should write in each section.

A more complex template includes input fields that prompt the author to supply required information. Confluence converts the input into content on the page when the author saves the page for the first time. Table 8.2 shows the types of input fields you can define on your template:

Table 8.2 – Input field syntax

Syntax for the input field	Description
`@FIELDNAME@`	A single line of text.
`@FIELDNAME|textarea(3,20)@`	A multi-line text box. The digits define the number of rows (in this case, 3) and width in characters (in this case, 20) for your input field.
`@FIELDNAME|list(V1,V2,Vn)@`	A dropdown list, which you can use to offer authors a choice of values. The values you define in the list will appear in a dropdown list when an author creates a page based on the template. The author can select one, and that value will appear on the page when they save it.

The field names must be unique. They appear on the form when an author adds a page based on the template. Apart from that, they are not of much significance to you as template designer. You cannot refer to them or use them as variables. Also note that you cannot have spaces anywhere in your variables, not even in the values in the dropdown list.

A note about the requirement to use wiki markup in templates in Confluence 4: This makes it impossible to draft your page as a normal page using the Confluence 4 rich text editor, then copy the page content into a template. Here is a workaround: Build the wiki markup in a text editor such as Notepad and paste it into the "Insert Wiki Markup" dialog of a page each time you want to test the template's content. When ready, copy the wiki markup from the text editor into the template.

An example of a page template

Ganache is working on the *ChocTactics* project, *Choc Ink*'s chocolate assessment tool. Some lucky people will taste the chocolate products and submit a report of their findings. Ganache adds a template for the reports that the chocolate assessors will write. Each report will be a page on the wiki containing the following information:

- A heading: ChocTactics Assessment Report.

- The assessor's identification number.

- The date of the assessment.

- The identification number of the sample being tested.

- The taste, selected from a list of predefined options: Sweet, Bitter, Bittersweet or Other.

- The texture, selected from a list of predefined options: Dry, Crumbly, Smooth, Creamy, Snappy or Gummy.

- Comments in free text form.

The above fields are contained in a table of two columns by six rows.

Example 8.1 shows the wiki markup for Ganache's template.

```
h2. ChocTactics Assessment Report

|| Assessor number || @AssessorNumber@ ||
|| Sample number || @SampleNumber@ ||
|| Date || @Date@ ||
| Taste | @Taste|list(Sweet,Bitter,Bittersweet,Other)@ |
| Texture | @Texture|list(Dry,Crumbly,Smooth,Creamy,Snappy,Gummy)@ |
| Comments | @Comments|textarea(3,20)@ |
```

Example 8.1 – Wiki markup for the ChocTactics Assessment Report

Would you like to see Ganache's template in action? It is up and running on the *Confluence, Tech Comm, Chocolate* wiki. Go to the *Chat and play* space at `https://wikitechcomm.onconflu-ence.com/display/CHAT/`. If you have not yet created your own tutorial playground there, follow the instructions on the page to do that now. Then add a new page within your playground and choose the *ChocTactics* template to base your page on.

Using a template as a basis for a new page

When wiki authors add new pages to the wiki, they can choose to base their pages on a template. The steps are slightly different in Confluence 3.5 and Confluence 4.

To use a template when adding a page in Confluence 3.5:

1. Go to the parent page, then choose "**Add**" > "**Page.**"

2. Click "**Select a page template to start from.**"

3. Select the template you want, then click "**Next.**"

4. If the template designer has included any variables (fields) in the template, Confluence will prompt you to supply values for those variables. Enter the information and click "**Insert Variables.**" If the template has no variables, you will not see this step.

5. The contents of the new page will appear in the Confluence editor. If you entered information in the variables, that information will now be part of the page. Add a page title and edit the content, if necessary.

6. Save the page.

To use a template when adding a page in Confluence 4:

1. Go to the parent page, then choose "**Add**" > "**Page from template.**"

2. Select the template you want, then click "**Next.**" From this point, the procedure is exactly the same as for Confluence 3.5, starting from step 4.

Note: The template is not associated with the page after the page has been created. In the standard Confluence setup, there is no way to make changes to a template and have those changes flow through to pages that were based on the template. Some plugins, however, do add this functionality. There is a list of useful plugins near the end of this chapter.

An example of adding a page based on a template

Ganache uses the *ChocTactics* template to add a new page. The screenshot in Figure 8.11 shows the step that prompts the author for the variable values, as described in step 4 of the procedure above.

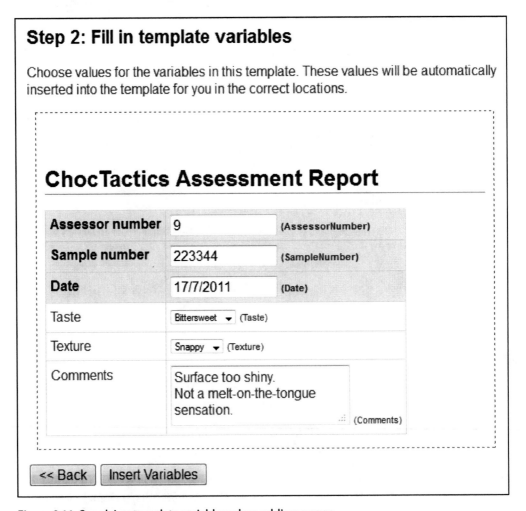

Figure 8.11. Supplying template variables when adding a page

Figure 8.12 shows the new page in the Confluence 3.5 rich text editor, which contains all the values supplied by the template. Ganache has not yet added a title or saved the page.

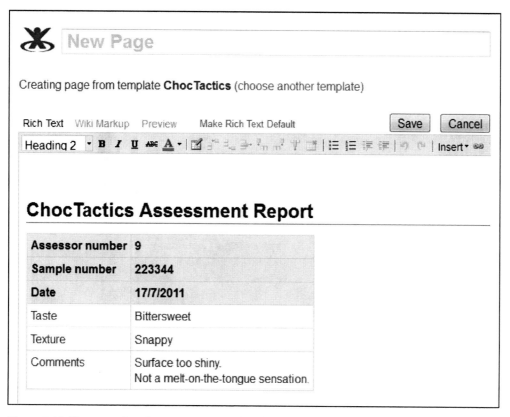

Figure 8.12. New page in editor containing values from template

Plugins for enhanced templates

In the standard Confluence setup, the template is not associated with the page after the page has been created. Using the *Scroll Wiki Forms plugin* or the *Scaffolding plugin*, you can create templates and forms that are retained after page creation. Details are in the section titled "References" at the end of this chapter.

Creating a space template

For technical writers on a wiki, it is useful to have a space containing a skeleton structure that you can copy again and again when you want to add a new set of documentation. Confluence does not have have the concept of a space template like this, but it is easy to create one yourself.

The idea is to develop your first documentation space to a stage where it meets your requirements, then copy it to another space and generalize it. Dub this second space your "template." When next you need to add a documentation space, just copy the template and add specific details.

Hint: Now that you have reached this point in the book, you have learned the most important techniques needed to set up your documentation space. If you have been creating a space as you follow the text, you can copy that space now and use it as a template.

To copy a space, you will need the *Copy Space plugin*. This functionality is not yet part of Confluence, though many think it should be! People interested in the debate can join the conversation in the Confluence issue tracker: `https://jira.atlassian.com/browse/CONF-14198`.

Installing the *Copy Space plugin* is the same as installing any other Confluence plugin. You need Confluence system administrator permission to do this.

To install the Copy Space plugin:

1. Log in as a Confluence system administrator.

2. Choose "**Browse**" > "**Confluence Admin**" > "**Plugins**" > "**Install**."

3. Type "**copy space**" into the search box and click "**Search**."

4. Click the name of the plugin in the list of search results. This will open the panel showing the plugin details.

5. Click "**Install Now**."

The *Confluence Administrator's Guide* has the details on installing plugins.[14]

Now that the plugin is installed, you can copy any space. The action will add a new space to the wiki, containing all the content from the original space. You need space administrator permissions on the space that you are copying and also permission to add a space to the wiki.

To copy a space: Go to a page in the space, then choose "**Browse**" > "**Space Admin**" > "**Copy Space**."

What you now know about content reuse and templates

With any luck, you are persuaded that content reuse is a good thing when done well. After following Ganache through the first steps of creating an inclusions library at the root of her documentation space, you then dived into two more complex examples.

The three macros, include, excerpt, and excerpt-include, no longer hold any mysteries for you. The interesting case of the hidden excerpt grabbed your attention. You dallied with the idea of hiding the reusable chunks from the search and making the search recognize the reused content where it appears. You tinkered with the thought of building a list of FAQs or glossary entries based on excerpts from pages using the children macro or the content-by-label macro.

Finally, you learned how to build page templates and space templates, thus ensuring that other authors follow the structure and format that you have designed for your documentation wiki.

Now that you have created such dazzling documentation, how do readers find their way to it and around it? That is the topic of Chapter 9.

References

These are the tools, websites, blog posts, and other references relevant to the content of this chapter.

Useful plugins

- *Copy Space plugin* by Atlassian, used to add a space by copying the content of another space: `https://plugins.atlassian.com/plugin/details/212`

- *Cache plugin* by Bob Swift, for caching part of the page, useful especially if the content is derived from an external source. A secondary use is to make the contents of the include macro searchable: `https://plugins.atlassian.com/plugin/details/152`

- *Multi Excerpt plugin* by Artemis Software, for defining more than one excerpt on a page and including excerpts from other spaces: `https://plugins.atlassian.com/plugin/details/169`

- *Builder Theme plugin* by Adaptavist, which includes the builder-show macro and the import macro for content reuse: `https://plugins.atlassian.com/plugin/details/274`

- *Scaffolding plugin* by CustomWare, for creating permanent templates and forms that are retained after page creation: `https://plugins.atlassian.com/plugin/details/190`

- *Scroll Office* by K15t Software, for exporting content from Confluence to Word: `https://plugins.atlassian.com/plugin/details/24982`

- *Scroll Versions*, a new plugin under development by K15t Software. This plugin enhances Confluence's content reuse features to support version control and provides cross referencing of reused content. (This functionality is not available in Confluence without the plugin.) When viewing a page that includes content from somewhere else, you can see where the included content comes from. When viewing a page that is included elsewhere, you can see all the pages that include it. See the section titled "K15t Software's focus on wiki documentation tools" (p. 436) for more details about the functionality in this plugin.

- *Scroll Wiki Forms plugin* by K15t Software, for extending the functionality of Confluence's page templates so that only the form fields can be edited when a user creates or edits the page. This plugin is available for Confluence 4 but not for Confluence 3.5. At time of writing, the plugin is in early access release: `https://plugins.atlassian.com/plugin/details/563990`

- Template bundles, for predefined page templates that you can download and import into your own Confluence site:

  - *CustomWare Page Templates*, by CustomWare: `https://plugins.atlassian.com/plugin/details/21408`

  - *Human Resources* template bundle by Atlassian: `https://plugins.atlassian.com/plugin/details/20493`

  - *Software Development* template bundle by Atlassian: `https://plugins.atlassian.com/plugin/details/20489`

- To find the full list of template plugins at the *Atlassian Plugin Exchange*: Go to `http://plugins.atlassian.com`, type **"template"** in the search box and select **"Confluence,"** then click **"Search."**

Further reading

Kai Weber explains the importance of consistent formatting in a technical document. He cites some academic studies that show how human intelligence can group related items faster and more confidently if the items are presented consistently. Conversely, inconsistent presentation leads to more errors and poorer performance: *Proving the benefit of consistency in tech comm*, a blog post on *Kai's Tech Writing Blog*, November 2011: `http://kaiweber.wordpress.com/2011/11/07/proving-benefit-of-consistency-in-technical-communications/`

Endnotes

[1] *Working with Macros*, in the *Confluence User's Guide*: `http://confluence.atlassian.com/display/DOC/Working+with+Macros`

[2] The *User management* (USERMAN) space on the Atlassian documentation wiki, used in the case study in this chapter: `http://confluence.atlassian.com/display/USERMAN`

[3] A page in the *Confluence Administrator's Guide*, used in the case study in this chapter: `http://confluence.atlassian.com/display/DOC/Configuring+User+Directories`

[4] A page in the *JIRA Administrator's Guide*, used in the case study in this chapter: `http://confluence.atlassian.com/display/JIRA/Configuring+User+Directories`

[5] *Content reuse on a wiki – a case study*, a blog post by Sarah Maddox on *ffeathers*, April 2011: `http://ffeathers.wordpress.com/2011/04/03/content-reuse-on-a-wiki-a-case-study/`

[6] *_LDAP Connection Pool Settings*, in the USERMAN space: `http://confluence.atlassian.com/display/USERMAN/_LDAP+Connection+Pool+Settings`

[7] *Configuring the LDAP Connection Pool*, in the JIRA documentation: `http://confluence.atlassian.com/display/JIRA043/Configuring+the+LDAP+Connection+Pool`

[8] *Configuring the LDAP Connection Pool*, in the Confluence documentation: `http://confluence.atlassian.com/display/DOC/Configuring+the+LDAP+Connection+Pool`

[9] *Making the include macro's content appear in the Confluence search results*, a blog post on *ffeathers* by Sarah Maddox, May 2011: `https://ffeathers.wordpress.com/2011/05/15/making-the-include-macros-content-appear-the-confluence-search-results/`

[10] *What part does chocolate play in your life as technical writer*, a discussion on *Technical Writing World*, started in May 2011: `http://technicalwritingworld.com/forum/topics/what-part-does-chocolate-play`

[11] A glossary built using the children macro with excerpts, in the *Crowd User's Guide*: `http://confluence.atlassian.com/display/CROWD/Crowd+User%27s+Glossary`

[12] The *"One Thing, in One Place, Once" Rule using Confluence*, a blog post by Jodie Miners on *JodieM.com.au*, August 2011: `http://jodiem.com.au/2011/08/13/the-one-thing-in-one-place-once-rule-using-confluence/`

[13] *Working with Templates*, in the *Confluence User's Guide*: `http://confluence.atlassian.com/display/DOC/Working+with+Templates`

[14] Guide to installing plugins, in the *Confluence Administrator's Guide*: `http://confluence.at-lassian.com/display/DOC/Installing+a+Plugin`

9

Findability and searchability

People are constantly in search of all sorts of things. *C'est la vie*. Documentation is what helps them in their search. (As technical communicators, that is our passionate belief.) What's more, searching for and within the documentation should be magic. Our readers should not have to think about the search tool itself.

Let's assume we, like Ganache, are writing a guide to Twitter. This chapter is all about how we can help people when they say things like this:

- Who can help me learn about Twitter?

- How do I Tweet?

- I know I saw something about a *schmilblick*,[1] but I can't for the life of me remember where or what or when!

- It's in the wiki. Somewhere.

Onwards and upwards. *Cherchez le schmilblick.*

Found in this chapter

This chapter covers searchability and findability under the following sections:

- **In search of findability.** Searchability is a measure of how easy it is to find a piece of information via a search engine. Findability is the measure of how easy it is to find the information, whether via a search engine or in some other way. This chapter looks at how we can optimize both searchability and findability in our wiki documentation.

- **The Confluence search.** As designers of documentation, we find it useful to understand the tools at our disposal. So we'll look at the elements of the Confluence search.

- **A quick guide to SEO.** Readers may come looking for information via the Confluence search or via an external search engine such as Google. This section gives guidelines on designing the documentation so that it returns the best search results to both sets of readers.

- **More about labels.** There is more to finding information than just the search. We'll see how to use Confluence labels to add similes to a page, to collect and display related topics, and even to build an index. Then we'll look at extended metadata options.

- **What you have discovered about findability and searchability.** A summary of this chapter.

- **References.** The tools, websites, blog posts, and other references relevant to the content of this chapter.

In search of findability

Technical communicators and SEO

It is interesting to see that many technical communicators market their knowledge of search engine optimization as part of their skill set. Try doing a Google search for "SEO technical communication."

There is no point in writing the best documentation in the universe if nobody can find it. Once people have found the documentation, they want one of two things. If they are new to the product, they want to know what it is all about. They will look for an overview and browse around a few topics to get the big picture. Other people are in search of a specific bit of information. They hope to dash in, grab a nugget of knowledge, and move on with their day.

The words "searchability" and "findability" may be painful to the ears. They may even bring a technical communicator close to tears. Is this a case where English completely disappears?[2] No, both words are in common use and they serve a useful purpose. Here is my working definition of the two terms:

- Searchability is a measure of how easy it is to find a piece of information via a search engine.

- Findability is the measure of how easy it is to find the information.

Searchability refers principally to search engine optimization (SEO), a set of techniques for making your pages appear high up in the search results when people search for a term that is important to you. SEO usually applies to global search engines such as Google, Bing, and Yahoo! Searchability also applies to the search tool provided by the documentation platform itself. Content management systems, wikis, and other online help systems provide search functionality of varying capabilities. It is up to us as technical communicators to learn the capabilities of our platform. Then we can design our documentation to take best advantage of the platform's search tool and help our readers find what they need.

Findability includes more than just the search. It is worth glancing back at Chapter 7, which examined some techniques for structuring and formatting wiki documentation. A good structure helps people find their way to the information they need, especially if they are browsing through the manual to get an idea of a product's features. Findability also includes the ability to find the appropriate section on the page once the search engine has sent you to the page. In web-based documentation, people can use the browser's search function to find a given word on the page. Good page structure, such as a table of contents at the top and consistency in the layout, also helps people know where they are and where to go next to find what they are looking for.

Let's start by getting to know the Confluence search, then move on to search engine optimization in general.

The Confluence search

Knowing how the Confluence search works, and how readers will use it, will affect the way we design the documentation.

Confluence has a big friendly search box at the top right of the screen, just where you would expect to see it. Remember that if your Confluence site uses a different theme from the default, the search box may be somewhere else. With any luck, it will be easy to find!

From the point of view of readers, the Confluence search just works. From our point of view as documentation designers, it is interesting to break the search down into its different components:

- **Quick navigation.** When you start typing a word into the search box, Confluence immediately starts looking for matching pages and other content types, offering them in a dropdown menu.

- **Full search.** If you type your search term and press Enter immediately, the full search swings into action. A new screen opens up, showing preliminary results and options to refine the results.

- **Search box in the Documentation theme.** If you type your search term into the search box in the left-hand navigation panel of the Documentation theme, the search looks for results in the current space only, starting from the level of the space's home page.

Let's look at each of these components in more detail.

The quick navigation feature

When you start typing a word into the search box at top right of the screen, Confluence immediately starts offering matching pages and other content types. The matching items appear in a dropdown menu, changing as you type. This quick navigation feature will pick up pages, blog posts, images and other attachments, spaces, and user profiles. It matches the search term against the title only, ignoring page content, labels, and other properties.

In Figure 9.1, Ganache has typed the word "twitter" into the search box. The screenshot shows the quick navigation dropdown menu under the search box, on the right of the screen.

Figure 9.1. The Confluence quick navigation feature

By default, the quick navigation feature looks for matches over the whole Confluence site. If you are using the Documentation theme, space administrators can configure the quick navigation to look in the current space only. Chapter 7 shows you how to configure the theme. Power users can override the restriction by entering "**all:**" and their search term. For example: "**all:tutorial**" will search all spaces for the word "tutorial."

Note that people can change the search parameters by going to the full search screen.

The full search

If you type your search term and press Enter immediately, the full search swings into action. You can also get to it by choosing the "Search for…" option at the bottom of the dropdown menu offered by the quick navigation feature.

This search matches the search term against the page content as well as the title. The position of the page, higher up or lower down in the results, depends on a ranking algorithm. You can refine the results by choosing a filter:

- Restrict the search to one or more spaces.

- Look for specific content types only, such as pages, blog posts, or attachments.

- Choose a date range.

- Specify an author.

The search also looks for matches in the content of blog posts, space descriptions, and attachments such as Microsoft Office documents, PDF files, and more. The *Confluence User's Guide*[3] has the details.

The results show any matching pages and other content types, as well as matching labels. You can click through to see the pages tagged with those labels (see Figure 9.2).

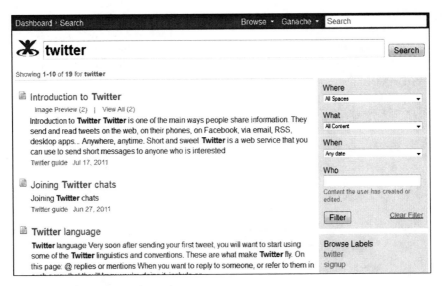

Figure 9.2. The Confluence search results

The full search includes a "**did you mean**" aid. When you enter a search term, Confluence checks to see if an alternative spelling occurs more often in the wiki content. If it finds a match, it offers the alternative search term, too. This is useful for correcting spelling mistakes or finding international differences. It is also useful if most of the people on the wiki habitually spell something incorrectly! We would hope, though, that such a thing does not happen too often on a technical documentation wiki.

Advanced readers will make use of the full search syntax that Confluence offers. People will expect to be able to use AND, OR, and NOT. They will use double quotes to indicate exact terminology. They will group phrases in brackets. They will even try wildcard searches. Confluence recognizes these tricks and more.[4] Power users will know that they can even search for specific content in specific fields.[5]

The left-hand search box in the Documentation theme

If your space uses the Documentation theme, there is a search box in the left-hand navigation panel (see Figure 9.3). The search triggered there will look for matches in the current space only.

In addition, the search will start at the level of the space's home page, looking at that page and its children only. This means that you can "hide" pages from the search, by putting them at the root level of the space. Why would you want to do that? One situation is for content reuse, when you want to define a library of information chunks that are used on various pages. Chapter 8 describes content reuse and how to add pages at the root of the space.

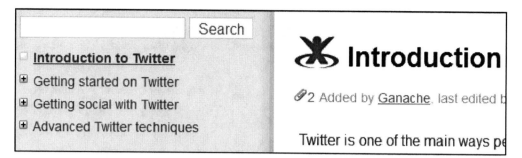

Figure 9.3. The Documentation theme's search box in the left-hand navigation panel

Ranking of Confluence search results

This is perhaps the most interesting aspect of the Confluence search. How does it decide which results to put at the top of the list? The *Confluence User's Guide* has a good description of the ranking algorithms used and of the philosophy behind them.[6]

- People come first. Confluence is a collaborative platform, so it gives a lot of weight to the content in people's user profiles on the wiki.

- Next up is the page title. If the match is found in the page title as opposed to the body of the page, the page will appear higher in the search results.

- Labels are important, too.

- Matches found in the body of the page are ranked higher if the matching term appears more than once on the page.

- Older content receives a lower ranking.

Let's look at the technicalities. Confluence applies weightings based on the following three properties of the matching content:

- The content type: user profile, page, blog post, attachment, comment, email message, and space description.

- The field in which the match was found: title, full name, label, content.

- The age of the content based on its creation date: 1 day, 2 days, 1 week, and so on.

A weighting is a number. The sequence of items in the lists above reflects the weighting, in descending order of importance. A user profile (weighting 9) ranks higher than a page (weighting 8). The title (9) ranks higher than a label (7). Content that is one day old ranks higher than two-day-old content. And so on.

Confluence determines the weightings for each of the three properties of a matching content item. It then multiplies the three weightings together and compares the result to the other matching items. The higher result will come higher in the list of search results.

The following example comes from the *Confluence User's Guide*. It offers a simplified scenario for illustrative purposes only.

Let's assume you are searching for a single term. Confluence finds a match in the title of an email message created today:

Weighting for the content type (email)	0.5
Weighting for field (title)	9
Weighting for age (today)	1.5
Total weighting	**6.75**

Confluence also finds a match in the content of a comment created three weeks ago:

Weighting for the content type (comment)	5
Weighting for field (content)	5
Weighting for age (up to 1 month ago)	1.2
Total weighting	**30**

Result: The comment (weighting 30) will be higher in the search results than the email message (6.75).

A quick guide to SEO

Here is a short list of practical tips for helping people find information on your documentation wiki, whether they are using the Confluence search or an external search engine like Google, Bing, or Yahoo!

- **Use good keywords, use them often, and use them consistently.** Spend some time deciding what the key terms are in your documentation and your product. Choose terms that you think people will naturally search for when trying to find the information you supply. Also choose terms that are important to your organization's marketing profile. Then, make sure that authors use these keywords all over the place: in titles, in headings, in the content of the page (especially near the top), and in labels. That said, you should use the keywords only where relevant to the content of the page. Do not let authors shoot keyword spam at the search engines and readers.

- **Make the page titles meaningful.** This will make Confluence's quick navigation feature more useful to readers. The page will appear higher up in the list of search results for both Confluence and external searches. It will also help the human eye find what it needs more quickly.

- **Keep the page titles short.** Keep your page names as short as possible, and put the significant words at the beginning of the name. Search engines like Google truncate page titles in the search results. The maximum advisable length is 60-65 characters. As far as external search engines are concerned, the page title is the content of the HTML `<title>` element inside the `<head>` element. Check the title that is concocted by your wiki. Confluence is pretty good on this score. It puts the page title first, followed by the space name and then the wiki site name. For example, Ganache has a page called *Introduction to Twitter* in her guide named *Using Twitter*. The page is on the *Confluence, Tech Comm, Chocolate* wiki.[7] The title in the HTML header looks like Example 9.1

```
<html>
<head>
<title>Introduction to Twitter - Using Twitter - Confluence, Tech Comm,
Chocolate</title>
</head>
<body>
Tweet Tweet.
</body>
</html>
```

Example 9.1 – Title in the HTML header

- **Avoid punctuation and special characters in page titles.** Search engines pay attention to the words in the URL of a page. For that reason, the URL is a good place to put your keywords. When using a wiki or a content management system, we do not have full control over the format of the URL, but we can control part of it. Confluence builds the URL of a page based on the page name along with the base URL of the wiki. If possible, it uses the exact words in the page name, joined together by plus signs. For example, if the page name is *Introduction to Twitter* then the last part of the URL is `Introduction+to+Twitter`. Here is the full URL of the page on the *Confluence, Tech Comm, Chocolate* wiki: `https://wikitechcomm.onconfluence.com/display/DOC/Introduction+to+Twitter`. This is a pretty URL, predictable and easy to read for people as well as machines. On the other hand, if the page name contains a character that Confluence cannot put directly into a URL, like a question mark, the URL looks something like this: `https://wikitechcomm.onconfluence.com/pages/viewpage.action?pageId=854392`. Ouch!

- **Add labels to the pages.** This is closely related to the point about keywords, but it is useful to highlight it separately. Labels, sometimes called tags, are significant in the weighting algorithms applied by the Confluence search. One factor that may be confusing is that, apparently, Google does not give much weight, if any, to the "keyword" meta tag in the HTML page header.[8] However, Confluence labels are not HTML meta tags. Instead, they form part of the body of the HTML page. Every label is also hyperlinked to another page, which lists all the pages tagged with that label. And, those page names are in turn hyperlinked back to the pages themselves. This cross linking gives the pages a higher weighting in Google and other search engines.

- **Add hyperlinks in and to your documentation.** Hyperlinks are like votes. Google and other search engines give a page a higher rating if there are links to it from other pages. The algorithms are complex and ever changing, and the incoming links must come from pages which are themselves "respected" – junk is not good. But on the whole, the more links the better. Add hyperlinks on your organization's website that point to the documentation. Write blog posts about the documentation with links into the relevant pages. Add related topics and other links within the documentation pages themselves. As technical communicators, we know that it is distracting to the reader to have too many links within the text. Instead, add links in a "related topics" section at the top or bottom of the page. This will be a help rather than a hindrance for the human readers. In Confluence, there are some handy macros that make this easy. We'll look at those later in this chapter.

Peg Mulligan[9] and Tom Johnson[10] have written some excellent posts about technical communication and SEO. The "References" section at the end of this chapter also lists some further reading about SEO in URLs and hyperlinks.

We technical writers can pat ourselves on the back. Technical documentation is by its very nature search-engine friendly. It is concise, choc-a-bloc with keywords, well structured, authoritative, and comprehensive. This makes a good argument when persuading management of the financial benefits provided by technical documentation, particularly online documentation.

More about labels

The previous section discussed using labels to increase the visibility of a page in the Confluence search results and in external search results. Now let's look at labels in more detail.

Using labels to add terminology to a page

You can use labels to add similes or obscure terminology to the page. This will ensure that the page appears in the search results, even if the word is not mentioned in the body of the page.

Perhaps you have no intention of mentioning the word "schmilblick" on your page, but you know that some people will search for it and expect to find the page. The doughty Ganache finds herself in just such a situation. In the hallowed halls of *Choc Ink*, where Ganache works, the term "schmilblick" has become inextricably linked with Twitter. The reasons for this association are obscure, having something to do with an elastic band, a time machine, and the documentation written by a French technical writer hired for a short duration. Nevertheless, to aid those hapless souls who search the *Choc Ink* documentation base for "schmilblick," Ganache as added the label to a few pages (see Figure 9.4):

Are you wondering what a "schmilblick" is?

A schmilblick is a nonsense, something that does not exist and can therefore be anything you want it to be. The term is most commonly used in the French language, although it has found its way into popular English and other cultures, too.[11]

Figure 9.4. Labels on a Confluence page

Using labels to collect related topics

Confluence's content-by-label macro[12] displays a list of pages that are tagged with a given label or labels.

Ganache has decided she needs a list of her schmilblick pages. She adds the content-by-label macro to a page (see Figure 9.5) called *Cherchez le schmilblick* on the *Confluence, Tech Comm, Chocolate* wiki.[13]

Figure 9.5. The content-by-label macro

The wiki markup for Ganache's content-by-label macro looks like this:

```
{contentbylabel:labels=schmilblick|showLabels=false|title=Pages labeled with
 "schmilblick"}
```

This is a useful way of collecting information. Instead of manually adding the links or making a hierarchical table of contents, you can use the content-by-label macro to display:

- Related topics at the top or bottom of a page.

- A group of frequently asked questions (FAQ) on a specific subject. In that way, one single FAQ entry can be used in more than one list. Just add extra labels to the page.

- The functions that a specific type of user, such as an administrator, can perform.

- Anything else you can dream up.

Let's look at an example of the last use case mentioned above, this time from the Bitbucket documentation.[14]

The wiki markup on the page in Figure 9.6 looks like this:

```
{contentbylabel:label=bb-admin|showLabels=false|
showSpace=false|maxResults=99|space=@self|sort=title}
```

The macro collects all the pages labeled with "bb-admin," restricting the list to the current space and sorting by page title.

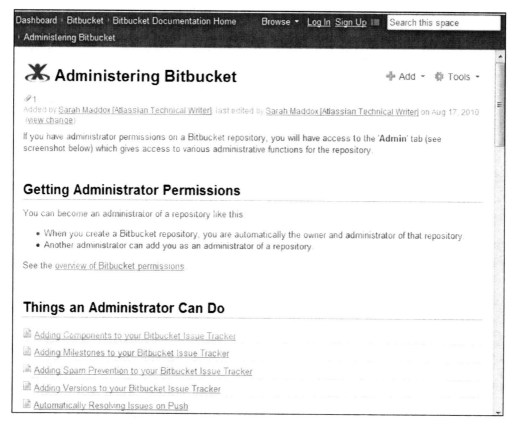

Figure 9.6. The content-by-label macro used to display administrative functions

Using labels to create an index or tag cloud

Isn't an index old hat these days? Do readers need an index as well as the online search? How can you create an index easily? Here is a suggestion. Maybe a tag cloud is the index *du jour*. Maybe labels can be the key words in the cloud.

There are a few different ways of viewing the labels applied to Confluence pages. One way is a visual representation called a tag cloud or heat map. Figure 9.7 shows the popular labels in the space containing Ganache's Twitter guide. Popular labels are the labels that are most frequently applied to pages in the space.

View: **Popular Labels** | All Labels

Below are the **16** most popular labels used in **Using Twitter**.
The bigger the text, the more popular the label. Click on a label
to see its associated content.
See also: global popular labels.

Order: **alphabetically** | popularity

apps chat documentation embed follow language
prepopulate quickstart schmilblick signup social
stream tweeting **twitter** widget wiki

Figure 9.7. Popular labels

To see the popular labels in a space: Go to a page in the space, then choose "**Browse**" > "**Labels**" >
"**Popular Labels.**"

As well as the popular labels, you can see the full list of labels in the space arranged alphabetically.
From either list of labels (popular or alphabetical) you can click through to see the pages tagged with
a specific label, then click through to a specific page.

Good, so that means that our readers can see a list of labels if they happen to follow the instructions
above. But, it would be nice if we could supply a link that says, "**Index.**" Clear and simple. Well, you
can do that too!

Adding a hyperlink that points to to the index

In Confluence, most things work via a URL. The URL for the alphabetical list of labels looks like this:

```
<CONFLUENCE-SITE>/labels/listlabels-alphaview.action?key=<SPACE-KEY>
```

Replace `<CONFLUENCE-SITE>` with the base URL of your Confluence site, and `<SPACE-KEY>` with
your space key.

So, in Ganache's Twitter guide, the URL is as follow (this URL, both here and below, is split after "labels/". If you copy the URL, join up the two lines without any spaces):

```
https://wikitechcomm.onconfluence.com/labels/
listlabels-alphaview.action?key=DOC
```

Given that information, it is easy to create a hyperlink on the word "index." Ganache decides to add an index link to the left-hand navigation panel in her space. She does that by configuring the Documentation theme (see Chapter 7) and adding the following wiki markup to the navigation section of the theme configuration:

```
[Index|https://wikitechcomm.onconfluence.com/labels/
listlabels-heatmap.action?key=DOC]
```

The square brackets are wiki markup for a hyperlink. The hyperlink consists of two parts, the linked text ("Index") and the URL.

Notice the URL in the above wiki markup. Ganache decided to link to the tag cloud, or "heat map," rather than the straight alphabetical list of labels.

Now the "**Index**" link appears in the left-hand navigation panel on all pages in the Twitter guide (Figure 9.8).

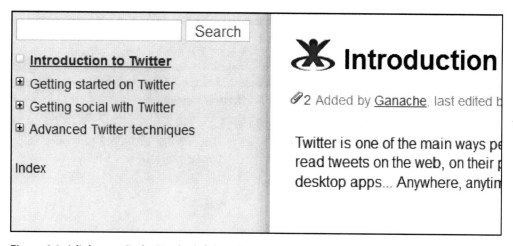

Figure 9.8. A link to an 'index' in the left-hand navigation panel

People can click the index link to jump straight to the tag cloud. You've got to believe we are magic!

Note: To be precise, Ganache had to do a little more than add the single line mentioned above to the configuration section of the Documentation theme. She also had to disable the default page tree and add it manually. The reason is that, by default, anything you add to the navigation panel will appear *above* the default page tree. If you want to put something below the page tree, you must remove the tick from the "**Page Tree**" check box and then add the wiki markup for the page tree yourself. This is what Ganache put into the "**Navigation**" text box:

```
{pagetree:searchBox=true}

[Index|https://wikitechcomm.onconfluence.com/labels/
listlabels-heatmap.action?key=DOC]
```

Extended use of labels and other metadata

Confluence supplies a number of macros that you can use to collect and display pages in logical groups by means of their labels. For example, there is a macro called the popular-labels macro. It displays a cloud, or heat map, of the most popular labels in the space. In fact, Ganache could use this macro for her index instead of the URL described in the previous section. The *Confluence User's Guide*[15] describes this macro and others. It also tells how to get an RSS feed filtered by labels, and how to view collections of labels in different ways.

Labels are a basic form of metadata. Does Confluence allow for more advanced manipulation of content via metadata?

- Core Confluence includes the metadata macro and the metadata-details-summary macro. Use the metadata macro to add properties to pages, then use the metadata-details-summary macro to display a list of those pages along with a tabulated list of the properties for each page.[16]

- The *Confluence Metadata Plugin 2* provides an extensive set of options for adding metadata to pages and spaces and for generating reports based on that metadata. It also lets you define a hierarchy of spaces by nominating one space as the parent of another. In core Confluence, it is not possible to nest spaces in this way.

What you have discovered about findability and searchability

Given that wiki documentation is online, most people will use a search tool to find the information they need. If the documentation is visible on the web, people will come in via Google and other external search tools. Once they have found a page, they may use the browser's page search to get to a specific piece of information on the page. They will also use the wiki search to find other pages within the documentation suite. If the documentation has a good structure, readers will happily move around the table of contents and click through to related topics. An index provides another way of accessing the information. Using labels and more complex forms of metadata, you can link related topics and provide alternative pathways into the documentation.

It is a good thing for a documentation designer to know the tools provided by the documentation platform. The Confluence search provides a quick navigation tool that matches on page titles only and a full search that matches on content as well as title. The full search will rank the results by applying a documented algorithm. Readers can filter the results by space, author, date, and content type. The Documentation theme adds a space-specific search box to the left-hand navigation panel, and also provides the option to make the quick navigation feature space-specific.

SEO (search engine optimization) is a big field. Yet there are some simple techniques you can apply when designing the documentation to optimize the search results, both for the internal search and for external search engines. In brief: give a lot of thought to keywords and titles.

The words findability and searchability may make you want to tear your hair out, but I cannot think of better ones!

References

These are the tools, websites, blog posts, and other references relevant to the content of this chapter.

Useful plugins

- *Confluence Metadata Plugin 2* by Andy Armstrong, for extended metadata and reporting functionality: `https://plugins.atlassian.com/plugin/details/5295`

- *Livesearch Macro plugin* by Atlassian, for adding a search box to your page. The search results appear in a dropdown list that changes to match the search term as the user types: `https://plugins.atlassian.com/plugin/details/168`

Further reading

- *SEO Cheat Sheet: Anatomy of A URL*, a blog post by Dr. Pete on *The Daily SEO Blog*, October 2009: `http://www.seomoz.org/blog/seo-cheat-sheet-anatomy-of-a-url`

- *Top 10 Fatal URL Design Mistakes*, a blog post by sam.d on *SEO Blog*, July 2008: `http://seo2.0.onreact.com/top-10-fatal-url-design-mistakes`

- *Building good quality SEO hyperlinks*, a blog post by Ben Morris on *Web and digital development stuff*, March 2008: `http://www.ben-morris.com/good_quality_seo_hyperlinks`

- Rhonda Bracey publishes an information-rich blog about technical writing. Every now and then she writes about Google search results that are amusing or that have provoked her curiosity. In *Suspicious Google search results* Rhonda describes a search for airline flight information that resulted in a list of people's names, all with the first name of Edward. Another post is titled *Google — you are too funny!* Rhonda had entered a search term of "anagram" and Google had responded, "Did you mean: *nag a ram*." Understandably, this made Rhonda laugh. On a more serious note, her post called *Search limitations* points out the folly of providing a less-than-helpful search engine on your site.

  - *Suspicious Google search results*, October 2011: `http://cybertext.wordpress.com/2011/10/06/suspicious-google-search-results/`

  - *Google — you are too funny!* August 2011: `http://cybertext.wordpress.com/2011/08/12/google-you-are-too-funny/`

  - *Search limitations*, June 2011: `http://cybertext.wordpress.com/2011/06/28/search-limitations/`

Endnotes

[1]   A *schmilblick* is a nonsense, something that does not exist and can therefore be anything you want it to be. The term is most commonly used in the French language, although it has found its way into popular English and other cultures too.

[2] The lyrics to one of my favorite songs, *Why Can't the English?* from *My Fair Lady*, 1964, a musical by Frederick Loewe with lyrics by Alan Jay Lerner: `http://www.stlyrics.com/lyrics/my-fairlady/whycanttheenglish.htm`

[3] *Searching Confluence*, in the *Confluence User's Guide*: `http://confluence.atlassian.com/display/DOC/Searching+Confluence`

[4] *Confluence Search Syntax*, in the *Confluence User's Guide*: `http://confluence.atlassian.com/display/DOC/Confluence+Search+Syntax`

[5] *Confluence Search Fields*, in the *Confluence User's Guide*: `http://confluence.atlassian.com/display/DOC/Confluence+Search+Fields`

[6] *Ranking of Search Results*, in the *Confluence User's Guide*: `http://confluence.atlassian.com/display/DOC/Ranking+of+Search+Results`

[7] *Introduction to Twitter* on the *Confluence, Tech Comm, Chocolate* wiki: `https://wikitech-comm.onconfluence.com/display/DOC/Introduction+to+Twitter`

[8] *Google does not use the keywords meta tag in web ranking*, a post by Matt Cutts on the *Google Webmaster Central Blog*, September 2009: `http://googlewebmastercentral.blogspot.com/2009/09/google-does-not-use-keywords-meta-tag.html`

[9] *Trends in Technical Communication: Exploring SEO-Friendly Authoring Tools*, a blog post by Peg Mulligan on *Content for a Convergent Word*, March 2011: `http://pegmulligan.com/2011/03/23/trends-in-technical-communication-exploring-seo-friendly-authoring-tools/`

[10] *Search Engine Optimizing Your Help Content for Google [Organizing Content 11]*, a blog post by Tom Johnson on *I'd Rather Be Writing*, May 2010: `http://idratherbewriting.com/2010/05/28/search-engine-optimizing-your-help-content-for-google-organizing-content-10/`

[11] Go to *Forvo* to see how to pronounce "schmilblick." `http://forvo.com/word/schmilblick/`

[12] Confluence's content-by-label macro, in the *Confluence User's Guide*: `http://confluence.atlassian.com/display/DOC/Content+by+Label+Macro`

[13] *Cherchez le schmilblick*, on the *Confluence, Tech Comm, Chocolate* wiki: `https://wikitech-comm.onconfluence.com/display/CHAT/Cherchez+le+schmilblick`

[14] Bitbucket administrator's guide: `http://confluence.atlassian.com/display/BITBUCKET/Administrating+a+repository`

[15] Guide to using Confluence labels, in the *Confluence User's Guide*: `http://confluence.atlassian.com/display/DOC/Working+with+Labels`

[16] Guide to the metadata macro, in the *Confluence User's Guide*: `http://confluence.atlassian.com/display/DOC/Metadata+Macro`

10

Workflow and release management

Workflow. It's what makes the world go round. We may rail against it as too restrictive, but most problems come when a workflow is too complex or does not fit the environment where it is used. What happens when you add a wiki to the mix?

The flow of this chapter

This chapter is about the life of a document from draft, through review, to publication and beyond. The chapter has the following sections:

- **Workflow: what's in the box.** Use the core Confluence functionality to take a page through the standard workflow: draft, review, and publish.

- **Release management.** Provide a complete set of documentation for each version of the product, using wiki spaces as the vehicle for release management.

- **Extended workflow and publishing capabilities.** Add sophisticated workflow and publishing tools to the core wiki functionality using plugins.

- **Sticky notes for review comments.** Attach movable comments to the content of the page using another plugin – useful and fun during the review process.

- **Going with the flow.** A summary of this chapter.

- **References.** The tools, websites, blog posts, and other references mentioned in this chapter.

Workflow: what's in the box

Keep it simple. That is the wiki philosophy. That is what's in the box. For many environments, that is all a technical communication team needs. This section is a guide to using the core Confluence functionality in a standard workflow, taking a document through the stages of draft, review, and publication. This will help you decide whether the core functionality meets your workflow requirements. Later in the chapter we will look at some plugins that provide extended workflow capabilities.

In this context, "draft" means a page that the author considers to be in the draft stage. It does not refer to a status or workflow stage in Confluence itself. In particular, it has nothing to do with the drafts that Confluence saves automatically every few seconds while you are editing a page.

Here is a summary of how to draft, review, and publish a page using the core functionality:

- **Drafting a page.**

  1. Add a hidden page to function as your draft:

     - If it is a new page, add the page as usual.

     - If you are revising an existing page, copy the page.

     - Set the page restrictions so that only you and other authors can see the page.

     - Save the page.

  2. Add the necessary content to the draft page, saving the page whenever you want to.

- **Organizing a review.**

  1. Set the page restrictions so that the reviewers can see the page. Also let them edit the page, if that is how you want the review to work.

  2. Make sure that the space permissions allow reviewers to add comments to the page.

3. Send a request asking people to review the page. You can use the page's "**Share**" option to send an email request.

4. Respond to and incorporate the review comments:

   ■ Reviewers will add comments to your page.

   ■ If you have allowed it, reviewers can also update the page directly.

   ■ Use the page history and "view change" options to see the changes made.

■ **Publishing the page.**

1. Remove all review comments.

2. Finalize the content of the page.

3. Make the content visible to all:

   ■ If this is a new page, remove the page restrictions so that everyone who has permission to see the space can also see the page.

   ■ If you are revising an existing page, copy the revised contents to the original page and delete the copy.

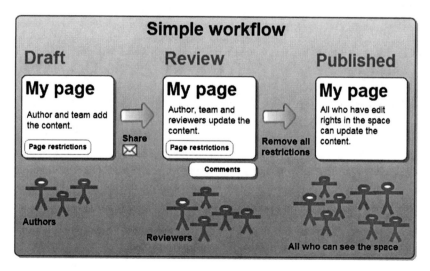

Figure 10.1. Simple workflow using core Confluence functionality

Let's take a closer look at each stage (see Figure 10.1 for a diagram of the workflow).

Drafting a page

The "page restrictions" in Confluence offer a handy way of hiding a page while it is under construction. If you are adding a new page, hide the page as soon as you add it. If you need to update an existing page, then it is useful to make a copy of that page and hide the copy so that you can work on it in peace.

Of course, it is not always necessary to make a copy of a page when you want to change it. If the change is simple and already decided, you can just go in there and edit the page. The quick edit-update-save workflow is the magic of a wiki!

When you add a page in Confluence, the page inherits the permissions (restrictions) of its parent page and of the space that the page is in. You can also add specific viewing and editing restrictions to the page itself. The way to do it is slightly different in Confluence 3.5 and Confluence 4.

To restrict viewing rights on a page:

1. Add, copy, or edit the page.

2. Choose the option for editing page restrictions, located under the editor panel:

   - In Confluence 3.5 it looks like this:

     Restrictions: Edit

   - In Confluence 4 it looks like this when there are no restrictions defined:

     Unrestricted

     and like this when there are restrictions on the page:

     Restricted

   - Choose whether you want to restrict viewing or editing the page. If you restrict viewing, then only people or groups that you choose will be able to see the page. Other people will not see it and will not be able to edit it, either. If you restrict editing, then other people can see the page but not edit it.

   - Specify the users or groups that should be able to see or edit the page. You can limit the rights to just yourself, to specific usernames, or to groups of people.

   - Save the restrictions, then save the page.

Now that the page is hidden, you have free rein. Use the usual technical writing techniques to outline the content of the page, add words and images, and get the draft to a stage where it is ready for review. Remember to save the page at regular intervals.

Confluence keeps a history of all your updates, and stores a separate version each time you save the page. If you want to retrieve some content that you removed during a too-enthusiastic cleanup of your draft, you can look at an earlier version of the page, copy the content, and paste it back into your most recent version. You can even revert to a previous version of the page. When you do that, Confluence will keep all the versions. The reversion becomes the latest version of the page. This means that you can undo the reversion itself, by reverting to the preceding version.

To see the history of a page: Go to the page, close the editor if you are editing the page, then choose "**Tools**" > "**Page History.**"

When viewing the page history you can select two versions and ask Confluence to compare them. They do not have to be sequential versions. Select version 1 and version 5 to see an aggregated view of the differences between them.

Confluence also stores a draft of your page at regular intervals. This is the auto-save draft, which is not the same thing as a page in draft stage. You will see a message appear above the editor panel at regular intervals, telling you when the last draft was saved. In case of a catastrophe, such as a laptop meltdown or an Internet tsunami, you can retrieve the last draft saved.

To retrieve an auto-saved draft: Claw your way back from whatever disaster caused the loss of the editing session and log back in to Confluence. Open the dropdown menu under your name at the top of the screen, then choose "**Drafts.**"

Organizing a review

Having team members and subject matter experts review a page is exciting, demanding, frustrating, and rewarding. The review stage is what makes great content awesome. It has saved me from many a howling mistake. Most technical communicators can tell tales of bloopers saved by peer review.

A wiki is built as a tool for collaboration, and it comes into its own in the review stage. You can choose whether reviewers should give their feedback as comments on the page or edit the page itself. Confluence has some fairly nifty code to handle concurrent editing. When you start editing a page that someone else is also editing, Confluence displays a message above the editor panel. You can choose to continue editing or cancel out and leave the other person alone.

The woman who thought phonographs taste like chocolate

Some people can taste words. This is a special form of synesthesia, which is what happens when two of your senses are intertwined. Some people see shapes when they hear a sound or associate colors with numbers. A report on *LiveScience.com* tells of a woman who tasted Dutch chocolate whenever she thought of a phonograph.[1]

If a number of people edit the same page at the same time, Confluence will do its best to work out if any of the changes conflict with each other, for example if two people edit the same paragraph. If possible, Confluence merges the changes when people save their versions of the page. If the changes conflict, then the last person to save will see a message that reports the conflict and asks him or her to resolve it.

Next, let people know that the page is ready for review. The easiest way is to use the "share" option.

When you are ready for the review, make sure that all of your reviewers can see the page. Adjust the settings for restricted viewing and editing of the page, and the permissions on the space, so that the reviewers can add comments and/or edit the page.

 Edit or comment?

Or both. I find that it works best to let people choose whether to add comments to the page or edit it directly. They have the freedom to fix small things quickly and to add comments for the bigger things. If someone makes a change that I disagree with, I can always undo it, and it's easy to monitor what's happening. We will look at that in greater detail in Chapter 14.

To share a page:

1. Go to the page, close the editor if you are editing the page, then choose "**Share.**"
2. Specify one or more people to send a message to. Start typing the person's name. Confluence will look for matching users on the wiki.
3. Enter a covering note.
4. Choose "**Share**" to send an email message to the person or people you have specified.

Set a watch for your page to be notified when comments are added or changes are made. Respond to the feedback and update the content in the usual way.

To watch a page: Go to the page and choose "**Tools**" > "**Watch.**"

Hint: Suggest to the reviewers that they watch the page, too. That will ensure that they keep abreast of events. By default, Confluence will add a person as a watcher of a page whenever that person makes changes to the page. But people can turn this functionality on and off.

Chapter 14 goes into detail about monitoring content on the wiki.

Publishing the page

It is publication day, the most exciting day in a technical writer's diary. Things need to happen fast. Make sure you have a list of the pages to be published and a good strong pen to tick them off with.

If the page is new:

1. Remove all the comments on the page.

2. Remove the page restrictions so that everyone who has permission to see the space can also see the page.

If the page is a revision of an existing page:

1. Check the original page to make sure that there have been no changes since you copied the page to make your draft. If there have been changes, apply them to the draft if they are still relevant.

2. Open the draft page in the editor and copy all its content into the clipboard.

3. Open the original page and paste the new content into it.

4. If necessary, attach any revised images or other files to the original page too.

5. Delete the draft page.

Drawbacks of the simple workflow

The above procedure works well for simple changes, but it becomes time consuming and error prone when there are many updates to be published. In addition, technical documentation in some industries is subject to strict regulation and therefore requires a more sophisticated workflow. Plugins to the rescue! More in the section titled "Extended workflow and publishing capabilities."

Ongoing updates

The story does not end when the page is published. People will need to update the content continually. A support engineer may add information about a known problem and workaround. A technical writer may refine the step-by-step instructions. A reader may fix a typographical error. When the next version of the product is released, the page may need extensive revision. Figure 19.1 in Chapter 19 illustrates the full life cycle of a page. All the techniques discussed above apply to these ongoing updates too.

Release management

It is quite likely that customers will need different versions of the documentation to match the versions of the product that they own. This is true if customers buy a specific version of the product and the developers then issue a later release with updated and new features. It is not true for hosted (software as a service) environments, where the developers can push out a new version to all customers at the same time.

If the documentation is shipped with the product, the technical writers make sure that the version-specific guides are packaged and installed in time to reach the customer with as the product. After shipping, there is no chance to change the documentation.

Things are a little different on a wiki. Customers come to the wiki to read the latest documentation. They get the benefit of all the updates made, issues discovered, and workarounds posted. Yet they still need documentation specific to their version.

Using spaces for release management

Enter the wiki space as the mechanism for providing version-specific documentation. Using spaces, you can match the release of documentation to the release cycle of the product.

The gallant Ganache has written a user's guide and an administrator's guide for *ChocTactics*, the chocolate assessment tool developed by *Choc Ink*. Now the developers are about to release a new version of *ChocTactics*. Ganache decides on the following structure for her documentation wiki:

- An overview space, holding the introductory material, quick start guide and release notes. Ganache gives this space a key of "TACT."

- A number of version-specific spaces, holding the detailed guides for each version of the product. These spaces have keys of "TACT010," "TACT011" and so on.

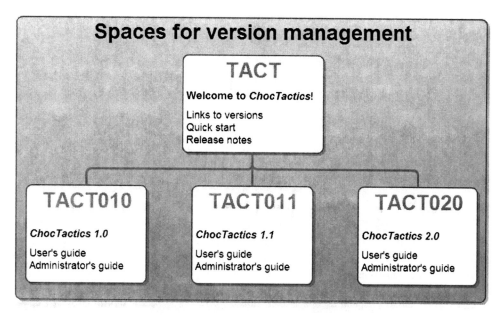

Figure 10.2. Spaces as mechanism for version control

A note: The diagram in Figure 10.2 does not imply a hierarchical structure of spaces. The spaces are not linked in any way except conceptually and via hyperlinks from the one to the other.

In software development, a product release cycle usually includes major releases, minor releases, and point releases. The naming of each type of release differs. Point releases are often called bug-fix releases. There may also be confusion in the naming of minor releases and point releases. For the purposes of this book, let's use the following convention:

- Major releases include significant changes and new features. The version number in front of the first point is incremented: 1.0, 2.0 and 3.0.

- Minor releases include improvements and less-significant new features. The version number after the first point is incremented: 1.1, 1.2 and 1.3.

- Point releases or bug-fix releases include fixes to reported problems. The version number after the second point is incremented: 1.1.1 and 1.1.2 and 1.1.3.

Both major and minor releases are likely to contain changes in the product that require significant changes in the documentation. If customers read the documentation for the wrong version, they will probably be confused.

Ganache's mantra is, "I see confused people." The documentation is there to help these people. Therefore, Ganache creates a new space for each major or minor release: 1.0, 1.1, 1.2, 2.0, 2.1, and so on.

The point releases, or bug-fix releases, should not include changes that affect the documentation. In principle they are merely fixing the product so that it behaves as it should. Therefore, there is no need to create a new space for a point release.

Creating the space for a new version

To create the space for a new version of the product, you can copy the space containing the latest version. We saw how to do that in Chapter 8, when we were copying a space to act as a template. You will need the *Copy Space plugin*, as described in that chapter. Here is a quick refresher, which assumes the plugin is already installed on your Confluence site.

To copy a space: Go to a page in the space, then choose "**Browse**" > "**Space Admin**" > "**Copy Space**."

An alternative structure for version-specific spaces

Figure 10.2 shows a simple structure with an overview space and a set of version-specific spaces.

An alternative structure looks like this:

- The main space (for example, TACT) always contains the documentation for the *latest* version of the product. The version documented in the space will therefore change over time.

- The version-specific spaces (TACT010, TACT011, TACT020 and so on) contain the documentation for *previous* versions of the product. These archive spaces come into existence only when the version is no longer the latest version.

Let's assume Ganache is about to release a new version of the documentation. At the moment, the latest version of *ChocTactics* is version 1.1, but the company is about to release version 2.0. In the alternative space structure, the TACT space currently documents version 1.1.

When *Choc Ink* releases *ChocTactics 2.0*, Ganache creates the archive space TACT011 to contain the documentation for version 1.1. She also updates the content of the main space TACT so that it documents version 2.0 (see Figure 10.3).

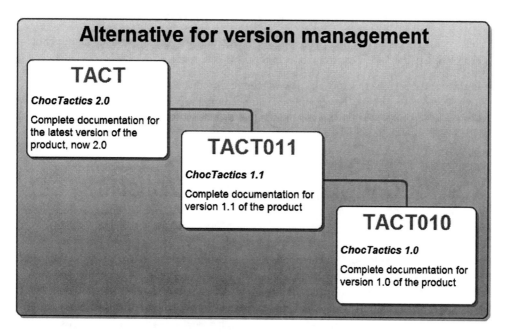

Figure 10.3. An alternative space structure for version control

The Atlassian documentation is currently structured using the approach illustrated in Figure 10.3. As I am writing this chapter, we are considering a change to the structure shown in Figure 10.2. There are pros and cons to both structures. As always, the choice depends on the requirements of the readers and the environment that the documentation serves.

These are some of the advantages of the second structure (Figure 10.3):

- Readers can monitor updates to the TACT space and know that they are always getting the latest news.

- External links pointing to pages in the TACT space will always point to the latest version of the documentation.

- People can add comments to the TACT space, using it as a central hub for conversation and community contributions.

On the other hand, the first structure (Figure 10.2) is simpler to maintain and simpler to understand.

- It is clearer to readers that the documentation is version-specific.

- Readers will always know where to find the documentation for a specific version. They do not have to look first in the TACT space and later in the TACTxxx space.

- Authors can create the space for the new version of a product as soon as development starts. The workflow is much simpler, because there is no need to hide pages that apply to the new version while working in the space that is still branded for the currently-released version.

Extended workflow and publishing capabilities

In many industries technical documentation is subject to strict regulation. Examples are the medical, aviation, military, and financial industries. Such documentation needs a strict review and approval workflow. Confluence does not provide any workflow tools as part of the core wiki functionality. The basic components of the draft, review, and publish process are available as described above. But if you want to mesh them together into a managed flow, you will probably want to extend the core functionality by installing a plugin.

There are a few plugins that add sophisticated workflow functionality to the wiki. The *Frevvo Live Forms and Workflows plugin* adds forms and associated workflows integrated with Confluence pages and with *Google Apps*. The *Ad hoc Workflows plugin* comes with a set of predefined workflows. You can also define your own workflows and share them with other workflow designers. The *Remote Publishing Plugin* adds one-click publishing to your wiki. In order to use it, you must have the *Ad hoc Workflows plugin* installed too. With the use of this plugin, you can develop your content in one space and then publish all the pages to a destination space on release date.

The section titled "Useful plugins and tools," at the end of this chapter, has more information about the above plugins. It also describes two more plugins where extended workflow functionality is currently under development: *Scroll Versions* and *Zen Foundation*.

To find the full list of workflow plugins at the Atlassian Plugin Exchange: Go to http://plugins.atlassian.com, type "**workflow**" in the search box and select "**Confluence**," then click "**Search**." The *Confluence CLI* is another tool worth mentioning here. It is very useful for bulk operations such as deletion of pages, setting of page restrictions, and much more. This is not a plugin, but rather an external tool that you can download and install on your own computer. It interacts with a Confluence server via the remote API. To use it, you specify the URL of the Confluence server, a username that has administrative access to that Confluence site, and the password of that username. Then you supply one or more commands, either in a batch file (.bat on Windows or .sh on Mac OS X) or on the command line. This may be very useful on release date, for example, when you want to delete or publish a number of pages all at the same time. The tool's documentation[2] includes some useful examples and a list of all the available actions and parameters.

Sticky notes for review comments

Reviewers can add comments to a page. That is part of the core Confluence functionality. If you install the *Spartez Kwik plugin*, reviewers will be able to add "sticky notes" to the page (see Figure 10.4). This is a really cute plugin that makes the review process fun and interesting. When you add a sticky note, you can make it visible to just yourself or to everyone. People can move the sticky notes around, change their color and dismiss them.

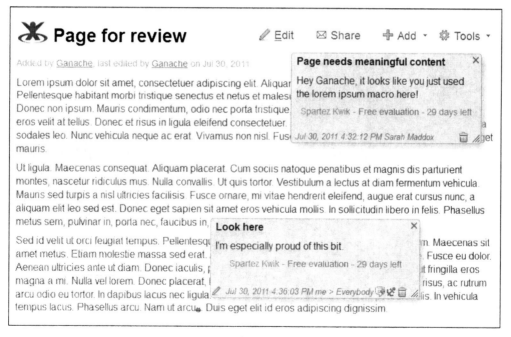

Figure 10.4. Sticky notes on a Confluence page, from the Spartez Kwik plugin

As with many tools, you can use this plugin free of charge for a thirty-day evaluation period. I did that when trying it out for this book. That is why the screenshot in Figure 10.4 shows the words, "Free evaluation - 29 days left."

Going with the flow

Having your content in a central location, where reviewers and editors can congregate and collaborate, saves a world of trouble. With the core wiki functionality, authors and reviewers can follow a simple workflow to draft, review, and publish a page. People can edit the page concurrently or consecutively, keep track of what other authors and reviewers are doing, and backtrack to a previous version of the page if necessary. When the time comes to publish, it is just a matter of removing page restrictions so that other people can see the page.

Spaces are a convenient mechanism for matching complete sets of documentation to specific product releases.

Some environments require a more sophisticated and comprehensive workflow, and will benefit from plugins that add extended workflow and publishing tools. Try a sticky notes plugin to add some fun as well as some very useful functionality during the review stage.

If possible, try to keep the flow as straightforward and untrammeled as possible. Simplicity is beauty. That's the wiki way.

References

These are the tools, websites, blog posts, and other references relevant to the content of this chapter.

Useful plugins and tools

- *Ad hoc Workflows plugin* by Comalatech, for adding workflow capabilities to Confluence: `https://-plugins.atlassian.com/plugin/details/142`

- *Arsenale Lockpoint plugin* by Arsenale, for adding checkin and checkout functionality to Confluence attachments. This means that you can lock an attachment for exclusive editing. Such attachments can be Microsoft Word, Excel and PowerPoint documents, Gliffy diagrams, Balsamiq mockups, and all other file types: `https://plugins.atlassian.com/plugin/details/36199`

- *Confluence CLI* by Bob Swift, a command-line client for interacting with a Confluence server: `https://plugins.atlassian.com/plugin/details/284`

- *Confluence Remote Publishing Plugin* by Comalatech, extending the *Ad hoc Workflows plugin* to publish content from a master Confluence space to a destination space: `https://plugins.atlassian.com/plugin/details/42115`

- *Confluence Workflows Utilities Plugin* by Comalatech, extending the *Ad hoc Workflows plugin* with utilities such as a workflow event logger and new comment trigger events: `https://plugins.atlassian.com/plugin/details/38741`

- *Copy Space plugin* by Atlassian, used to add a space by copying the content of another space: `https://plugins.atlassian.com/plugin/details/212`

- *Frevvo Live Forms and Workflows for Confluence*, for creating input forms and associated workflows: `https://plugins.atlassian.com/plugin/details/8965`

- *Scroll Versions*, a new plugin under development by K15t Software. When this plugin is installed and activated in a Confluence space, you will be able to define a number of versions of your documentation. The plugin offers flexible options for publishing a version at any time, to the same space or a different space. See the section titled "K15t Software's focus on wiki documentation tools" (p. 436) for more details about the functionality in this plugin.

- *Spartez Kwik plugin*, for adding sticky notes to Confluence pages: `https://plugins.atlassian.com/plugin/details/37075`

- *Sticky Notes and Snip-Edit Plugin* by Comalatech, for adding sticky notes and editing sections of pages. With this plugin, sticky notes are not automatically available on every page. To enable sticky notes on a page, add the stickynotes macro at the beginning of the page: `https://plugins.atlassian.com/plugin/details/193`

- The *Zen Foundation* theme for Confluence, by Stepstone Technologies, adds the option to save a page as a draft. The author can then work on the page, saving the draft as often as needed, and publish it when ready. At the time of writing this book, the drafts are visible to the author only. The next version of *Zen Foundation*, due for release by the time this book comes off the press, will add the ability to share drafts with all users who have write access to the page. It will also offer bulk publishing of draft pages: : `https://plugins.atlassian.com/plugin/details/32965`

Endnotes

[1] *New Insight into People Who Taste Words*, an article on *LiveScience.com*, November 2006: `http://www.livescience.com/1141-insight-people-taste-words.html`

[2] Documentation for the *Confluence CLI*: `https://studio.plugins.atlassian.com/wiki/display/CSOAP/Confluence+Command+Line+Interface`

## 11

Importing and exporting content

Content cannot be stuck on a wiki. We must be able to move it around. At times we need to peel it off the wiki, ball it up, and throw it at people. In turn, we may need to catch a ball of content lobbed at us, straighten it out, and put it on the wiki.

In other words, we need to get content out of the wiki into a standard, portable format. When pulling content into the wiki, we want to accept as many formats as possible. Four years ago, when I first became acquainted with a wiki, options were decidedly limited. Since then plugin developers have worked wonders. They have started building the technology that technical communicators get excited about.

Bouncing into this chapter

This chapter is about getting content into and out of a wiki. The chapter has the following sections:

- **Why do we need to get content into and out of the wiki?** This section examines some of the reasons why people need to convert content from one format to another. It also discusses single source publishing and the thorny question of round tripping.

- **Import and export options summarized.** A table shows the file formats that can be converted into Confluence pages and a list of the tools available to do the job. There is a similar table for export formats.

- **Details of the import and export options.** Next comes a section for each of the import options: Other wikis, Microsoft Word, Adobe FrameMaker, Confluence XML, DITA XML, and DocBook XML. Then follows a section for each of the export options: Print, PDF, Microsoft Word, Confluence XML, DocBook XML, HTML, Eclipse Help, JavaHelp, and EPUB.

- **Rolling on out of this chapter.** This chapter's summary includes a short discussion of a wiki as a tool for technical communicators.

- **References.** The tools, websites, blog posts, and other references mentioned in this chapter.

Why do we need to get content into and out of the wiki?

The thesis of this book is that a wiki is the platform extraordinaire for technical communication. By the time you get this far in the book, you have invested a fair amount of time learning how to set up a documentation wiki. It is easy to understand that you might now want to move some existing content onto the wiki. But why, oh why, would you want to think about moving content off the wiki?

Some scenarios

Here are some of the circumstances where documentation developers or readers want to move content onto or off the wiki.

- **Printed output.** People may need to print the content for reading when they are offline, for sharing with other people who are offline, or just because they prefer reading printed pages. Printed hard copy is often required for legal and contractual reasons.

- **Moving to a different platform.** A documentation development team may decide to develop, store, and publish the content on a new platform. With any luck, this will mean a move onto the wiki rather than off it!

- **Single source publishing.** Sometimes called single sourcing, this is a common scenario in the technical communication world. We develop the documentation on one platform, but readers need it in a number of different formats and platforms. I have devoted an entire section to this topic below.

■ **Translation and localization.** Let's assume that we employ an external company to translate our content into other languages. We need to give the translators a copy of the documentation in a format that they can use. Many translators require Microsoft Word documents. Others ask for an XML format to feed into their translation tools. If we can produce a standard XML format, we can run XSLT tasks to transform the output to the specific format required by the translation tool.

■ **Backups.** Confluence stores all its content in a database. Regular database backups, rather than XML exports, will protect us from a site-wide catastrophe. It is also useful to back up the contents of each space individually at times, such as at each major release date. The Confluence XML export and import provide a useful way of backing up and restoring the content of a space.

■ **Moving or copying a space to another Confluence site.** Perhaps you have developed your documentation on a staging site and now want to move it to production. Or perhaps you want to transfer a copy of the documentation to a testing server. The XML export is the tool to use here as well. Export the content to XML, then import the XML into the test site. *A reminder:* The *Copy Space plugin* is useful for copying a space within the same Confluence site. See Chapter 8 for an overview of this plugin.

■ **Giving customers an offline version of the documentation.** Some organizations block access to the Internet, making it difficult for their staff members to come to an online wiki outside their firewall. It is useful to give such customers a copy of the documentation that they can install on their own Confluence site. The Confluence XML export and import are the tools for the job. Alternatively, you can give people an HTML or PDF version of the documentation.

■ **Giving readers a portable version of the documentation.** Why would we want to export documentation to an epublishing format? A common use case is the manuals for service representatives who work at a client's site. For example, technical writers will compile the printer repair manual on the wiki and export it to a mobile device. The service representatives take the manual with them and can use it even when offline.

Single source publishing

Single source publishing, also known as single sourcing, is a methodology for developing content on a single platform and then publishing it to a number of different mediums and platforms.

As technical writers, we always keep our readers' needs at the forefront of our minds. One of the most basic requirements is that our readers have access to the documentation. No matter what tool we use to write the documents, we must be able to publish them in various output formats depending on where our readers are and the tools they have at their disposal.

A quick clarification. This book makes a distinction between single source publishing and content reuse. By content reuse, I mean the technique of splitting information into small chunks that you can use in a number of different locations within the documentation suite. This is covered in Chapter 8.

Let's consider single source publishing from two points of view.

- **Developing the documentation on a wiki, then publishing it in other formats.** Some customers may not be able to read the online documentation, perhaps because their work environment offers no Internet access or because they are constantly on the move. They need a downloadable version of the documentation that they can access behind a firewall or on a mobile device. People may want to integrate parts of the documentation into their own environment, such as a developer's IDE (integrated development environment).

- **Developing the documentation with some other tool, then publishing it on a wiki.** After a detailed analysis of the requirements, audience, and environment, we may decide to develop the document-ation in a format such as DITA XML, Microsoft Word, or Adobe FrameMaker. By publishing it on the wiki, we can take advantage of the wiki's collaborative and social features too.

The crux of single source publishing is that we need to publish again and again, with as little manual intervention as possible. We need to develop the content on the wiki and export it regularly to other formats, or *vice versa*, preferably via an automated job. Some of the tools discussed in this chapter are well on the way to providing the functionality we need.

Round tripping

Here's a howdy do! Does round tripping work, and do we even want to attempt it? The term "round tripping" means the repeated conversion of content from one format to another and back again. For example, consider this scenario:

1. Develop a topic in DITA XML, using an XML editor, and store the XML code in a source repository.
2. Publish the topic as a Confluence page.
3. Allow collaborators to update the Confluence page.
4. Export the page from Confluence to XML and import it into the source repository.
5. Update the topic in the source repository. For example, you may need to update it for the next version of the product.
6. Publish the updated topic in Confluence.
7. Go back to step 3, again and again.

That sounds fairly simple. Now add the fact that people will be updating the content in the source repository and on the wiki at the same time. Also, since the two media are so different, the topic will probably lose meaning and structure in the conversion. This is particularly true of the conversion from DITA XML, a semantically rich structure, to wiki format, which is less structured.

Given those complexities, I am not convinced of the value of round tripping on an ongoing basis. Instead, it is simpler and safer to write the documentation on one platform and then convert it to one or more other platforms when publishing the content. Chapter 19 has some ideas about how to structure the wiki so that people can interact with and collaborate on content even if it is generated outside the wiki.

Round tripping is technically possible from Confluence to DocBook XML and back again. There is a DocBook importer (from Red Hat) and a DocBook exporter (from K15t Software), both available as plugins for Confluence. This is an exciting development, whether we decide to make full use of round tripping or whether we stick with one-way conversion. Details of the plugins are below.

Importing chocolate into Australia in your luggage

Australian Customs authorities are pretty strict about foodstuffs, as many visitors to this fair land can attest. As a rule of thumb, do not try to bring any food, plant, or animal material in your suitcase. You will have to throw it away before entering the country.

A blogger named TBJ was planning to visit Australia and wanted to bring in some Belgian chocolates as a gift to his host. TBJ wrote to the Australian Quarantine and Inspection Service asking for advice. A day later, he received a helpful and comprehensive reply. The long and the short of it is: **Chocolate is OK!**[1]

Import and export options summarized

The two tables below summarize the tools available for importing and exporting content into and out of Confluence wiki. The rest of this chapter gives more detail about each tool. At the end of the chapter is a list of all the plugins and tools, with links to the websites where you can download the software and read the guides supplied by the developers.

Summary of import options

In Table 11.1, the first column shows a documentation format that you may want to import into Confluence wiki. The second column describes the tools available.

Table 11.1 – Import options

Import format	Tool
Another Confluence wiki	The Confluence XML export and import tools are useful for downloading the content of a space, or an entire site, and uploading it into another Confluence site.
Other wikis	The Universal Wiki Converter is a standalone tool that assists in converting pages to Confluence format from other wiki formats.
Microsoft Word	Confluence's built-in Office Connector imports a Word document and converts it to one or more wiki pages based on the criteria you define. It processes one Word document at a time. WebWorks ePublisher provides a set of standalone tools for converting Word documents to Confluence, based on custom styles and formats. Batch processing and scheduling are available.
Adobe FrameMaker	WebWorks ePublisher converts FrameMaker documents to Confluence.
DITA XML	DITA2wiki is a standalone tool that converts DITA XML documents to Confluence format and uploads them into the wiki via a set of Ant commands. WebWorks ePublisher offers a set of standalone tools for conversion and import from DITA XML to Confluence wiki.
DocBook XML	The *DocBook import plugin* is a Confluence plugin that converts DocBook XML to wiki format and imports the pages into a Confluence space.

Summary of export options

In Table 11.2, the first column shows a documentation format that you may want to convert your Confluence content into. The second column describes the tools available.

Table 11.2 – Export options

Export format	Tool
Another Confluence wiki	The Confluence XML export and import tools are useful for downloading the content of a space or an entire site and uploading it into another Confluence site.
Print	The browser's print option will print a Confluence page. Export to PDF for sophisticated printing options.
PDF	Confluence's built-in PDF export converts a selected page or pages into a single PDF file and provides customized layout and stylesheets. *Scroll Wiki PDF Exporter*, a Confluence plugin, offers a PDF export with flexible themes for configuring layout and styles.
Microsoft Word	Confluence's built-in Word export performs a basic conversion of wiki content to HTML and applies some predefined Word CSS stylesheets. It processes just one page at a time. *Scroll Office* is a Confluence plugin that exports pages to Word based on custom templates. You can export a single page or a hierarchy of pages.
DocBook XML	*Scroll Wiki DocBook Exporter* is a Confluence plugin that converts a selected page or pages to DocBook XML.
HTML	Confluence's built-in HTML export converts a single page, a selection of pages, or an entire space to HTML. The *Scroll Wiki HTML Exporter* is a Confluence plugin that exports content from Confluence pages to styled HTML files.
Eclipse Help	The *Scroll Wiki EclipseHelp Exporter* is a Confluence plugin that exports content from Confluence to Eclipse Help format.
JavaHelp	The *Scroll Wiki JavaHelp Exporter* is a Confluence plugin that exports content from Confluence to JavaHelp format.
EPUB	The *Scroll Wiki EPUB Exporter* is a Confluence plugin that exports Confluence pages to EPUB format.

Now let's look at each of the options in detail.

Importing from other wikis

The Universal Wiki Converter (UWC) is a standalone tool that you can download and install. It assists in automating wiki conversions and migrations from other wikis to Confluence. This is a one-way conversion only. Below is a list of the wiki formats that the UWC will import into Confluence. Note that many of the conversions are in alpha release, which means that they are fairly experimental:

- DokuWiki
- Jive Clearspace
- JSPWiki (alpha release)
- MediaWiki
- MindTouch (alpha release)
- MoinMoin
- PmWiki
- SharePoint (alpha release)
- SMF Wiki (alpha release)
- Socialtext (alpha release)
- SWiki
- TikiWiki
- Trac
- TWiki
- VQWiki
- XWiki (alpha release)

The UWC is an extensible framework. People can develop their own converters and make them available to others. Some of the conversions listed above are contributed by community developers.

AppFusions has recently taken over development and support of the UWC. They have published a *Wiki Migration Checklist*[2] with advice on how to manage the migration project. At time of writing, the checklist covers three areas of planning:

- **User community planning:** Communicating with all stakeholders, both within the organization and in the wider community of wiki users.

- **Systems planning:** Technical environment, syntax elements within the content, metadata, attachments, extensions and plugins, and more.

- **Information architecture:** A look at the existing structure of the content and a suggestion that a migration is the ideal time to plan for a new structure on the new wiki.

Importing from Microsoft Word

A Word import is available in the built-in Confluence Office Connector and via a standalone platform from WebWorks.

Built-in Office Connector

The Office Connector is part of the core Confluence functionality.[3] Confluence can import a Word document and convert it to one or more wiki pages based on the criteria you define. It processes one Word document at a time. You can choose to split the document into more than one Confluence page based on a given heading level within the Word document.

To import a Word document: Choose "**Tools**" > "**Import Word Document.**"

WebWorks ePublisher

WebWorks ePublisher provides a set of standalone tools for converting Word documents to Confluence. Note that ePublisher is not a plugin, but rather an independent set of tools that interact with the Word documents and with Confluence. You can design templates to define the styles and format of the Confluence pages resulting from the import. Best of all, you can automate the conversion using batch processing and scheduling. A while ago I played with ePublisher and Confluence and wrote a blog post about the experience with step-by-step instructions and screenshots.[4] There is some good information in the comments on that post, too. Another post on *Atlassian blogs* has more up-to-date screenshots.[5]

Importing from Adobe FrameMaker

WebWorks ePublisher, mentioned above, also converts Adobe FrameMaker documents to Confluence.

Importing Confluence's proprietary XML

A section below, called "Exporting and importing Confluence XML," covers both import and export of content to and from Confluence XML.

Importing from DITA XML

DITA (Darwin Information Typing Architecture) is a popular XML standard for technical document-ation. DITA was originated by IBM but is now maintained by OASIS. In a DITA-based design, a book or other publication is a collection of topics. Procedures, help systems, and other types of documentation can be broken down into topics. DITA is designed to make content reuse work. The DITA website, at `http://dita.xml.org/`, invites contributions from anyone interested in developing the standard.

WebWorks ePublisher

WebWorks ePublisher is a star player here, too, offering conversion and import from DITA XML to Confluence wiki.

DITA2wiki

DITA2wiki is a standalone tool that converts DITA XML documents to Confluence format and uploads them into the wiki via a set of Ant commands. It is an open source project on *SourceForge*. The project was started by Lisa Dyer, who has also written a number of blog posts and presentations about the use of DITA2wiki in the *Lombardi Wikis*.

For an overview of a real-life implementation of DITA2wiki, take a look at Lisa's presentation on *SlideShare* titled *Lombardi Wikis, a model for collaborative information development (with DITA XML in the mix)*.[6] Anne Gentle, champion of wikis and open source, wrote an excellent post announcing the DITA2wiki project.[7] I have had fun experimenting with DITA2wiki too. There is some excellent discussion in the comments on the resulting blog post.[8]

Importing from DocBook XML

DocBook is an XML standard, developed by a group of companies, including HAL Computer Systems and O'Reilly & Associates, as a means of writing and exchanging documentation. It is now an open standard maintained by OASIS (`http://oasis-open.org`). Where DITA is designed for topic-based development, DocBook is document-based and designed for large chunks of information such as chapters. The *DocBook Project* website is at `http://docbook.sourceforge.net/`.

The *DocBook import plugin* is a Confluence plugin that accepts a zipped archived of DocBook source files and converts them to Confluence pages. Vlastimil Eliáš developed the plugin for the JBoss Community's *Project Documentation Editor* at `http://docs.jboss.org/author`. JBoss community members use the site to provide documentation for their open source projects.

Printing pages and manuals

There is no print option in the Confluence menus. To print a single page, use the web browser's print option or export the page to PDF. To print more than one page at once, use the PDF export function.

When you print a page using the browser's print option, Confluence jumps in and gives you a pretty printed page that includes only the page content and comments, without navigational elements and the browser framework.

A six pack and a tablette de chocolat

To the French, a finely muscled stomach looks like a bar of chocolate. A *tablette de chocolat* is the Gallic equivalent of the British six pack.[9]

Exporting to PDF

Adobe's Portable Document Format (PDF) is an open standard for creating and delivering documents that can be read on almost any electronic medium, and printed too. The layout of a PDF document is fixed, meaning that you can define the page size, header, footer, font styles, and other layout options, and know that your reader will see the document exactly as you designed it.

Confluence offers built-in PDF export options, and plugins provide extended functionality.

Built-in PDF export

People reading Confluence pages can export a single page to PDF.

To export a single page to PDF: Go to the page, then choose "**Tools**" > "**Export to PDF.**"

In addition, if you have the space export permission, you can export a set of pages, selected pages, or an entire space to a single PDF file. For technical communicators, this is a good way to give customers a downloadable version of a manual.

To export a set of pages or a space to PDF: Go to a page in the space, then choose "**Browse**" > "**Advanced**" > "**PDF Export**."

You can customize the PDF layout. It is fairly simple to add a cover page with a corporate logo, title, and other branding elements. Page headers and footers are also customizable. Depending on your level of expertise, you can adjust the PDF stylesheet, too, by adding CSS rules for more sophisticated effects. The *Confluence User's Guide* offers some examples.[10] Only space administrators can change the PDF layout and stylesheet for a given space.

To change the PDF layout for a space: Go to a page in the space, then choose "**Browse**" > "**Space Admin**" > "**PDF Layout**."

To change CSS rules affecting the PDF export: Go to a page in the space, then choose "**Browse**" > "**Space Admin**" > "**PDF Stylesheet**."

Scroll Wiki PDF Exporter plugin

Scroll Wiki PDF Exporter, a Confluence plugin, provides flexible themes for configuring PDF layout. You can select one of Scroll's built-in themes and configure your table of contents, header, and title pages. For even more flexibility, you can add your own theme plugins to interact with the Scroll plugin. Scroll supports themes based on the DocBook XSL stylesheets. See the Scroll developer's guide.[11]

Exporting to Microsoft Word

Confluence offers a basic Word export option, and plugins provide extended functionality.

Built-in Word export

People reading a Confluence page can export the page to Microsoft Word. This option performs a basic conversion of wiki content to HTML and applies some Word CSS stylesheets. It processes just one page at a time.

To export a page to Word: Choose "**Tools**" > "**Export to Word**."

Scroll Office plugin

Scroll Office is a Confluence plugin. Once installed, the plugin replaces Confluence's built-in "Export to Word" functionality. You can define your templates in Word in the usual way, then upload them to Confluence as global templates or space templates. When someone exports the Confluence pages to Word, Scroll Office will use those templates to build Word documents from the wiki pages. You can export a single page or a hierarchy of pages.

Scroll Office provides additional features such as enforcing page-breaks, setting the page orientation to landscape or portrait, and excluding specific content from the export. The *Scroll Office User Guide*[12] describes the techniques and available macros for tailoring the Word exports. Scroll Office also offers a REST-style API for automated exports.[13]

Exporting and importing Confluence XML

Confluence provides XML export and import capabilities for individual spaces and for the whole site. The XML format is proprietary and is intended for backups or for transferring a space from one Confluence site to another. If you write your documentation on Confluence and your customers have their own Confluence site, too, then you can export your manuals to XML and customers can upload them onto their own Confluence site. This is useful for customers who protect their internal systems from the Internet by putting them behind a firewall. In such environments, people will not be able to get beyond the firewall and come to your wiki to read the online documentation.

Another use for the XML export and import is simply to back up your documentation space so that you can restore it after some catastrophe. You may also want to transfer documentation from one site to another, such as from a staging site to a production site, or from a production site to a test server.

You need the space export permission to be able to export content to XML. It is a good idea to limit the number of people who can do exports, because exports can put a heavy load on the Confluence server. This is true especially if the spaces contain a large amount of content.

To export a space to Confluence XML: Go to a page in the space, then choose "**Browse**" > "**Advanced**" > "**XML Export.**"

You need the Confluence system administrator permission to be able import content from XML.

To import a space from Confluence XML: Choose "**Browse**" > "**Confluence Admin**" > "**Backup & Restore.**"

A note about matching Confluence versions: The Confluence site that you are importing the XML into must be running the same version of Confluence as the site that you exported the XML from. Confluence cannot import XML from a different version. To get around this problem, install a temporary Confluence site running the Confluence version that matches the exported XML. Import the XML into the temporary site, then upgrade the site to a version that matches the destination site. Export the XML and import it into its final destination.

Exporting to DocBook XML

This is where we technical writers prick up our ears and start looking excited! When offered an XML export to a recognized XML standard we see worlds of possibility opening up, including portability to other document formats via existing XSL transformations. Earlier in this chapter we discussed DocBook XML and the *DocBook import plugin*. Now let's look at a plugin that exports wiki content to DocBook XML: *Scroll Wiki DocBook Exporter*.

Once the *Scroll Wiki DocBook Exporter* plugin is installed into the Confluence site, a new option appears in the "Tools" menu: "Export to DocBook." The export offers a number of configuration options, and you can choose to export the page you are currently viewing or a tree of pages starting from the current page (see Figure 11.1).

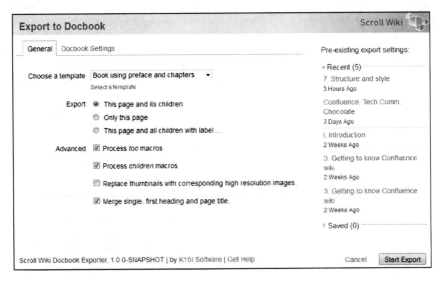

Figure 11.1. DocBook export options

A note of interest: We used Confluence and *Scroll Wiki DocBook Exporter* to publish this book. I wrote the content on Confluence, and the reviewers joined me there to add their comments. After incorporating the feedback from the technical review, we exported the pages to DocBook format for further processing by the publisher and printer.

Exporting to HTML

You can use the built-in Confluence HTML export or add a plugin for extended options.

Built-in HTML export

Using the built-in Confluence HTML export, you can export a single page, a selection of pages, or an entire space to HTML. Confluence supplies the HTML and associated files in a zip file. You need the space export permission to use this option.

To export a page or pages to HTML: Go to a page in the space, then choose "**Browse**" > "**Advanced**" > "**HTML Export.**"

Scroll Wiki HTML Exporter plugin

The *Scroll Wiki HTML Exporter plugin* provides a styled export of Confluence content to HTML. It also interacts with the *Ad hoc Workflows plugin* (see Chapter 10) so that you can choose to export the latest published version of your documentation.

Exporting to Eclipse Help

How cool, to write your documentation and online help content in Confluence wiki and convert it to Eclipse Help format! Eclipse[14] is a platform that developers use for writing code. It is an open source IDE (integrated development environment) that aims to provide a universal toolset for development. Developers can extend Eclipse, adding extra functionality by writing plugins for the Eclipse platform, much the same way as people write plugins for Confluence. The driver for creating the Eclipse Help format was to allow people to write documentation for Eclipse plugins.

An *IBM developerWorks* article, titled *Documenting your project using the Eclipse help system*, gives an excellent overview of how to build and use Eclipse Help for "easy-to-use and searchable help documentation."[15] Another article on *KeyContent.org* gives detailed information about writing documentation and help for Eclipse projects and plugging the documentation into the Eclipse platform.[16] The Eclipse documentation guidelines are good, too.[17]

Now, people who are developing a plugin for the Eclipse development platform can use Confluence to write the topics, then export them to Eclipse Help format. *Scroll Wiki EclipseHelp Exporter* is a Confluence plugin that does just that.

Here is the cool thing from our point of view as technical communicators: you can use a cut-down version of the Eclipse Help platform to provide online documentation for any system. It does not have to be an Eclipse tool that you are documenting. This is a nice, lightweight solution for standalone help, even if your customers are not using Eclipse.

Exporting to JavaHelp

JavaHelp is a part of the Java platform. It is a system that developers and authors can use to present online help, primarily for people using Java applications. Computers running the help system need to have the JRE (Java Runtime Environment) installed.

Scroll Wiki JavaHelp Exporter is a Confluence plugin that converts Confluence pages to JavaHelp format.

Exporting to epublishing formats

The EPUB format is an open ebook (electronic book) format developed and maintained by the International Digital Publishing Forum (http://idpf.org/). The format is designed to support reflowable content, meaning that the content's layout changes to suit the device on which the content is being read. HTML is another example of a reflowable format. The *Wikipedia* page on EPUB[18] lists the devices that support the EPUB format, including the iPhone, iPad, Android devices, and a number of other ebook readers and tablets.

Scroll Wiki EPUB Exporter is a Confluence plugin that converts Confluence pages to EPUB format. At the time of writing this chapter, the plugin is in beta release.

Not in the least bit daunted by the word "beta," our hero Ganache has installed the *Scroll Wiki EPUB Exporter plugin* onto her Confluence site. Now she can export her Twitter guide in EPUB format. She clicks the "Export to EPUB" option that the plugin has added to the Confluence "Tools" menu and accepts the default configuration options, as shown in Figure 11.2.

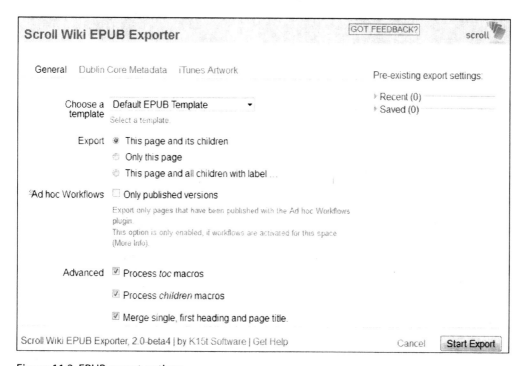

Figure 11.2. EPUB export options

The plugin produces a zipped file with a .epub extension. Ganache's next step is to get the EPUB file onto her iPad. She emails the file to herself, opens the email message on her iPad, taps the attached file, and chooses "Open in iBooks." Figure 11.2 shows the result. Indubitably, we are magic!

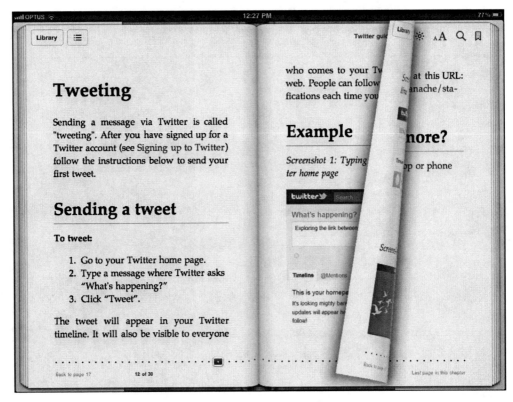

Figure 11.3. Ganache's Twitter guide in EPUB format on the iPad

iBooks is a built-in app (application) on Apple's iPad. It is also available for download at the *App Store* (*Apple Store*).

Rolling on out of this chapter

Technical writers take heart! There is no need to worry about documentation getting stuck on a wiki. There are already a number of options for getting content into and out of Confluence. Wikis are the cool kids on the block, despite having been around since 1995. They are undergoing constant and enthusiastic development, both by the wiki vendors and by the ecosystem of plugin developers. Even if a particular export or import option is not available right now, there is a good chance that it will be soon. As a case in point, *Scroll Wiki EPUB Exporter* is brand new at time of writing this chapter. If there is something we need, we can put our hands up and ask for it.

Should a technical communicator use Confluence as the primary tool for developing technical documentation? Here is my opinion as a self-confessed wiki hugger. As with all tools, the answer depends on the environment, audience, and requirements. If our readers will view the documentation primarily online, if we want to build a supportive and active community around the documentation, and if we want agile development and publishing of good-looking and functional documentation, then a wiki is the tool for us. What about a technical writer's experience of the wiki? There will be more about this later in the book. For now, I will just say that I have found the wiki a lot of fun and very rewarding to work with.

If the result of the requirements analysis is that another tool is better as the primary tool for developing the documentation, so be it. It will be less fun, though!

Before you roll on out of this chapter, remember to keep checking for more plugins.[19] Quite a few of the developers in the plugin community are fired up about the possibilities of technical documentation on a wiki. Let's keep them interested by telling them what we think of their tools and by asking for more.

References

These are the tools, websites, blog posts, and other references relevant to the content of this chapter.

Useful plugins

- *DocBook import plugin* by Red Hat, for importing a zipped archived of DocBook source files into Confluence: `https://plugins.atlassian.com/plugin/details/43179`

- *Scroll Office plugin* by K15t Software, for exporting wiki content to Microsoft Word: `https://plugins.atlassian.com/plugin/details/24982`

- *Scroll Wiki DocBook Exporter plugin* by K15t Software, for exporting wiki content to DocBook XML: `https://plugins.atlassian.com/plugin/details/725767`

- *Scroll Wiki EclipseHelp Exporter plugin* by K15t Software, for exporting wiki content to Eclipse Help format: `https://plugins.atlassian.com/plugin/details/420619`

- *Scroll Wiki EPUB Exporter plugin* by K15t Software, for exporting wiki content to EPUB format: `https://plugins.atlassian.com/plugin/details/44371`

- *Scroll Wiki HTML Exporter plugin* by K15t Software, for exporting wiki content to HTML: `https://plugins.atlassian.com/plugin/details/420604`

- *Scroll Wiki JavaHelp Exporter plugin* by K15t Software, for exporting wiki content to JavaHelp format: `http://www.k15t.com`

- *Scroll Wiki PDF Exporter plugin* by K15t Software, for exporting wiki content to PDF: `https://plugins.atlassian.com/plugin/details/7019`

Standalone platforms and tools

- *DITA2wiki*, including *DITA2Confluence*, an open source tool that converts DITA XML documents to Confluence format and uploads them into the wiki via a set of Ant commands: `http://source-forge.net/projects/dita2wiki/`

- *Universal Wiki Converter*, a standalone tool that converts pages to Confluence format from other wiki formats: `https://studio.plugins.atlassian.com/wiki/display/UWC/Universal+Wiki+Converter`

- *WebWorks ePublisher*, a set of tools for converting documents from a number of formats (Adobe FrameMaker, Microsoft Word, DITA XML) into other formats (Confluence wiki, PDF, HTML and others): `http://www.webworks.com/`

Further reading

- Guide to installing plugins, in the *Confluence Administrator's Guide*: `http://confluence.atlassian.com/display/DOC/Installing+a+Plugin`

- *Styled Word Documents from Your Wiki with Scroll Office*, a blog post by Stefan Kleineikenscheidt on *Atlassian blogs*, August 2010: `http://blogs.atlassian.com/confluence/2010/08/technical-project-documentation-wiki-export-microsoft-word-templates.html`

Endnotes

[1] *Australian Customs: biscuits and chocolate are OK*, a blog post by TBJ on *To Australia, Over The Rainbow*, January 2011: `http://austour.wordpress.com/2011/01/17/australian-customs-biscuits-and-chocolate-are-ok/`

[2] *Wiki Migration Checklist*, published by AppFusions: `https://www.appfusions.com/display/Dashboard/Wiki+Migration+Checklist`

[3] *Working with the Office Connector*, in the *Confluence User's Guide*: `http://confluence.atlassian.com/display/DOC/Working+with+the+Office+Connector`

[4] *WebWorks ePublisher for converting documents to Confluence wiki*, a blog post by Sarah Maddox on *ffeathers*, April 2009: `http://ffeathers.wordpress.com/2009/10/04/epublisher-for-converting-documents-to-confluence-wiki/`

[5] *Technical Writing in a Wiki - Single Source Publishing*, a blog post by Sarah Maddox on *Atlassian blogs*, November 2010: `http://blogs.atlassian.com/confluence/2010/11/technical-writing-wiki-single-source-publishing.html`

[6] *Lombardi Wikis, a model for collaborative information development (with DITA XML in the mix)*, a presentation by Lisa Dyer on *SlideShare*, February 2008: `http://www.slideshare.net/lisa.dyer/lombardi-wikis-model`

[7] *DITA Meets Wiki – Output DITA to Wikitext*, a blog post by Anne Gentle on *Just Write Click*, October 2008: `http://justwriteclick.com/2008/10/16/dita-meets-wiki-output-dita-to-wikitext/`

[8] *Playing with DITA2Confluence*, a blog post by Sarah Maddox on *ffeathers*, October 2008: `http://ffeathers.wordpress.com/2008/10/26/playing-with-dita2confluence/`

[9] The French idiom *tablette de chocolat* is described in *Pardon My French*, a book by Charles Timoney, published by Penguin Books, 2007.

[10] Guide to exporting Confluence pages and spaces to PDF, in the *Confluence User's Guide*: `http://confluence.atlassian.com/display/DOC/Exporting+Confluence+Pages+and+Spaces+to+PDF`

[11] *Scroll Developer Guide*: `http://www.k15t.com/display/SCROLLDOC/Developer+Guide`

[12] *Scroll Office User Guide*: `http://k15t.com/display/OFCE/User+Guide`

[13] Scroll Office guide to the REST API: `http://k15t.com/display/OFCE/Automating+Scroll+using+the+REST+API`

[14] Eclipse Foundation open source community website: `http://www.eclipse.org/`

[15] *Documenting your project using the Eclipse help system*, an article by Arthur Barr on *IBM developerWorks*, January 2004: `http://www.ibm.com/developerworks/opensource/library/os-echelp/`

[16] *Eclipse Help System, Writing Documentation and Help for Eclipse Projects and Plug-ins*, an article by Rick Sapir, Bill Albing and Jeremy H. Griffith on *KeyContent.org*: `http://www.keycontent.org/Eclipse+Help+System`

[17] *DocumentationGuidelines* on the Eclipse wiki: `http://wiki.eclipse.org/Documenta-tionGuidelines`

[18] A page about the EPUB format on *Wikipedia*: `http://en.wikipedia.org/wiki/EPUB`

[19] The *Atlassian Plugin Exchange*: `https://plugins.atlassian.com`

12

Building online help

"Help!" Don't panic. Wikis understand a call for help too.

A technical documentation wiki is a hive of activity. Developers, support engineers, and technical writers are continually updating and enhancing the pages. Visitors to the site add comments day and night. People swap ideas and techniques using the documentation as their communication hub. Can this possibly be the right destination for context-sensitive help links? Is it safe to point a link from a software application directly into a wiki page? Yes. In fact, the currency and interactive nature of the information make the wiki a good solution for online help.

A quick guide to this chapter

Let's see how we can use a wiki as an online help platform. The chapter has the following sections:

- **Introduction to online help.** Clarifying the terminology: web-based, context-sensitive, and embedded help. This chapter describes how to use a wiki to provide web-based and context-sensitive help. It also suggests how a wiki can be the source of the information displayed in embedded help.

- **A wiki as candidate for online help.** The attributes of a wiki that are useful in a help system.

- **A simple online help solution.** How to make sure people get just the help they need when they ask for it.

- **Tool tips and embedded help.** A quick guide to using a wiki as an in-application help solution.

- **A case study.** How the online help works in the Atlassian Crowd application, using an Apache redirect to change the destination of the incoming help requests mid flight.

- **How this chapter was helpful.** A summary of the chapter.

- **References.** The tools, websites, blog posts, and other references mentioned in this chapter.

Introduction to online help

What do people mean when they talk about online help? They may mean a web page that they go to when they have a question or a problem, or a standalone help system such as those provided by WinHelp or HTML Help. They may mean context-sensitive help, tool tips, or even help text that appears on the application screen itself. This chapter deals with the following types of help:

- **Web-based help.** This is documentation hosted on a website. People come to the site to read it.

- **Context-sensitive, or contextual, help.** When someone clicks a help link on an application screen, they get the instructions directly related to that particular screen. At a deeper level, when someone clicks a help icon next to a field they get help about that specific field.

- **Tool tips and embedded help.** Another type of help is the content that appears when someone hovers their mouse over a field and a short description pops up in a bubble. That is often called a tool tip. Or perhaps the user clicks a control that expands a panel on the screen containing the help text. That is embedded help or in-application help.

Technical communicators use a number of tools to develop online help. Examples are Author-it, Adobe RoboHelp, MadCap Flare, Help & Manual, and many more. Then there are the wikis. This chapter describes a solution that Atlassian uses to provide web-based and context-sensitive help for a number of its own software applications. The online help platform is Confluence wiki. The applications, such as Crowd, JIRA, FishEye, and Confluence itself, include code that responds when a user clicks a help link by displaying the appropriate page in the documentation. As in most online help systems, the mapping of application screen to help topic is held in a simple text file in the form of key-value pairs.

Towards the end of the chapter we examine how a wiki can be the source of the information displayed in embedded help.

A wiki as candidate for online help

How is a wiki useful for online help?

- A wiki page is accessible via a URL. The application screen can link directly to the page.

- Wiki pages are continuously being updated and enhanced by technical writers, support engineers, and developers. When someone clicks the help link, they get the most up-to-the-minute information possible.

- If wiki permissions allow it, readers can add their own tips as comments on the wiki page.

- A good wiki has a search engine that people can use to find more information if the linked page does not have what they need.

- Wiki pages can include multimedia content such as images, movies, and videos.

- A wiki provides tools for runtime integration with other software. This means that a page can display information directly from another platform. For example, a wiki page can show a list of bug fixes drawn dynamically from an issue tracking system.

- Some wikis include blogging as part of the wiki platform. A documentation page can embed a list of the latest blog posts related to the topic of the page.

- Developers are continually adding features and plugins that technical communicators can use to enrich the help content.

A simple online help solution

Let's take a look at a simple way of providing web-based and context-sensitive online help for a software application. This is a cut-down version of the way Atlassian does it. The case study later in this chapter explains why Atlassian's solution is slightly more complex.

The indomitable Ganache has decided to add context-sensitive help links to the *ChocTactics* application (see Figure 12.1). Ganache writes and publishes the help topics on the wiki. Readers come directly to the documentation via Google search or via links from websites, blogs, and more. In addition, they come by clicking the help links on the *ChocTactics* screens.

When someone clicks a help link, exactly the right page appears. The content is up to date, to the minute. If there is a technical problem with that part of *ChocTactics*, other people may have encountered the same problem and added helpful comments on the page too. Support engineers will suggest workarounds and add a link to the report in the issue tracker so people can keep track of the fix.

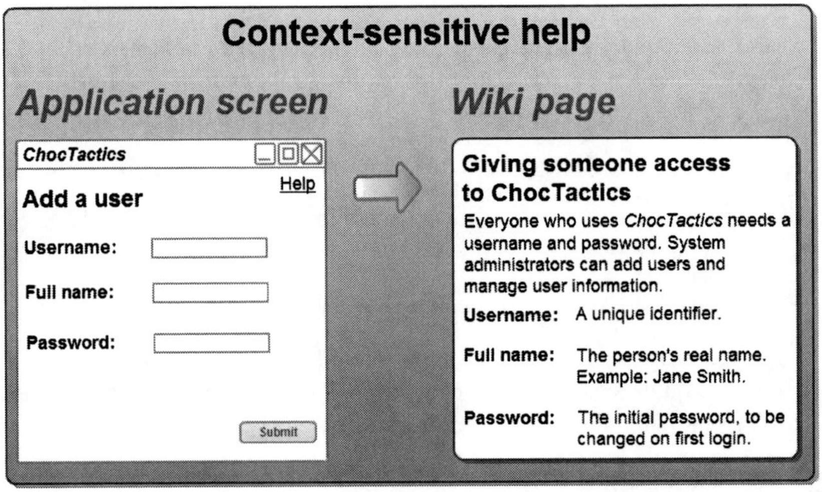

Figure 12.1. An application screen with a help link pointing to the appropriate wiki page

Technical requirements for the application

The *ChocTactics* application includes a properties file, which Ganache has named `help-paths.properties`. This file has three functions:

- Provide the stem (meaning the first part) of the URL, containing the location of the wiki server.

- Indicate the version of *ChocTactics* that the user is running.

- Map the *ChocTactics* screens to the help topics.

The `help-paths.properties` file consists of a list of key-value pairs, also called attribute-value pairs. The key is understood by the *ChocTactics* application to indicate either a URL stem or a specific screen or control. The value is a variable that Ganache and other *ChocTactics* administrators can configure. By putting this data mapping into a text file, Ganache has made it possible to adjust the application's behavior without changing the application code.

Example 12.1 shows the properties file for *ChocTactics*:

```
help.prefix            http://confluence.chocink.com/display/TACT020
help.login             Logging+in+to+ChocTactics
help.home              About+ChocTactics
help.sample.browse     Managing+chocolate+samples
help.sample.add        Adding+a+chocolate+sample
help.sample.assess     Assessing+a+chocolate+sample
help.user.add          Giving+someone+access+to+ChocTactics
```

Example 12.1 – Properties file for *ChocTactics*

Row 1 contains a key (`help.prefix`) and the first part of the URL, where the help topics are found. The URL has these parts:

- `http://confluence.chocink.com` – The location of the wiki server.

- `/display/` – A standard part of the URLs used in Confluence to display a page.

- `TACT020` – The space key containing the help topics for a specific version of the application, in this case *ChocTactics 2.0*. Ganache uses wiki spaces to hold the documentation for each released version of the application. (See Chapter 10.)

The rows from 2 onwards define a list of locations inside the *ChocTactics* application (the keys on the left) and the corresponding wiki pages (the values on the right). The keys are defined by the developers inside the application. The page names include the plus signs and other URL encoding as applied by Confluence when forming a URL.

For example:

- The key `help.login` refers to the *ChocTactics* login screen. The value `Logging+in+to+Choc-Tactics` refers to the wiki page containing the help text for the login screen.

- The key `help.user.add` refers to the *ChocTactics* screen for adding a user. The value `Giving+someone+access+to+ChocTactics` refers to the wiki page that gives instructions on adding usernames and related information to the *ChocTactics* user database.

The above example of a help properties file has just a few key-value pairs. In reality, the file would contain many more rows.

Let's assume that a system administrator called Gianduja wants to add a user to the *ChocTactics* application. Gianduja finds her way to the *ChocTactics* screen for adding users, but then she needs some help, so she clicks the help link at the top of the screen.

To show Gianduja the help topic, the *ChocTactics* code performs the following steps:

1. Look in the code or in the application data store to find the help key for the screen: `help.user.add`.

2. Look up the corresponding page name in the `help-paths.properties` file: `Giving+someone+access+to+ChocTactics`.

3. Splice the page name onto the URL stem found at the top of the same properties file, to construct the full URL: `http://confluence.chocink.com/display/TACT020/Giving+someone+access+to+ChocTactics`.

4. Open the URL in a new browser tab.

An alternative to using page names in the key-value pairs

The `help-paths.properties` file in the above example uses the full page names, such as `Giving+someone+access+to+ChocTactics`. This means that the solution breaks when someone changes the page title on the wiki. The technical writing team needs to put procedures in place to ensure that no-one changes a page title if that page is referenced from the online help.

An alternative is to use the tiny links that Confluence supplies for each page. They look something like this: `http://confluence.chocink.com/x/GgUN`. The URL has these parts:

- `http://confluence.chocink.com` – The location of the wiki server.

- `/x/GgUN` – A value, generated by Confluence, that uniquely identifies the page within the site. No part of the URL reflects the space in which the page is located.

To find the tiny link for a Confluence page: Go to the page, then choose "**Tools**" > "**Link to this Page**."

The advantage of this link is that the tiny URL remains the same even if someone changes the page name. The disadvantage is that the space key is not part of the URL. Therefore, you cannot use the space key to indicate the version of documentation for a specific release of the product. For every release, the technical writers will need to update all the URLs in the properties file. There is also not much point in specifying the first part of the URL (the `help.prefix`) separately, as the page identifiers will be different on every wiki server. In effect, the help properties file will contain the full URL for every single key-value pair.

Example 12.2 shows how the properties file might look.

```
help.login              http://confluence.chocink.com/x/GgUN
help.home               http://confluence.chocink.com/x/FAY
help.sample.browse      http://confluence.chocink.com/x/B4H6BQ
help.sample.add         http://confluence.chocink.com/x/D
help.sample.assess      http://confluence.chocink.com/x/GwHo
help.user.add           http://confluence.chocink.com/x/CHOC
```

Example 12.2 – Help properties file

What happens if people cannot get to the wiki?

Some corporate environments are protected by a firewall, meaning that people cannot go to an external website. The above solution gives customers a way to work around this problem. Since the help mappings are held in a text-based configuration file, customers can edit the file and point the help links to a different server. They will download a copy of the *ChocTactics* documentation (see Chapter 11) and install it on their own server. Then they will adjust the help.prefix in the help-paths.properties file to point the help links at their own version of the documentation.

Note that if you use tiny URLs, the customer will need to change all the URLs in the properties file.

Providing help files in another language

The same solution works for customers who need to replace the documentation with a translated version. They can adjust the help.prefix and page names to point to the relevant wiki and translated pages.

Tool tips and embedded help

First, a quick reminder about the terminology as used in this book. A tool tip is content that appears when someone hovers their mouse over a control and a short description pops up in a bubble. The user may click a control that expands or opens a panel containing the help text. This is called embedded or in-application help.

This section contains a suggestion of how a wiki can be the source of the information displayed in tool tips and embedded help. It is not something I have tried myself. In practice, it may not be worth the effort for tool tips. But for embedded help, having the content easily editable on the wiki may be very useful.

The technical solution described above provides part of the solution needed here: the mapping of application screens and controls to wiki pages. In addition, the application needs to supply the following functionality:

- Instead of linking to the appropriate topic on the wiki, the application will pull the associated piece of content off the wiki. I suggest that the technical writers modularize the content so that the application can pull in an entire page at a time. (More on this in the next paragraph.) The developers can use the Confluence APIs[1] to retrieve the content of a page.

- The application will display the content appropriately in a popup, sidebar or panel on the application screen.

Another difference is in the design and development of the help content itself.

- The content must be modularized and tailored for the small amount of space available in the tool tip, popup, or panel. Technical writers will structure their content into small chunks, designed specifically for use in the embedded help. The simplest design is to have a wiki page for each popup or panel. The page contains only the content required in the popup or panel. That will make it easy for the application to pull in the content it needs. If the content of these pages is useful elsewhere, such as in the wiki-based documentation, then the content reuse techniques described in Chapter 8 are applicable here too.

- Content developers can take advantage of the fact that the help appears on the screen and in context. It does not need much contextual framework.

- Links to other topics should be few and far between. At most, a single link will point to the wiki page that contains the full help topic.

- Screenshots are unnecessary since the help appears on the screen itself. They would also take up too much space.

A case study of web-based, context-sensitive online help

Atlassian uses the documentation wiki to provide online help for the applications it develops, including JIRA, FishEye, Crucible, Bamboo, Crowd, and Confluence itself. Let's look at the online help for Atlassian Crowd as an example.

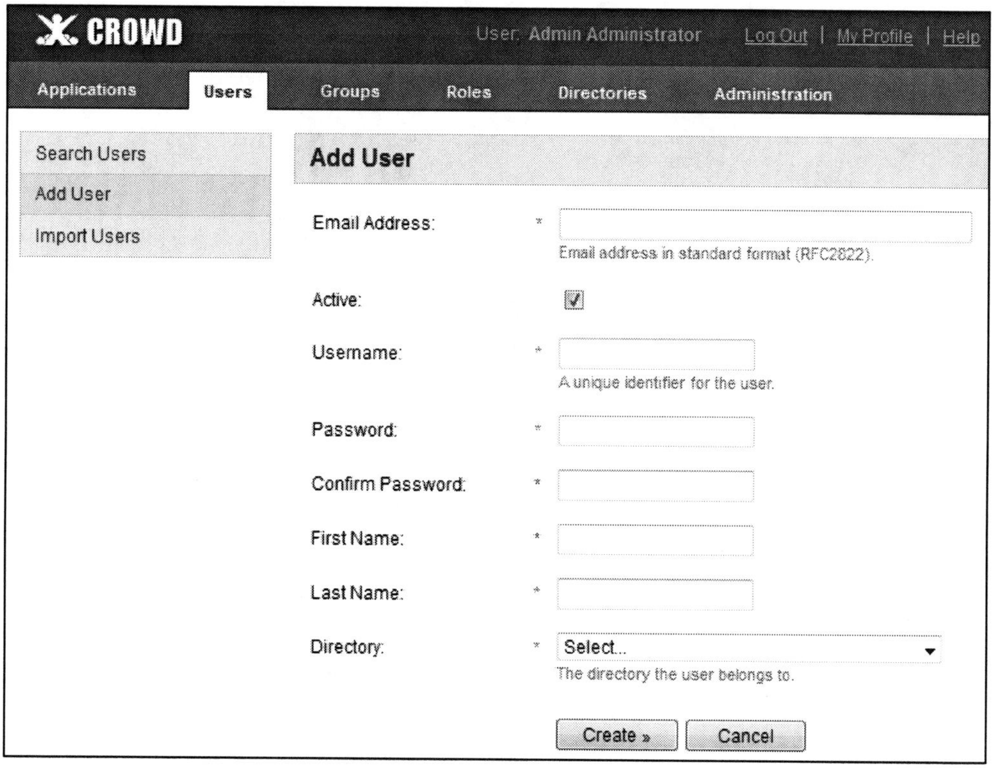

Figure 12.2. Adding a user in the Crowd application

When a Crowd administrator clicks "Help" at the top right of the screen (Figure 12.2), a new browser window opens showing the guidelines on adding a user (Figure 12.3).

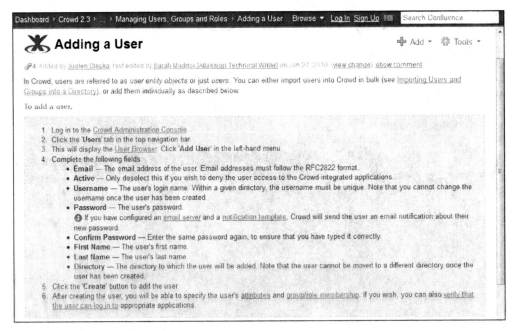

Figure 12.3. The Crowd topic on adding a user

The help comes directly from the wiki page. The information is up to date, to the minute.

When designing the documentation and the application screens, the technical writers and developers work closely together to decide where the help links are needed. The technical writers structure the documentation pages to provide a good landing place for each help link. They also make sure that there is a separate set of documentation for each release of the application.

Redirecting the help links to archive spaces for earlier releases

There is one slight difference between the simple help solution described for *ChocTactics* and the solution used by Atlassian. The difference is a result of the way the technical writers use spaces for release management.

Chapter 10 shows how to store each version of the documentation in a separate wiki space. The chapter also mentions that Atlassian always stores the current version in the same space with a constant space key (see the section titled "An alternative structure for version-specific spaces" (p. 208)).

This means that the online help links must change to a new destination at some point after the application has been shipped to the customers.

Let's assume that the latest release of Crowd is version 2.3. The documentation spaces are shown in the table below.

Space key	Documentation version
CROWD	Crowd 2.3
CROWD022	Crowd 2.2
CROWD021	Crowd 2.1
CROWD020	Crowd 2.0

If Gianduja clicks the help link in a Crowd 2.3 system, she must see the documentation in the CROWD space.

Now let's assume that Crowd 2.4 is released. The documentation for Crowd 2.3 will move to a new space CROWD023, and the CROWD space will hold the documentation for Crowd 2.4.

Space key	Documentation version
CROWD	Crowd 2.4
CROWD023	Crowd 2.3
CROWD022	Crowd 2.2
CROWD021	Crowd 2.1
CROWD020	Crowd 2.0

When using Crowd 2.3 Gianduja now needs the documentation in the CROWD023 space, not the CROWD space. Somehow, magically, the online help link must swap its destination without any change to the Crowd 2.3 application or its `help-paths.properties` file.

How it works

The `help-paths.properties` file shipped with the Crowd application has the same format as the *ChocTactics* file, discussed above. The difference lies in the first line: the destination URL.

Example 12.3 shows the first few lines of the `help-paths.properties` file for Crowd 2.3:

```
help.prefix http://docs.atlassian.com/crowd/docs-023/

help.login                         Logging+in+to+Crowd
help.home                          About+the+Crowd+Administration+Console

help.application.browse            Using+the+Application+Browser
help.application.add               Adding+an+Application
help.application.view.details      Using+the+Application+Browser
help.application.view.directories  Mapping+a+Directory+to+an+Application
help.application.remove            Deleting+or+Deactivating+an+Application

help.user.browse                   Using+the+User+Browser
help.user.add                      Adding+a+User
```

Example 12.3 – help-paths.properties file for Crowd 2.3

The destination URL, in the first line of the file, does not point to the wiki. Instead, it points to a different location, `docs.atlassian.com`, which runs an Apache server that redirects the incoming requests to the relevant documentation space.

The magic is in the redirect

The Apache server contains a directory-level configuration file named `.htaccess`. The Apache documentation[2] has the details on the various ways to use such a file. In the Atlassian Crowd solution for online help, the `.htaccess` file is used to rewrite the incoming URL.

The incoming URL for a Crowd help request looks something like this: `http://docs.atlassian.com/crowd/docs-023/Adding+a+User`.

Example 12.4 shows the contents of the `.htaccess file` for Crowd:

```
RewriteEngine on

# eg docs.atlassian.com/crowd/docs-023/Crowd+Documentation
RewriteRule ^/?docs-023/(.*)$ http://conf.atlas.com/display/CROWD/$1 [L]
RewriteRule ^/?docs-(\d+)/(.*)$ http://conf.atlas.com/display/CROWD$1/$2 [L]
```

Example 12.4 – .htaccess file for Crowd

(I have adjusted the URL of the Atlassian documentation wiki in these examples, because it was too long to fit into the fixed-with format above. In real life, `conf.atlas.com` is `confluence.atlassian.com`. If you noticed that, have a chocolate.)

The file in Example 12.4 redirects requests of the form `http://docs.atlassian.com/crowd/docs-023/Adding+a+User` to `http://conf.atlas.com/display/CROWD/Crowd/Adding+a+User`. That takes care of the space for the latest release. All other versions are directed to spaces with version numbers in the space key. For example, requests of the form `http://docs.atlassian.com/crowd/-docs-022/Adding+a+User` are redirected to `http://conf.atlas.com/display/CROWD022/-Adding+a+User`.

Crowd 2.4 will have its own `help-paths.properties` file, with a first line like this:

`help.prefix http://docs.atlassian.com/crowd/docs-024/`

Of course, Crowd 2.3 will still have its original `help-paths.properties` file, which cannot be changed because the application is out there in the wild. Instead, the Atlassian engineers adjust the `.htaccess file` on the Apache server.

How? Alter the `docs-xxx` in the first RewriteRule to match the version in the `help-paths.properties` of the new release. So, on the release of Crowd 2.4, the engineers change the rule to:

`RewriteRule  ^/?docs-024/(.*)$  http://conf.atlas.com/display/CROWD/$1 [L]`

That's it! Simple, once you know how.

How this chapter was helpful

Now you know how to use a wiki as an online help platform. A wiki publishes pages that you can address via unique URLs. It is easy to hook up the help links from an application screen to the matching topic on the wiki. If the mapping of help links to wiki pages is held in a configuration file, then customers can adjust the mappings and destination of the help links to suit their needs. They can hide behind a firewall, add their own procedures, or swap to a translated version of the documentation.

A well-maintained and interactive wiki page becomes a valuable resource for anyone seeking help, whether they come to the page from a web-based search or by clicking a help link. People come to the page, swap hints, and help each other. That is taking online help to the next level. Help 2.0, anyone?

References

These are the tools, websites, blog posts, and other references relevant to the content of this chapter.

Useful plugins

Scroll Versions is a new plugin under development by K15t Software. This plugin supplies stable page keys across different versions of the same page, for use in online help solutions that support document-ation under version control. See the section titled "K15t Software's focus on wiki documentation tools" (p. 436) for more details about the functionality in this plugin.

Endnotes

[1] Guide to the Confluence APIs in the Confluence developer documentation: `https://developer.atlassian.com/display/CONFDEV/Confluence+Remote+API+Reference`

[2] Guide to `.htaccess` files in the Apache HTTP Server documentation: `http://httpd.apache.org/docs/current/howto/htaccess.html`

Life on a wiki

The wiki is up and running, the documents are under development, the procedures are in place. Now, what is it like to work on a wiki day to day?

13

A day in the life

A regular day on a wiki. It's fast-paced. It's rewarding. It can be funny and frustrating in turn. It's everything that makes us love our role as technical communicators. It is the full monty.[1]

Good morning wiki!

Let's see what a day on a wiki looks like, from a technical communicator's point of view. The readers are there, too, including customers, support engineers, developers, and product managers. What does their wiki day feel like? This chapter has the following sections:

- **A day in the life of a wiki technical writer.** A typical day when nothing special happens, but a lot happens nonetheless. Checking the wiki notifications, attending a standup, reviewing wiki updates, working on new documentation, and restructuring part of the wiki.

- **A few days in the life of a wiki reader.** A customer's experience of wiki documentation.

- **A day in the life of a support engineer.** How a support engineer uses the wiki to solve customer problems.

- **A product manager alive on a wiki.** Getting early feedback from customers. Collaborating on high-profile documents such as release notes.

- **A developer walks into a wiki.** The technical writer says, "I pwn your typos."[2]

- **The sun never sets on a wiki.** A technical communicator in a global environment.

- **References.** The tools, websites, blog posts, and other references mentioned in this chapter.

A day in the life of a wiki technical writer

Ganache, technical communicator extraordinaire, is our hero in this section. She spends most of her day on the wiki.

9 am. Check the wiki notifications

The sweet, spicy aroma of nutmeg and cocoa heralds the arrival of Ganache and her mug of hot chocolate. It is time to see what has happened on the wiki overnight. Ganache opens her email inbox and scrolls through the notifications sent by the *Choc Ink* documentation wiki.

What part does chocolate play in your life as technical writer?

From Andrew Lui on Technical Writing World:[3] "I think this quote sums it up quite nicely ;-)

'Chocolate is nature's way of making up for Mondays' – Anonymous"

The first shows an update to a documentation page made by one of the developers. Ganache sips her chocolate while she assesses the update, then nods her agreement and deletes the notification. No further action required.

On to the next notification. Ganache bursts into giggles, spraying chocolate froth onto her monitor. A reader has commented about a spelling error on page, ending his comment with this quip: "It was a little *ChocInk* to find a typo on such otherwise awesome documentation!"

Ganache fixes the error on the page and adds a comment of her own. "Thank you for pointing out the error. Fixed. Your *ChocTactics* worked, by the way – I laughed out loud when I read your comment."

She also adds a reminder to her task manager, so that she will come back in a week and delete both comments once the reader has had time to see her reply.

A support engineer named Roald has added a new troubleshooting page, which describes a known issue with the latest software release and provides a workaround that customers can use until the problem is fixed. A quick glance at the notification tells Ganache that the page could do with some technical writer love. She marks it for later attention and moves on to the next item in her inbox.

The next notification requires immediate action. A spammer has created a login account on the wiki and added a couple of comments advertising free holidays. Ganache deletes each comment, wondering absently why anyone would think such an advertisement on a documentation wiki is likely to generate sales. *Ah well,* she thinks, *it's good that the wiki is attracting enough traffic to make this spammer think it's worthwhile.*

A product manager has added some screenshots to the release notes for the next version of *ChocTactics*. Ganache skips through those notifications, knowing that she already has a task in her issue tracker to review the release notes later today. She is pleased to see that the product manager has found time to work on them already.

9:45 am. Attend the team standup

All around Ganache, the *ChocTactics* developers are getting up from their desks, stretching, yawning, and chatting. They group together to form a rough circle. One of them grabs the team mascot, a cuddly but slightly moth-eaten toy koala bear, from its habitual perch on the fridge. Ganache drifts over to join them, adroitly catching the koala bear, which is suddenly flying towards her. Evidently she will be first to speak in the standup today.

A standup is a short team meeting, held standing up. It is a phenomenon of agile development, discussed in Chapter 15. Standups usually happen daily, usually in the morning, and are usually very short. The number of occurrences of the word "usually" indicates that a standup is whatever the team makes of it. The aim is for team members to report what they did yesterday, what they plan to do today, and what roadblocks, if any, are hindering their progress. All team members listen to the reports, and if anyone says anything that affects somebody else's work, the people concerned get together for a discussion after the standup.

Ganache kicks off the proceedings. "Yesterday I started the draft release notes for ChocTactics 2.0 and finalized the updates to the administrator's guide. Today I will create a new documentation space for ChocSpice, finish the first draft of the ChocTactics 2.0 release notes, and restructure the ChocTactics user's guide. One of the support engineers has reported a bug in ChocTactics 1.1: the assessment report shows a glitch if the date is the 29th of February. I will work with him to finalize the troubleshooting page."

She tosses the koala bear to the next person in the circle and listens carefully as each person speaks. One of the developers, Mick, has just committed a fix for the leap year problem addressed by the new troubleshooting page. Ganache makes a mental note to talk to Mick after standup. Another team member mentions his work on a new feature for *ChocTactics 2.0*. Ganache notes that down for when she starts work on the related documentation.

After the standup, Ganache has a chat with Mick about the status of the fix for the leap year problem. She writes down the issue key so that she can direct readers to the issue tracker from the documentation and arranges to send the wiki page to Mick for technical review when it is ready. She and the team return to their desks to continue work.

10 am. Review the wiki updates

Ganache does a quick review of the troubleshooting page that Roald, the support engineer, has created to address the leap year bug. The page is currently visible to the world, because the Roald judged it important to get the information out to customers as soon as possible. Ganache agrees with his decision and is confident that the information supplied is correct.

She fixes the spelling, standardizes the page structure, and adds contextual information at the top of the page and to related topics at the bottom. She publishes her updates immediately, being similarly confident that she has added value and not added anything technically incorrect. The simple wiki workflow is good for this use case. Edit, update, save, and the updates are immediately visible to everyone who has permission to see the page.

Now Ganache forwards the page to Mick for technical review. He will add any extra information that the customers will find useful. Ganache will see the changes appear in her inbox and make any adjustments, or talk to Mick and Roald, if necessary.

10:15 am. Work on new projects

Choc Ink has formed a development team to work on a new product called *ChocSpice*. Ganache needs to create a new space on the wiki for the *ChocSpice* documentation so she and the team will have a framework for the manuals and release notes.

To create the new space, Ganache uses the *Copy Space plugin* and her space template. (See Chapter 8.) The template contains the skeleton for a user's guide, administrator's guide, installation guide, and the other bits and pieces needed in a technical documentation space. Because the product is new and the documentation will not yet contain any information that is useful to customers, Ganache sets the space permissions to make the space visible to *Choc Ink* employees only. She will adjust the permissions when the product manager wants to publish the documentation.

All in all, it takes Ganache about half an hour to copy the template, rebrand it for *ChocSpice*, and set up the release notes for the first beta release. Then she sends a notification to the product manager letting him know that he can start work on the release notes immediately.

Next, Ganache turns to the documentation for a new release of *ChocTactics*. She works through the set of features and improvements marked for this release in the project issue tracker, noting those that require a mention in the release notes and those that require a change in the documentation. The page for the *ChocTactics 2.0 release notes* currently has page restrictions applied so that it is visible to *Choc Ink* employees only. Ganache adds and updates items in the release notes, then sends a notification to the product manager and development team asking for feedback. The document will undergo a number of reviews before release date and will be ready for publication with the software release. When the product manager gives the go-ahead, Ganache will remove the page restrictions so that the release notes are visible to everyone who has permission to see the space.

What part does chocolate play in your life as technical writer?

From Shweta Hardikar on Technical Writing World:[4] "Reading this post on a Monday morning cheers me up. I always keep a bar of chocolate or candies for people in my drawer. Team mates take that much needed break and share a minute or two chatting under the excuse of eating chocolates or candies. In short, we bond over such treats and get to know each other well.

"We also have a custom of getting sweets for sharing with the team on all good events, birthdays, anniversaries, baby births, new home, new bike. In India, it's always Cadbury's.

"One more benefit: people are always ready to help the 'chocolate-lady'. As a technical writer, forging some kind of bond with the team helps me a lot."

1:30 pm. Do some wiki restructuring

After lunch Ganache tackles a reorganization of the table of contents in the *ChocTactics* user's guide. When writing some new pages for *ChocTactics 2.0*, which is as yet unpublished, Ganache realized that the guide itself could do with some restructuring. She wants to add a new section to the table of contents, move some existing pages into it, and add some new pages specifically for *ChocTactics 2.0*. The new structure will benefit customers using the current version of the product, as well as being essential when the new version is released. Therefore, Ganache has decided to make the changes on the live wiki.

Last week, Ganache drew a diagram illustrating the revised table of contents and got approval from the product team. Now it is time for the fun part. First, she adds a new page to hold the new section. Then she moves the existing pages into it by making them children of the new page. Where necessary, she adjusts the contextual information at the top and bottom of the pages to reflect their new position.

Wiki readers will immediately see the pages in their new locations in the table of contents. The web addresses of the pages are not affected by the move, because the pages have remained in the same space. In Confluence, a page's URL includes the space key and the page name, but not the position of the page within the space. This means that links to these pages are unaffected.

Next up for Ganache's attention are the new pages for *ChocTactics 2.0*. She originally created these pages in a holding area until she had time for the restructuring exercise. They are visible to *Choc Ink* staff only, because Ganache added page restrictions to each page. Now she moves them into the new section, too. They remain visible to *Choc Ink* staff only, because the move does not affect the page restrictions. Ganache will publish those pages when the product manager gives the go-ahead.

5 pm. Check the wiki notifications again

Ganache checks her wiki feed again to see what has happened since this morning. A customer has added a comment to the new troubleshooting page, expressing disappointment that the bug exists at all. Roald has already replied, apologizing on behalf of *Choc Ink* for the inconvenience caused and saying that the fix is under development.

Another customer has added a comment in praise of the new troubleshooting page.

"Thank you for the workaround," the customer enthuses, "You saved my life! I would just like to point out to other people that this glitch happens only in a leap year, and only on the 29th of February. That's today. I am sure *Choc Ink* will release a fix before the next time this bug can affect us, so there's no need to apply the workaround unless you need the report today."

Good point, thinks Ganache. She adds a similar note to the wiki page and thanks the customer for the comment. On that happy note, Ganache logs out of the wiki and heads off home.

A few days in the life of a wiki reader

People come to a technical documentation wiki for many reasons. The technical communicators who write the documentation are themselves also wiki readers. So are the support engineers, developers, quality assurance engineers, and product managers who contribute to the documentation. In this section let's confine ourselves to the people who come looking for information on how to use the product. These are the readers mentioned in the title of this section, and Gianduja is our hero today.

Gianduja is a customer of *Choc Ink* and an administrator of the *ChocTactics* application. She first encountered the documentation wiki by clicking an online help link on a *ChocTactics* screen. (See Chapter 12.) She drops in every now and then to check up on some facts and procedures. Unlike

Ganache, Gianduja does not spend every waking hour on the wiki. But she has become quite closely involved with it and the community of people who use it.

One day, wondering whether *ChocTactics* could display its reports in a funky purple instead of the usual sedate hues, Gianduja goes to Google and searches for "ChocTactics report customization." The search results lead her to a section of the *ChocTactics administrator's guide* that lists the community-contributed plugins. One of the plugins delivers just what she needs.

Gianduja remembers seeing a comment from another customer, named Caraque, on the page describing the *ChocTactics* reports. Caraque was asking for color-coded reports too. Gianduja goes back to the page, finds the comment and adds a reply. "Hey Caraque, there's a plugin that may help." She adds a link and goes back to her day job.

A couple of days later Gianduja receives a grateful reply from Caraque, "Thanks, just what I needed." A glow of satisfaction warms the cockles of Gianduja's heart.

How did Gianduja receive that reply? She has set her notification preferences on Confluence so that the wiki adds her as a watcher on any page where she makes an update or adds a comment. When Caraque replied to her comment, Confluence sent her an email notification. See Chapter 14 for more about notification options.

Gianduja returns often to the community part of the documentation space to see whether anyone has added an interesting new plugin. She has subscribed to an RSS feed on the documentation space so she does not have to keep going to the wiki to see if anything has changed. When people add comments or pages, the updates appear in her RSS reader. Confident that she has useful knowledge to share with other readers, she has signed up for a wiki username. When she spots a mistake in a document or has some useful information to add, she edits the page directly.

Gianduja is getting to know Caraque and other community developers, swapping tips and tricks, sharing her new-found knowledge with colleagues, and brightening their day job, too. What's more, Ganache knows Gianduja by name from seeing her updates and comments in the wiki notifications. They are even starting to recognize each other's profile photographs.

A day in the life of a support engineer

Roald is a *Choc Ink* support engineer. He works in the *ChocTactics* team and will shortly move over to support the new product, *ChocSpice*. He monitors the support queue and helps customers with their problems or questions. Roald is also occasionally to be found on the wiki when needs must and the devil drives.

8 am. The first support request

Roald's day has not started very well. Early in the morning, very early by Roald's clock, a customer lodges a support request about a glitch in the *ChocTactics* reporting functionality. For some reason the reports are not displaying the date. Roald does a quick search of all bugs reported in the issue tracker. There are no similar complaints. *Great,* Roald groans, *a whole new bug.*

8:07 am. The second support request

Another support request pops up in the queue. The same problem. Roald runs the reporting program on the *ChocTactics* test site and sure enough, empty dates. A quick investigation reveals that the reporting program does not handle leap years correctly. *And today's date? You guessed it, 29th of February. Easy enough to fix.*

8:30 am. The troubleshooting page

Roald hops over to the knowledge base area on the *ChocTactics* wiki and adds a troubleshooting page describing the known issue. He promises to post a workaround soon. Then he goes back to the growing queue of support requests and responds to each one, pointing the customers at the wiki page. Roald puts together a patch and uploads it onto the wiki page for customers to download. He also creates a bug report in the *ChocTactics* issue tracker, so that the developers will fix the problem in the next release of the software.

Hopping back to the troubleshooting page on the wiki, Roald adds a macro that draws information directly from the bug report onto the wiki page. That way, he can be sure that the page will have the latest information about the developers' progress on the issue, without having to make any further wiki updates.

Back to the busy queue.

10:15 am. The technical writer review

Some time later, Roald receives a wiki notification that Ganache has polished up his wiki page. Roald smiles. Just a little. And goes back to the queue.

10:30 am. Two customer comments

Within a few minutes, two more notifications come in when two customers add comments to the page. One of them is a little upset, so Roald apologizes on behalf of *Choc Ink*. The other customer thanks Roald for the workaround. Roald smiles. A little wider. And goes back to the queue.

2 pm. A development project

After lunch, with a wide grin, Roald hands the support queue over to a colleague and turns his attention to a development project he is working on. He is building a Confluence plugin that will add functionality especially useful to support teams. He wants to be able to customize a wiki space so that it works more like a knowledge base:

- Add a feedback form on every page, asking readers simple questions about the usefulness of the information.

- Generate reports that aggregate the responses received, both for a page and across the entire space.

- Adjust the way the Confluence search returns its results. Pages that readers judge to be more useful should appear higher in the search results than the less popular pages. The search should also be able to target specific spaces, instead of just one space or the entire wiki.

- Provide a macro, for insertion on wiki pages, that displays the most popular pages.

After a couple of hours on his project, Roald checks that there are no urgent issues still waiting in the queue. Then he heads out for a beer and a barrel of laughs with his colleagues.

Note: Although Roald is fictitious, the plugin that he is working on does exist and is available for download. It is the *Content Survey and Reporting plugin* created by Jeremy Largman and Ulrich Kuhnhardt, support engineers at Atlassian. Details are at the end of this chapter.

A product manager alive on a wiki

Charlie is the *ChocTactics* product manager. He has many channels for getting feedback from *Choc Ink's* customers. The wiki is one of those channels. Charlie watches for comments on the documentation and responds when a customer asks something about the direction of the product. In particular, Charlie keeps a close eye on the release notes, the supported platforms page, and the security advisories. He is also working closely with Ganache on the change management pages for *ChocTactics 2.0.*

9 am. Check the wiki notifications

Like Ganache, Charlie checks his wiki feeds early each morning. A customer has commented on the release notes for the previous *ChocTactics*, asking when a certain feature will be available. Charlie replies that *ChocTactics 2.0* is planned for release in the next month and will include the requested feature.

9:45 am. Attend the team standup

Charlie pokes his head into the standup to hear the reports from the technical writers and developers and to check that there are no roadblocks impeding progress.

10 am. Add to the FAQ entries for the new release

The new release is just around the corner and includes some significant new features that will have an impact on the way customers currently use the product. Charlie is keen to let the customers know about the changes, get feedback directly from anyone who is concerned about the impact, and work with people to help them manage the change within their own organization.

For a couple of months now, Charlie and Ganache have worked together on a change management guide that is publicly visible on the wiki. The guide informs people of the new features, gives them guidelines on changes they may need to make to their internal procedures, and provides training materials.

In addition, customers can add comments and feedback on the FAQ (frequently asked questions) page. Charlie responds to all such comments and adds an FAQ entry to the page when relevant. Today he amalgamates a couple of points made by interested customers and adds them to the content of the page.

Note: The Confluence documentation itself provides a real-life example of how product managers can get customer feedback before a major product release. Confluence 4 brings significant changes to the way people use the wiki. In particular, the editor is quite different from earlier versions of Confluence. See Chapter 3 for a discussion of the changes. The Confluence technical writers and product managers worked together to produce a set of pages aimed specifically at helping customers manage the change to Confluence 4.[5]

One of the pages is a list of frequently asked questions.[6] Readers add comments to the page, asking questions and voicing their concerns. The product managers monitor the page, respond to the comments, and add information to the page itself when relevant. In this way, customers and product managers have a central location for exchanging ideas, concerns and responses.

11 am. Take a break from the wiki

Product managers do have other work to do. Charlie goes off and does it.

4 pm. Fix the release notes

Charlie receives Ganache's message that the release notes are ready for review. He drops in on the wiki and tweaks the text here and there. Release notes have a hybrid purpose, part technical, part marketing, and part instructional. It is important for product managers to work closely with technical writers and developers to craft the perfect document. Charlie adds a couple of screenshots, complete with rounded corners.

A developer walks into a wiki

Mick is a developer at *ChocTactics*. He works in the source code and on the issue tracker, with occasional forays into the documentation wiki.

In mid-morning, Mick receives a request from Ganache to review the troubleshooting page. He knows what this is about, since she mentioned it at the standup. He runs an expert eye over the document, adds an explanatory sentence about the fix he is working on, then dives back into the code. Ganache sees his update and is happy.

That afternoon, Mick is back on the wiki. He is currently working on the new application, *ChocSpice*, and he needs to use the *ChocTactics* API to develop an integration between *ChocSpice* and *ChocTactics*. Mick spots an area in the API documentation that could be improved. He pops into the wiki page, makes the change, copies some example code into the page, then goes back to coding.

A couple of days later, a developer who works outside *Choc Ink* makes use of Mick's example code. She adds a comment to the page:

"Thanks for the awesome example, Mick. I've used your code and written a test for it too. Here's a link to the test source. Anyone is free to use it."

Cool, thinks Mick.

The sun never sets on a wiki

It is evening in Sydney, Australia. Ganache and the *Choc Ink* team say their good nights and head off into the sunset. As Ganache peels her attention away from the absorbing world of the wiki, a thought flits through her head, *Where does the documentation go when I'm asleep?* The sun is rising in Rio de Janeiro and Londoners are already trickling into work. While Ganache is off duty other people will be reading and contributing to the documentation. The results will be waiting in her notification inbox tomorrow morning.

Many people work in a global environment. Customers never sleep. Some organizations have teams all over the world, ensuring that they offer support in all time zones. A number of tools help people to communicate in such environments. A wiki is one of those tools. Ganache can put up a draft, notify the reviewers on different continents, and carry on with other tasks. Wherever they may be and whenever they are at work, the reviewers congregate in a central virtual location – the wiki page. A couple of comments later, after some collaborative updates have been completed, the page is ready for publishing.

One fine day, Ganache finds herself traveling from Sydney to San Francisco to train a new technical writer. All she needs is her mobile phone, some fog-proof clothes, and an emergency chocolate cache. All her writing tools are on the wiki. She can log in from any computer, fire up a web browser, and update the documentation.

The wiki never sleeps.

References

These are the tools, websites, blog posts, and other references relevant to the content of this chapter.

Useful plugins

Content Survey and Reporting plugin by Atlassian, especially useful for knowledge bases: `https://-plugins.atlassian.com/plugin/details/24440`

Further reading

A support engineer's guide to creating a knowledge base space, in the *Confluence User's Guide*: `http://confluence.atlassian.com/display/DOC/Creating+Your+Knowledge+Base+Space`

Endnotes

[1] The phrase "the full monty" means something that includes everything relevant to the subject, or something that is complete and sufficient for the purpose in mind. *The Phrase Finder* gives some suggested origins of the phrase: `http://www.phrases.org.uk/meanings/-full%20monty.html`. There is also a film, made in 1997, titled *The Full Monty* which tells the story of a group of British men who found themselves without a job and decided to do a striptease act to make money, taking off all their clothes on stage and thus "going the full monty."

[2] The word "pwn" is techno slang for "conquer" or "dominate." It probably came from mistyping the word "own." See the definition on the *urban dictionary*: `http://www.urbandictionary.com/define.php?term=pwn`. The word "typos" is a common abbreviation of "typographical errors."

[3] *What part does chocolate play in your life as technical writer*, a discussion on *Technical Writing World*, started in May 2011: `http://technicalwritingworld.com/forum/topics/what-part-does-chocolate-play`

[4] *What part does chocolate play in your life as technical writer*, a discussion on *Technical Writing World*, started in May 2011: `http://technicalwritingworld.com/forum/topics/what-part-does-chocolate-play`

[5] Change management guide for Confluence 4.0: `http://confluence.atlassian.com/display/DOC/Planning+for+Confluence+4.0`

[6] FAQ page in the change management guide for Confluence 4.0: `http://confluence.atlassian.com/display/DOC/Confluence+4.0+Editor+FAQ`

14

What's happening on the wiki

Technical writers are guardians of the documentation. It is crucial that they know what is happening on the wiki. Other people also want to know when things change. A customer awaits a software release with bated breath. Administrators keep an eye on security advisories and supported platforms. Employees need a reliable way of receiving notification when procedures change. Developers like to know when the new kid on the block posts a shiny code sample.

How do people see where the action is on a wiki? Smoke signals? Bush telegraph? No, it all happens via activity streams, wiki watches, and RSS feeds. And, to get the big picture, wiki administrators dabble in web analytics, too.

What happens in this chapter

This chapter describes the tools available for monitoring wiki activity. The chapter has the following sections:

- **Activity streams.** Drop in on the Confluence dashboard to see the latest updates. Or visit someone's profile page to see their most recent activity on the wiki.

- **Wiki watches and notifications.** To monitor activity on a specific area of the wiki, add yourself as a watcher, and Confluence will send an email notification whenever something happens to that page or space.

- **Monitoring updates via RSS feeds.** RSS feeds provide another way of getting targeted information on wiki activity and sharing it with other people. Upon request, Confluence will send an XML file containing the most recent updates matching a set of selection criteria. You can read the response in an RSS feed reader, such as Sage or Google Reader, or display the feed on a Confluence page.

- **Web analytics and usage statistics.** Web analytics tools, such as Google Analytics and Webtrends, give valuable insight into the number of people on the wiki and where they are spending their time.

- **Summary of this chapter.** See what you have learned in the chapter.

- **References.** The tools, websites, blog posts, and other references mentioned in this chapter.

Activity streams

An activity stream is a list of recent events on a given platform or platforms, in this case the wiki. The stream shows the name of the person involved, the date and time of the activity, a summary of what happened, and a link to the related page. In Confluence, the most obvious activity stream is on the dashboard. By default it shows the most recent updates across the entire wiki. Each person can configure their own view of the dashboard so that they see only the updates of interest to them.

Figure 14.1. The activity stream that appears on the Confluence dashboard

There is another activity stream on every user's profile page, showing the most recent activity.

Dropping in on the dashboard is a great way to see what is happening right now. It often gives an alternative view of the organization and provides some light relief. While you would use wiki watches and RSS feeds to keep track of the content that is important to your daily job, you may find it rewarding to hop over to the dashboard and see what other people are up to.

Wiki watches and notifications

It is easy enough to look at the history of a specific page (see the section titled "Drafting a page" (p. 202)). But how can you find out that a change has been made to that page or to any other content in your documentation space? It is not feasible to keep visiting the page on the off chance that someone has updated it since you were last there.

One way is to watch a page or a space so that you get a notification whenever something happens.

Watching a page or a space

Anyone who has logged in to the wiki can watch a page. Whenever someone makes a change to the page or adds a comment, Confluence will send the watchers an email notification.

To watch a page: Go to the page, then choose "**Tools**" > "**Watch**."

You can watch a whole space too. Confluence will send an email notification whenever someone adds, updates, or deletes any content in that space.

To watch a space: Go to a page in the space, then choose "**Browse**" > "**Advanced**" > "**Start watching this space**."

Setting notification options

Perhaps you want to prevent Confluence from sending a notification every time you update the wiki yourself. Perhaps you want Confluence to add you as a watcher of every page that you add or update so that you get a notification of all subsequent updates that other people make. You can set these and other options in your user profile.

To set your notification options: Open the menu under your username at the top of the screen, then choose "**Settings**" > "**Email**." Click "**Edit**" to update the settings.

The *Confluence User's Guide*[1] describes the options available.

To set your email address: Open the menu under your username at the top of the screen, then choose "**Profile**." Click "**Edit**" to update your email address and other details.

Monitoring updates via RSS feeds

Wiki watches, as described above, send notifications to your email inbox. Another way to monitor wiki updates is via an RSS feed.

On request, Confluence will send you an XML file containing information about the most recent updates that people have made to Confluence pages or other content. This is called an RSS feed. People often say that RSS stands for "really simple syndication," but *Wikipedia* states that it originally stood for "RDF site summary."[4] The XML file conforms to a standard (RSS or Atom) that various RSS readers can interpret.

These are the stages involved:

1. Define the RSS request, using the Confluence feed builder to build a query. The resulting query takes the form of a URL with parameters that Confluence understands.

The sweetest RSS feed ever

The *SCANDYBARS* blog[2] has the clearest, most delicious-looking pictures of chocolate I have ever seen. The author makes the pictures by cutting the chocolates in half and then scanning them. That's right, on a digital scanner. It must get a bit sticky, but the results are perfection. One of the best is the picture of the Duplo bar from Italy.[3] Try the sweetest RSS feed ever: `http://-scandybars.tumblr.com/rss`.

2. Send the request to Confluence whenever you want to see the most recent updates. The easiest way to send the request is via a feed reader.

3. Read the response. A feed reader will parse the response and present it to you in readable form.

Defining the request

The RSS request is a URL, pointing to your Confluence server and passing it various parameters that it will interpret when choosing the information to send back to you.

The URL looks something like this (that's all on one line):

```
https://wikitechcomm.onconfluence.com/spaces/createrssfeed.action?types=page&
types=comment&spaces=DOC&maxResults=15&title=Twitter+guide&publicFeed=true
```

There are a number of predefined RSS feeds available in Confluence, or you can build your own.

To get the predefined RSS feeds for a space: Go to a page in the space, then choose "**Browse**" > "**Advanced**" > "**RSS Feeds**."

The feed builder offers a more flexible solution. It displays a form that you can use to tailor the information in the feed, by specifying the events that you want to know about:

- One or more content types – pages, blog posts, attachments, comments or mail archives.

- One or more spaces.

- Pages tagged with a given label or labels.

- Updates made within a given period of time.

To build a feed in Confluence 3.5: Go to the dashboard then choose "**Feed Builder**."

To build a feed in Confluence 4: Choose "**Browse**" > "**Feed Builder**."

Reading the RSS feed

It is an interesting exercise to paste the URL of the RSS request into a web browser and see the XML that Confluence returns. Once that novelty has passed, a feed reader is much more people-friendly. The feed reader will send the RSS request to the Confluence site and display the response in an easy-to-read format.

Here are a few RSS feed readers, also called feed aggregators:

- Sage, an add-on for Firefox web browser.

- Google Reader, a web-based feed reader.

- NetNewsWire for Mac OS X, iPad and iPhone.

- SharpReader for Windows.

- Liferea for UNIX and Linux.

- Mozilla Thunderbird, an email client that can read RSS feeds and display them in a format similar to email messages.

Wikipedia[5] has an extensive list of feed readers, and *RSS Info*[6] has reviews of RSS readers.

A closer look at Sage

I recommend Sage as a light-weight, well-designed feed reader. For me, the best thing is the way Sage helps me scroll through the wiki updates quickly and drill down to the ones that need attention. The Sage window has three panels.

- At top left is the list of all my feeds. I just click the feed that I want to read.

- At bottom left are one-liners for all the items in the chosen feed

- The main panel on the right is a two-column detailed display of all the items in the feed.

In Figure 14.2, Sage is displaying a feed from the Twitter guide on the *Confluence, Tech Comm, Chocolate* wiki. As you can see, Ganache has been busy updating pages in the guide.

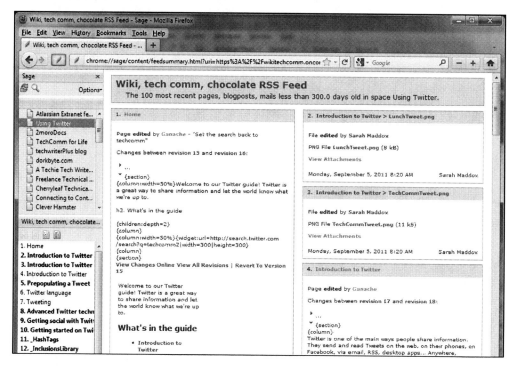

Figure 14.2. Sage feed reader in Firefox

I can scroll down quickly, scan the list, right-click the items I am interested in, and open them in a separate tab for closer attention. In the bottom left panel, items that I have not yet read are in bold text. When I have reviewed all the items, I click the option that marks all the items as read so the next time I ping Confluence I will know where I left off.

Sage is an add-on for Mozilla Firefox. To install Sage:

1. Start Firefox.

2. Go to this web page: `https://addons.mozilla.org/en-US/firefox/addon/77`.

3. Click the "**Add to Firefox**" button.

Note that Firefox treats feeds as bookmarks. To add a feed to Sage:

1. Copy the feed URL from the source, such as the Confluence feed builder or an RSS link on a website or blog.

2. Click the Sage icon (a green leaf) in the Firefox toolbar.

3. The Sage panel will appear on the left. Right-click in the top left Sage panel and choose "**New Bookmark**."

4. Paste the URL of the RSS feed into the "**Location**" text box and click "**Add**."

To read the RSS feeds:

1. Click the Sage icon in the Firefox toolbar.

2. The Sage panel will appear on the left. Select one of your feeds in the top left-hand panel.

3. The page titles will appear in the bottom left-hand panel, and the content in the right-hand panel.

Displaying an RSS feed on a Confluence page

Another way of reading an RSS feed is to embed the feed on a Confluence page. In this way, you can share the news items with colleagues. The items in the feed may come from Confluence itself or from any other feed. To display an RSS feed on a Confluence page, use the RSS macro.

This is a little risky, as the Confluence documentation so eloquently warns,[7] because an RSS feed is essentially raw HTML. It may therefore contain JavaScript elements, which can introduce malicious code into the Confluence page. Before embedding an RSS feed onto a Confluence page, it is a good idea to make sure you trust the source of the feed. For that reason Confluence system administrators can choose to enable or disable the RSS macro and to configure a whitelist of trusted URLs. (A whitelist is the opposite of a blacklist. In other words, it is a list of goodies rather than baddies.)

Web analytics and usage statistics

Website administrators and technical communicators can use web analytics to gain valuable insight into the number of people using the wiki and where they are spending their time. Let's take a look at Google Analytics and a couple of useful Confluence plugins.

Google Analytics

Google Analytics[8] is a free service provided by Google. It analyzes the traffic on a website and displays graphs and figures showing how people found the site and what they did when they got there.

As an example, let's look at some Google Analytics data about the Atlassian documentation wiki.[9] This wiki hosts the documentation for all the Atlassian products: JIRA, Confluence, Crowd, FishEye, Crucible, Bamboo, Atlassian OnDemand, and more. Each product has its own space on the wiki. In fact, most products have a number of spaces, one for each major release of the product.

Figure 14.3 shows an overview of visitors to the site in the month from 22 July to 21 August 2011.

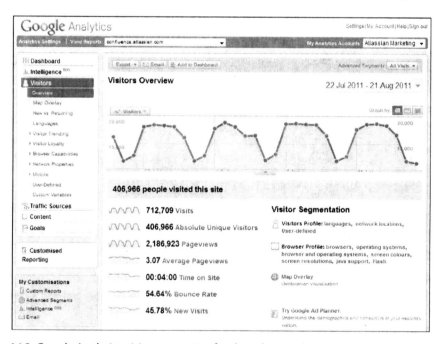

Figure 14.3. Google Analytics visitors overview for the Atlassian documentation wiki

The visitors overview shows the number of visits (712,709), the number of unique visitors (406,966), and the number of pages that they looked at (2,186,923). It also gives other interesting figures such as the percentage of new visits (45.78%) and the average amount of time each person spent on the site (4 minutes). The bounce rate (54.64%) indicates the proportion of people who left the site after viewing just one page. Google says that a high bounce rate generally indicates that the page the visitor landed on was not relevant to the visitor. Another interpretation, though, is that the visitor found an answer immediately and left satisfied.

The screen shown in Figure 14.3 is just the start. You can drill deeper to get more figures on just about every section of the screen.

Google Analytics also gives a detailed analysis of the most popular content on the site. Figure 14.4 shows an overview of the top content in the Atlassian documentation wiki for the stated period.

Figure 14.4. Google Analytics top content overview for the Atlassian documentation wiki

Unsurprisingly, the documentation landing pages for Atlassian's two flagship products are in the top three pages visited: JIRA and Confluence. The most popular page is the wiki dashboard. Again, Figure 14.4 is just the tip of the iceberg.

The traffic sources overview (see Figure 14.4) shows that most of the visitors come via a search engine. Drilling deeper (not shown in Figure 14.4) reveals that Google is far and away the biggest source of search engine visits (397,822), with Bing second in line (4,533), followed by Yahoo (1,137).

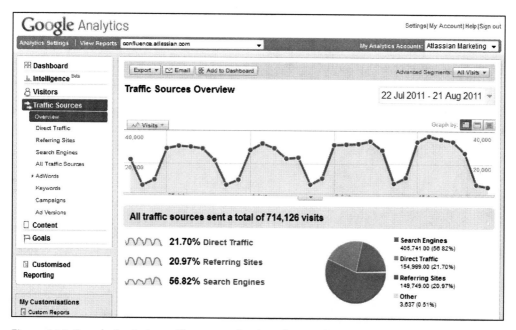

Figure 14.5. Google Analytics traffic sources for the Atlassian documentation wiki

There is much information to derive from the data that Google Analytics provides. In the section titled "Further reading" at the end of this chapter are some links to guidelines on making the most of Google Analytics.

Adding analytics to Confluence pages

AppFusions has developed two plugins for Confluence that add analytical data to the wiki:

- The *Analytics Plugin* displays Google Analytics data about a given space, such as top content, depth of visit, traffic sources, search engine keywords, and more.

- The *Webtrends Analytics Connector* provides a Confluence macro that integrates site-level data from Webtrends into a Confluence page.

Summary of this chapter

A wiki is a collaborative tool. People are updating content all the time. Technical writers, as guardians and curators of the documentation, need an efficient way to monitor activity. Wiki watches and RSS feeds are good tools for monitoring specific pages and spaces. Activity streams show the current activity on the wiki. Google Analytics and other web analytics tools are powerful, flexible platforms for analyzing and reporting website traffic. Since a wiki is essentially a website, technical communicators can draw a lot of information from analytics tools. The graphs and drill-downs are fun and pretty, too.

Ours is a world of information overload. It is all about staying on top of the news but not being swamped by it. I hope the information in this chapter gives you some guidelines on how to do that.

References

These are the tools, websites, blog posts, and other references relevant to the content of this chapter.

Useful plugins

- *Analytics Plugin for Confluence* by AppFusions, for reporting Google Analytics data about a given space: `https://plugins.atlassian.com/plugin/details/16722`

- *Webtrends Analytics Connector for Confluence* by AppFusions, for integrating Webtrends Analytics 9 Insight data into a Confluence page. Plugin information: `https://plugins.atlassian.com/-plugin/details/39438`. Webtrends information: `http://www.webtrends.com/`

Some RSS feed readers

- Sage feed reader for Firefox: `https://addons.mozilla.org/en-US/firefox/addon/sage/`

- Google Reader, a web-based RSS feed reader: `http://www.google.com/reader/`

- NetNewsWire, an RSS feed reader for Mac OS X and iOS: `http://netnewswireapp.com/`

- SharpReader, an RSS feed reader for Windows: `http://www.sharpreader.net/`

- Liferea, an RSS feed reader for UNIX and Linux: `http://liferea.sourceforge.net/`

- Mozilla Thunderbird email client: `http://www.mozilla.org/en-US/thunderbird/`

Further reading

- *Tracking Updates to Confluence Pages and Spaces*, in the *Confluence User's Guide*: `http://confluence.atlassian.com/display/DOC/Watching+Changes`

- *Analytics Help*, the official Google Analytics help site: `http://www.google.com/support/-googleanalytics/`

- Some excellent introductory information in *Understanding Google Analytics*: `http://understand-inggoogleanalytics.com/`

- *Understanding Google Analytics Reports*, a blog post by Eldon Beard on *eldon beard*, August 2011: `http://eldonbeard.com/understanding-google-analytics-reports/`

- The official *Google Analytics* blog: `http://analytics.blogspot.com/`

Endnotes

[1] Guidelines on email notifications and options, in the *Confluence User's Guide*: `http://confluence.atlassian.com/display/DOC/Subscribing+to+Email+Notifications+of+Updates+to+Confluence+Content`

[2] The *SCANDYBARS* blog, showing the clearest, most delicious-looking pictures of chocolate ever spotted by a chocolate lover: `http://scandybars.tumblr.com/`

[3] A cross section of a Duplo chocolate bar on *SCANDYBARS*: `http://scandybars.tumblr.com/-post/8895254403/duplo-italy`

[4] The *Wikipedia* page about RSS: `http://en.wikipedia.org/wiki/RSS`

[5] A comparison of feed aggregators on *Wikipedia*: `http://en.wikipedia.org/wiki/Comparison_of_feed_aggregators`

[6] Information about RSS and RSS readers, on *RSS Info*: `http://blogspace.com/rss/readers`

[7] A guide to the RSS feed macro, in the *Confluence User's Guide*: `http://confluence.atlassian.com/display/DOC/RSS+Feed+Macro`

[8] Google Analytics: `http://www.google.com/analytics/`

[9] The Atlassian documentation wiki is a Confluence site: `http://confluence.atlassian.com`

15

Going agile

When Ganache first learned that she was to work in an agile environment, she did not know whether to say "yikes" or "yaayyy!" Either way, a number of questions sprang to mind. What is "agile," and how can technical communicators use a wiki to good advantage in such an environment?

In this chapter

This chapter describes technical communication and a wiki in an agile environment. The chapter has the following sections:

- **Introduction to agile methodology.** Agile methodology is a collection of guidelines on how to organize a development team.

- **The Agile Manifesto.** The document that introduces the principles of agile methodology.

- **Three agile methodologies.** A quick look at Extreme Programming (XP), Scrum, and Kanban.

- **Technical writing in an agile environment.** How technical writers can work with an agile development team.

- **Wiki in an agile environment.** The features of a wiki that make it useful for agile technical writing.

- **Special techniques for agile teams.** Innovation sprints, documentation blitz tests, and code reviews.

- **Summary of this chapter.** Some conclusions.

- **References.** The websites, blog posts, and other references mentioned in this chapter.

Introduction to agile methodology

Agile methodology is a collection of guidelines on how to organize a development team. It is based on a set of philosophical principles about the best way to produce software systems. These principles are encapsulated in a fascinating and admirable document called the *Agile Manifesto*, quoted below.

Although people talk about "the agile methodology," there are actually a few different methodologies that have arisen as various groups of people have developed procedures to put the agile principles into practice. The most well-known agile methodologies are Extreme Programming (XP), Scrum, and Kanban. More about them later. It is also important to note that very few teams follow one methodology exclusively. Instead, teams combine ideas and practices from two or three methodologies, adapting the procedures to suit their needs

A prominent goal of agile methodology is to focus on what the customer wants and to discover those requirements by personal interaction and iterative development. The most revolutionary concept in the methodology is to put the control of the development process squarely in the hands of the engineers. The development team is self-organizing, collaborates with customers, and examines the results of its work regularly so that it can improve its own processes.

Many of the ideas in agile methodology come from the way Toyota works. In particular, Toyota's teams regularly stop and examine their work to see how they can improve its quality. There is a good summary on the *Wikipedia* page about *The Toyota Way*.[1]

The Agile Manifesto

This is the full text of the *Agile Manifesto* at `http://agilemanifesto.org/`:

Manifesto for Agile Software Development

We are uncovering better ways of developing software by doing it and helping others do it. Through this work we have come to value:

- **Individuals and interactions** over processes and tools

- **Working software** over comprehensive documentation

- **Customer collaboration** over contract negotiation

- **Responding to change** over following a plan

That is, while there is value in the items on the right, we value the items on the left more.

Kent Beck	James Grenning	Robert C. Martin
Mike Beedle	Jim Highsmith	Steve Mellor
Arie van Bennekum	Andrew Hunt	Ken Schwaber
Alistair Cockburn	Ron Jeffries	Jeff Sutherland
Ward Cunningham	Jon Kern	Dave Thomas
Martin Fowler	Brian Marick	

© 2001, the above authors.

this declaration may be freely copied in any form, but only in its entirety through this notice.

The *Agile Manifesto* links to another document, the *Principles behind the Agile Manifesto*, which describes the twelve principles of agile software:[2]

Principles behind the Agile Manifesto

We follow these principles:

- Our highest priority is to satisfy the customer through early and continuous delivery of valuable software.

- Welcome changing requirements, even late in development. Agile processes harness change for the customer's competitive advantage.

- Deliver working software frequently, from a couple of weeks to a couple of months, with a preference to the shorter timescale.

- Business people and developers must work together daily throughout the project.

- Build projects around motivated individuals. Give them the environment and support they need, and trust them to get the job done.

- The most efficient and effective method of conveying information to and within a development team is face-to-face conversation.

- Working software is the primary measure of progress.

- Agile processes promote sustainable development. The sponsors, developers, and users should be able to maintain a constant pace indefinitely.

- Continuous attention to technical excellence and good design enhances agility.

- Simplicity – the art of maximizing the amount of work not done – is essential.

- The best architectures, requirements, and designs emerge from self-organizing teams.

- At regular intervals, the team reflects on how to become more effective, then tunes and adjusts its behavior accordingly.

Many of these principles fit very nicely with the aims of technical communication. Listening and responding to customers is what technical communicators have been advocating ever since Ada Lovelace first picked up her pen.[3] The manifesto's emphasis on communication, collaboration, and simplicity tie in with a technical writer's role and convictions.

One of the principles does present a bit of a hiccup: "We have come to value… Working software over *comprehensive documentation*." But the manifesto goes on to clarify, "That is, while there is value in the items on the right, we value the items on the left more." This principle is a rebellion against, and a safeguard against, those projects that become so bogged down in feasibility studies, requirements analyses, functional and technical specifications, multiple sign offs and risk analyses, that the engineering work takes a back seat and sometimes falls right off the donkey cart.

Three agile methodologies

A number of distinct methodologies have arisen, offering practical ways of embodying the agile principles. Generally, agile practitioners pick and mix techniques from these methodologies to suit their needs. To get a better idea of how a development team can put the agile principles into practice, let's take a look at three of the agile methodologies: Extreme Programming (XP), Scrum, and Kanban.

Extreme Programming (XP)

Extreme Programming is the father of all agile variants, supplying a well-developed set of principles and procedures. The Extreme Programming organization offers a "gentle introduction" to XP.[4] It includes a diagram showing how an XP project moves from the creation of user stories and an architectural spike, through a number of development iterations to the release of the software.[5]

XP, and agile methodology in general, introduces a number of new terms.

- A **spike** is a quick foray into the development of a feature. It takes a predefined period of time and has the aim of testing the feature's feasibility and estimating the time it will take to develop the complete feature.

- A **user story** is a task-based specification of what the user needs to do.

- **Pairing**, more prosaically known as "pair programming," is one of the most endearing XP terms. It refers to two developers working closely together on a story.

The *XP Rules* page[6] has a useful list of terminology.

Scrum

Scrum is a lightweight framework designed to be tailored by the development team. These are the basics of the Scrum process:

- The product owner writes feature requests as **stories**. A story includes the specific persona and the reason that persona wants to perform the task specified.

- The **product backlog** is the list of stories in order of priority, with the most important at the top. Stories at the top of the list should be small. In other words, the stories themselves have been through iterative analysis and development until they reach the stage where development can start. The stories at the bottom of the backlog tend to be just ideas and often still represent big chunks of work. Some projects use the concept of an "epic" to group related stories and keep them together as they approach the top of the list.

- The **sprint backlog** is the list of stories to be tackled in a particular sprint.

- A **sprint** is a period (often two weeks) during which the development happens. There is a planning meeting at the beginning of the sprint and a retrospective at the end. The idea is that each sprint delivers a potentially shippable product.

- Scrum defines three roles. The **product owner** creates and prioritizes the backlog, defines the stories, and decides what goes into the next sprint. The **scrum master** works with the team day to day and is focused on the current sprint, fixing anything that is preventing the team from working. The **team** does the development work.

- When estimating the time required for a story, the team uses the idea of **story points** rather than time. The story points are expressed as a number, which is determined by estimating the relative complexity of the work required.

- The team calculates its **velocity** based on the number of story points that it has managed to complete during the sprint. For example, if the developers had an estimate of 10 points for the stories in a sprint, and completed 8 of them during the sprint, then the velocity is 8. The team will work out an average over a number of sprints, and use this to compile estimates for the product backlog.

See the section titled "Further reading" at the end of this chapter for sources of information about Scrum.

Kanban

The Kanban methodology focuses on organizing items of development work in a queue. Each team member pulls an item of work, or **card**, from the queue when he or she is ready to work on it. Kanban is designed to prevent any part of the development team from being overloaded with work and so forming a bottleneck that could stop the rest of the team from progressing.

The first step is to map the workflow of the development team. The workflow stages are usually depicted as columns drawn on a board, a wall, or using a software package. This is called the **Kanban board**. For example, the columns on the board may be:

- Scheduled
- In development
- Ready for QA (quality assurance)
- In QA
- Ready for deployment
- Documentation
- Done

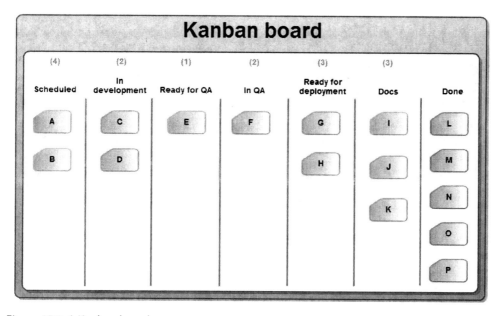

Figure 15.1. A Kanban board

The team sets a limit on each column, determining the maximum number of cards (tasks or issues) that can exist in a column at the same time. For example, the columns in Figure 15.1 have these limits: 4, 2, 1, 2, 3, 3.

When ready for more work, a developer takes the top card from the schedule and puts it into the next column: "In development." When development is finished, the card moves into the next column: "Ready for QA."

The crux of the matter is this: You cannot move a card into the next column if the column is already full. In other words, if the number of cards in the column has reached the predefined maximum, no-one can add another card until someone removes one. This often means that the person who wants to move the card must first tackle one of the cards in the column that is blocking the flow. In this way, the work flows through to the last column. The aim of the process is to end up with a number of cards in the "Done" column.

Technical writing in an agile environment

Two mainstays of agile methodology are continual adaptation to change and constant review of the team's processes to ensure ongoing improvement. From a technical writer's point of view, this is an excellent situation. It means that the development team will be able to adapt to include the documentation and other technical communication tasks in the workflow, provided that the team is convinced of the usefulness of and necessity for doing so.

Another agile goal that has significant implications for technical communicators is this: At the end of each sprint, the team should produce a shippable product. Ideally, the documentation should be ready for shipping too.

A number of teams are investigating and experimenting with the best ways to integrate technical communication into the development team's workflow. Many have already achieved a deep level of integration, while others find lighter integration more efficient. Deep integration means that the technical communication tasks are part of the development team's workflow. The documentation issues reside on the same issue tracker as the engineering tasks, and the technical writers attend all the planning meetings and retrospectives. Light integration means that the technical writers follow their own workflow. They do attend the team standups (daily short meetings held standing up) and keep an eye on the development issue tracker, but essentially they work alongside the engineering team.

Nirvana is getting the documentation tasks included in the "definition of done." In other words, no task can be declared done until the documentation is complete. If the entire team is committed to ensuring that the documentation is ready for deployment at the same time as the software, there is less chance technical writers will experience the dreaded crunch time towards the end of the release cycle. Another advantage is that the development team will be more aware of the work the technical writers are doing and of what needs doing.

There is a risk of increased tension between the engineers and the technical writers, especially in the early days of integrating the documentation tasks into the workflow. A good strategy is to make sure management and the team leads are on board and ready to emphasize the importance of the documentation in the delivery of the product to the customers. Technical writers can also demonstrate the added value that they can bring to the design and development stages of the workflow.

One of the big debates is whether the documentation tasks should lag one sprint behind the engineering tasks or be part of the same sprint. Let's say the team is developing this story around the development of a Twitter widget for a web application:

"As a web page designer, I would like to add a Twitter stream to a page, showing the latest Tweets from a given person."

In a two-week sprint the designers map out the user experience for the story, the engineers write the code, and the quality assurance team tests the results. Should the technical writer also finish documenting the story in this sprint, or should the documentation task be part of the next two-week period? There are advantages to both methods. It is often difficult for the documentation to be written until very near the end of the sprint, because the feature is not finalized until then. Therefore, it seems practical to schedule the documentation tasks for the next sprint. On the other hand, to be true to the ideal of ending a sprint with a shippable product, the documentation must be ready at the same time as the software. Also, the engineers will move on to new stories in the next sprint. It is counterproductive if they have to work with the technical writers on a story from the previous sprint.

Agile technical writers

A number of technical writers working in agile environments have shared their experiences and lessons learned.

- Mary Connor has written an excellent series of posts on her blog, *Clever Hamster*, about her team's move to an agile environment. The first post is titled *How does Agile affect documentation?*.[7]

- Edwin Dawson's three-part series on *Technical Writing In Agile Software Development* appears on the *Atlassian blogs*. In the first part, *Technical Writing in Space*,[8] Edwin mixes a merry story of mayhem in with the serious discipline and strategy required to work within agile teams.

- Diana Ost's post, *Sprinting for the Finish Line*, has good tips on how to work in a sprint. Her blog is aptly named *The Agile Technical Writer*.[9]

- Ben Minson, over on *Gryphon Mountain Journals*, writes regularly about agile practices. A good example is his post called *Suggestions for Survival in an Agile Environment as a Technical Communicator*.[10]

Social sites and online forums are an excellent place to ask questions and see what agile practitioners are experimenting with. Try the *Agile Technical Writers* group on *LinkedIn*[11] and various discussions on *Technical Writing World*.[12]

Agile tips

Here are a few practical tips on getting started in an agile environment:

- Attend the development team standups. This is a good way to get advance notice of new features, patch releases and changing deadlines.

- In your standup report, let the development team know what you are working on. Your work will dovetail with theirs quite neatly. A technical writer tackling a problem area in the documentation will often find that the developers are at the same time completing bug fixes and the support engineers are weathering a storm of problem reports around the same topic.

- Keep the contributions short at the standups. Just report on the work done yesterday, the work planned for today, and any problems that are hindering the work.

- Make sure the documentation is seen as part of the product, and the technical writing timeline is factored into the product release.

- When documenting a new feature, evaluate the user's experience. Such feedback is valuable to the engineers and designers during the development sprint. Often the technical communicator is the first end user to experience the software. The technical communicator also frequently has more of a task-oriented and big-picture view than the developer who writes the code.

- Apply the principle of iterative development to the documentation as well as to the software. Contribute to the quality assurance process and be ready to adapt the documents to reflect any resulting code changes.

- Subscribe to blogs, wiki watches and any other information source that is available. Information overload is a valid concern, but most people quickly learn how to scan the items coming in and pick up on the relevant ones. It is important to stay ahead of the news in an agile environment.

- Seek even more input. Keep your eyes open for impromptu and scheduled training sessions and any other sources of information.

- Monitor the bug reports and enhancement requests coming in for the product. Take note of any that will affect the documentation.

For a deeper level of integration into the agile team's processes:

- Attend the development team's planning sessions at the beginning of each sprint and include the planning of the documentation tasks as part of each session.

- Log the documentation issues in the development team's issue tracker. Make sure that the issues are marked for completion in the relevant sprint. Depending on the team's choice, that may be the same sprint in which the engineering and quality assurance is done or the following sprint. Either way, the documentation tasks will be obvious to the scrum master or development team lead.

- Help define the stories so that each one represents a complete unit of work that can be documented.

- Make it clear to the team that the developers need to complete their stories in a sprint, because otherwise it makes it difficult for the technical writers to do their work.

- Take part in the retrospectives at the end of each sprint. This is a good opportunity to make the team aware of the documentation team's requirements. Report on what went well from the point of view of the documentation and what needs improving.

- If you run into persistent problems in working with the team, sit down with the scrum master or development team lead, explain the challenges, and work together to find solutions.

- Seat the technical writers with the engineers. If there is a team of technical writers, assign each technical writer to a specific product rather than sharing the products amongst the writers.

We have come to value chocolate over extensive verbal compliments

My favorite principle in the *Agile Manifesto* is that you get good people, put them in a good environment and then trust them to do a good job. A good environment – that is where chocolate comes into it. At the office we have cupboards full of the stuff.

Undeterred by such abundance, the technical writing team also goes on regular "doc hot choc" outings. A doc hot choc is a team visit to a restaurant where we order mugs of hot chocolate and talk about documentation. Sometimes we even allow one or two hangers on, such as a developer who has written a Confluence plugin just for us.[13]

A wiki in an agile environment

Ward Cunningham, inventor of the very first wiki,[14] is also one of the authors of the *Agile Manifesto*. In his keynote presentation at WikiSym 2005, titled *Crucible of Creativity*, Ward talked about the importance of accepting change as an integral part of software development. The technique of refactoring a program to allow for ongoing change is a key practice in agile methodology. Ward mentioned software programs, documents, and the wiki in the same breath:

 Computer programs and documents get very brittle very quickly. What wiki's done is say, let's accept that things are evolving.[15]

To Ward, wikis and agile development are intimately intertwined.

In an agile environment, more emphasis is placed on team work and less on the idea that a team member is a specialist. Each developer can work on any story. Some say there should be no specialists at all in an agile team. That is a bit extreme, but it is true to say that there will come a time when the whole team needs to work on a particular task to get it across the line. That task may very well be a documentation task.

Technical writers can give good input into the design of a feature and how it will affect other parts of the product. Similarly, developers can write the initial drafts of the documentation to ensure it is all finished within the sprint. If the team is using a task board, such as the Kanban board shown in Figure 15.1, team members can see quite clearly the stages that the various stories have reached. If a number of stories get stuck in the documentation stage, that will be clear to the entire team. Therefore people will be more likely to jump in and grab a documentation task off the board.

Agile methodology is based on the assertion that people are very good at organizing themselves if they all understand the goal of the team and if they frequently discuss the best way to get there. A wiki works well for people who have a common aim. It provides the medium for the discussion on how to achieve that aim. If the aim is technical documentation, the wiki is simultaneously the platform that hosts the results.

When compared to other documentation development tools, a wiki offers these advantages to an agile team:

- **Simplicity and accessibility.** Some documentation tools require in depth technical knowledge that makes it counterproductive for people other than the technical writers to update the documents. A wiki provides a simple editing environment. This is useful for people who need to write the odd document or make an occasional update but do not want to immerse themselves in a specialized documentation technology.

- **Short path to publication.** The simple wiki workflow fits well into the agile ideal of producing a shippable product at the end of every sprint.

- **Geekiness.** Despite the fact that they have been around for more than fifteen years, wikis are still the tool of choice for the cool kids on the block. Developers feel an affinity for wikis. They are more likely to hop in and update the documentation when required.

- **Central, online location.** Many agile teams work in an increasingly global environment. It is useful to have a central location with good version control and monitoring tools where people can make updates as and when required.

- **Collaboration.** Wikis are designed to make collaboration work. Collaboration makes the agile team work. Take a look at Chapter 19, which is about managing updates by multiple authors. Equally important, Chapter 14 describes the tools a wiki provides for monitoring changes.

- **Extensibility.** Wikis make it easy to develop add-ons. A development team can customize the wiki to suit the ever-changing needs of the agile environment. If it makes sense to include a live display of issues from the team's issue tracker in the release notes, the developers can whip up a plugin to do that. If it makes sense to generate the change log for a minor release automatically so the technical writers can focus on major releases, that is possible too.

One of the ways to help the agile team feel that it owns the documentation is to help the developers feel they own the documentation platform. A wiki is a platform that engineers and product managers as well as technical writers can own.

Special techniques for agile teams

This section includes some innovative ideas that will appeal to developers, support engineers, and technical communicators in an agile team. These techniques work especially well if a wiki is available. Let's watch Ganache, our technical communicator extraordinaire, as she walks the agile walk and talks the agile talk.

Innovation sprint

An innovation sprint, also known as "technical writer funny hat day," is a short period of time set aside for the technical writers to try something new.

Ganache and the other writers at *Choc Ink* are embedded in the agile development teams. They churn out documentation updates quickly and efficiently, producing a shippable product and high-quality documentation at the end of every sprint. This is good, but there is little time to step outside the box and try something new. The technical writers need some time to think about innovative techniques to improve the team's procedures and the documentation itself. Ganache puts on her thinking cap and comes up with an idea.

"Let's hold an innovation sprint!"

She decides to do it once a quarter, at the beginning of each major release cycle. At that time there are fewer technical documentation tasks in the sprint because the engineers are building the infrastructure for the new features.

To prepare for the innovation sprint, Ganache sets a date and asks the writers to start thinking about what they will do during the sprint. The sprint itself will be very short: just one day. It is essential that each person has a good plan and has done the necessary preparation beforehand to make the best use of the time available.

Ganache starts an "innovation register," where the team can collect their ideas. She also asks the product managers if they have any input about things they would like to see in the documentation. A week before the sprint, the team gets together to decide which items are feasible and who would like to tackle what. They swap advice about some of the items, sharing ideas from other projects they have worked on.

The big day dawns. The technical writers arrive bright and early, all wearing hats. The funnier the better. Why the funny hats? It is a tactful way of letting people know that the technical writers are otherwise engaged for the day. It also builds team spirit and makes a good photograph.

A few days later, the team holds a demo session where each person demonstrates what they achieved during the sprint. Ganache has cleverly put a weekend between the sprint and the demo session. A couple of the team members are so enthusiastic about their projects that they have worked extra hours to have something special to show. Is this cheating? No, because the innovation sprint is all about the ideas and the team spirit, not about time spent.

Note: The Atlassian technical writers hold innovation sprints every now and then. A blog post, *Technical writers hold an innovation sprint,*[16] describes the fun the team had in the first one. A product of that sprint was the "Tips via Twitter" initiative described in Chapter 21. Out of another innovation sprint came the "Busted Stuff Report" described in Chapter 23.

Documentation blitz test

Ganache has been working on the documentation for a major product release. The release cycle has included a number of two-week sprints, and the documentation updates have gone through technical review at the end of each sprint. Nevertheless, Ganache would like a review of the documentation as a whole, instead of piecemeal.

Therefore, Ganache has a chat with the development team leads and gets their agreement to hold a "documentation blitz test."

In a blitz test, people spend a short period of time, usually an hour, doing concerted and concentrated testing of the documentation. The testers are developers, technical writers, product managers, QA engineers, and anyone else useful. It is a good a way of viewing the documentation as whole rather than the disparate bits that have previously been reviewed individually. Best of all, it is a way of working together to improve the documentation.

Ganache creates a test plan with the following information:

- Date and time of the blitz test.

- Aims and scope: A list of all the new and updated documents that need testing, grouped by functional area.

- Location of the kickoff meeting. After the initial briefing, the testers will work at their own desks.

- People who will be doing the testing, as agreed with the team leads and the testers themselves.

- The test environment: The documentation wiki and a test installation of the software release.

- Where and how people can record their test results and feedback. Ganache organizes an online chat room, which she and the other technical writers will monitor. The testers can also add comments to the documentation pages on the wiki.

At the appointed date and time, the testers congregate in the briefing room. Ganache allocates specific functional areas and documents to each person and checks that everyone knows where to put their results. Then she gives the go-ahead. People rush back to their desks, the office takes on an air of frenzied quiet, and the chat room and wiki explode into action.

Ganache is very satisfied with the results. People report a few documents that need fixing. In some cases the engineers jump right in and update the wiki rather than just logging the fault. One engineer finds an issue that he has fixed in the code, but has forgotten to mark as "fixed" in the issue tracker, with the result that Ganache has not yet updated the documentation. Another engineer finds some screenshots that are out of date because the code was changed after the documents were updated. As an unexpected bonus, a developer finds a bug in the application itself when she tests the procedure against the step-by-step guide.

An intangible benefit of blitz testing is that it increases the engineering team's sense of ownership in and responsibility for the documentation. It also bonds the technical writers and developers into a closer team.

Note: The Atlassian quality assurance (QA) team first introduced blitz tests to test the applications themselves rather than the documentation. The technical writers decided to use the idea for the documentation, too. A blog post, *A blitz test of the documentation*,[17] describes one such event. Anne Gentle and the OpenStack team heard about the idea and held a blitz in time for the Diablo release in September 2011. See Anne's post, *OpenStack Documentation Blitz*.[18]

Doc sprints

Doc sprints are a special case and have a chapter all of their own, Chapter 22.

Code reviews

Ganache asks the engineers to include her in the code reviews for major updates, especially those which affect the user interface. This is fairly easy because the development team uses an online code review tool. They can share the review comments with any number of people. Ganache reviews the wording on the new application screens, giving feedback about spelling, grammar, consistency, and usability. She also picks up useful information just by watching the review comments whiz by.

Summary of this chapter

The core requirement of agile methodology is that the team must be able to respond to change. That is why there is no fixed methodology. Each team studies the practices of other teams and adapts them to suit its own needs.

An agile team needs a toolset that is compatible with constant change. Everything is flexible, including the roles within the team. A wiki makes it easy for team members to work together to get the documentation tasks across the line. People do not need to learn a complex documentation tool. An engineer can hop onto the page, apply a change, and move on to the next task in the sprint. A product owner updates the release notes. Technical writers monitor the updates and follow up if necessary.

The collaborative features of the wiki encourage the team to try innovative techniques for developing and testing the documentation. An example is documentation blitz testing, which involves the entire team in actively reviewing the documentation against the product.

There are aspects of a wiki that are still clumsy for the purposes of technical documentation. But wikis are built to be extensible. Developers are keen to experiment with ways of improving agile processes, including technology changes. Technical communicators strive to create fit-for-purpose documentation, and to do it in ways that are sustainable and efficient. In a later chapter we will look at how technical writers can help to drive wiki development, tailoring the tool for their needs.

For now, Ganache has reached the end of another eventful day. She sits back with a satisfied smile and muses about agility and making her wiki fly.

References

These are the websites, blog posts, and other references relevant to the content of this chapter.

Further reading

- *What is Extreme Programming?*, on *XProgramming.com, An Agile Software Development Resource*: `http://xprogramming.com/book/whatisxp/`

- A visual presentation about Scrum methodology, from *Mountain Goat Software*: `http://www.mountaingoatsoftware.com/presentations/30-an-overview-of-scrum`

- A Scrum reference site, *ScrumAlliance*: `http://scrumalliance.org/`

- *What is Scrum?*: `http://scrumalliance.org/pages/what_is_scrum`

- Jeff Sutherland's blog about Scrum: `http://scrum.jeffsutherland.com/`

- A presentation on *Scrum Metrics for Hyperproductive Teams*, by Jeff Sutherland, October 2010: `http://scrum.jeffsutherland.com/2010/10/scrum-metrics-for-hyperproductive-teams.html`.

- *Kanban 101, Intro to kanban for software development*: `http://www.kanban101.com/`

- *Agile in a Flash*, speed-learning cards for agile methodologies by Jeff Langr and Tim Ottinger, published by Pragmatic Bookshelf, January 2011: `http://oreilly.com/catalog/-9781934356715`

- *Agile in a Flash*, a blog about the above cards and about agile principles: `http://agileina-flash.blogspot.com/`

Endnotes

[1] *The Toyota Way* on *Wikipedia*: `http://en.wikipedia.org/wiki/The_Toyota_Way`

[2] *Principles behind the Agile Manifesto*: `http://agilemanifesto.org/principles.html`

[3] Ada Lovelace (1815-1852) is credited with being the first computer programmer. She translated a document titled *Sketch of the Analytical Engine Invented by Charles Babbage* and wrote a detailed set of notes that included an algorithm, or program, that the hypothetical engine could process. Also in her notes, Ada wrote about the importance of the engine, and of mathematics itself, in aiding people to communicate our understanding of the world around us. See the "Notes by the Translator" at the end of the document: `http://www.fourmilab.ch/babbage/sketch.html`

[4] *Extreme Programming: A gentle introduction*: `http://www.extremeprogramming.org/index.html`

[5] Diagram of an Extreme Programming project: `http://www.extremeprogramming.org/map/-project.html`

[6] *XP Rules*, from the Extreme Programming organization: `http://www.extremeprogramming.org/rules.html`

[7] *How does Agile affect documentation?*, a blog post by Mary Connor on *Clever Hamster*, February 2011: `http://www.cleverhamster.com/clever_hamster/2011/02/how-does-agile-affect-documentation.html`

[8] *Technical Writing In Agile Software Development - Part 1*, a post by Edwin Dawson on the *Atlassian blogs*, August 2009: `http://blogs.atlassian.com/developer/2009/08/technical-writing-in-agile-software-development-part-1.html`

[9] *Sprinting for the Finish Line*, a blog post by Diana Ost on *The Agile Technical Writer*, December 2009: {{http://theagiletechnicalwriter.com/2009/12/02/sprinting-for-the-finish-line}

[10] *Suggestions for Survival in an Agile Environment as a Technical Communicator*, a blog post by Ben Minson on *Gryphon Mountain Journals*, April 2009: `http://www.gryphonmountain.net/-2009/04/suggestions-for-survival-in-an-agile-environment-as-a-technical-communicator/`

[11] The *Agile Technical Writers* group on *LinkedIn*: `http://www.linkedin.com/groups/Agile-Technical-Writers-1115987`

[12] *Technical Writing World*: `http://technicalwritingworld.com`

[13] *Why chocolate is essential to agile tech writing*, a blog post by Sarah Maddox on *ffeathers*, February 2009: `http://ffeathers.wordpress.com/2009/02/03/why-chocolate-is-essential-to-agile-tech-writing/`

[14] Ward Cunningham developed the word's first wiki, *WikiWikiWeb*, at address `http://c2.com/-cgi/wiki`. See Chapter 2 of this book.

[15] Seb Paquet's notes about Ward Cunningham's keynote presentation at *WikiSym 2005*, on *The Conference Wiki of the 2005 International Symposium on Wikis*: `http://www.wikisym.org/-ws2005/wiki/space/Ward+Cunningham%2C+Seb%27s+notes`

[16] *Technical writers hold an innovation sprint*, a blog post by Sarah Maddox on *ffeathers*, July 2010: `http://ffeathers.wordpress.com/2010/07/03/technical-writers-hold-an-innovation-sprint/`

[17] *A blitz test of the documentation*, a blog post by Sarah Maddox on *ffeathers*, December 2010: `https://ffeathers.wordpress.com/2010/12/04/a-blitz-test-of-the-documentation/`

[18] *OpenStack Documentation Blitz*, a blog post by Anne Gentle on *The OpenStack Blog*, September 2011: `http://www.openstack.org/blog/2011/09/openstack-documentation-blitz/`

16

Wiki best practices and advice

Now that you have reached this point in the book, you have a solid grounding in the technicalities of developing technical documentation on a wiki. Even so, questions will arise in your day-to-day work about the best way to do things. Other people may ask you about best practices too. In fact, I have found that people ask all sorts of questions, all the time. Wiki technical writers are breaking new ground, and people are keen to know how we do it.

In this chapter

This chapter is a handy reference for day-to-day life on a wiki, including tips about wiki tech comm best practices and answers to common questions:

- **Tips and best practices.** Lessons learned by a wiki technical writer during four years on the job.

- **Answers to some common questions.** Procedures and philosophies.

- **Getting technical.** The "how to."

- **Confluence and SharePoint.** A comparison of the two.

- **Kay's chocolate cake: part 2.** The second part of the recipe started in Chapter 5.

- **Summary of this chapter.** The crux of the chapter in just one sentence.

- **References.** The tools, websites, blog posts, and other references relevant to the content of this chapter.

Tips and best practices

These are the lessons I have learned during the four years I have worked as technical writer on a wiki.

Simple is best

It is tempting to use all that shiny wiki technology to build complex structures with recursive content reuse, hyper-smart metadata, and tamper-proof workflow. But simplicity is the virtue of a wiki. It is what makes collaboration possible.

Achieving a simple structure means spending a fair chunk of time in the design phase. In that respect, using a wiki is just like using any other documentation platform. Each documentation suite has its own requirements. With any luck, it will not need all the add-ons and sophistication that wiki technology has to offer. Instead, you can cherry pick the features you need. Simplicity takes time. Simplicity sings.

Chapter 8 includes a case study on designing a content reuse solution. For ideas on structuring the wiki, take a look at Chapter 7. Workflow options, both simple and sophisticated, are the topic of Chapter 10.

Use plugins wisely

Plugins make the wiki go round. The best advice is: Use the plugins you need, and use them wisely.

Check the support policy for each plugin before installing it. Does the wiki vendor support the plugin? If not, does the plugin developer offer support? For Confluence plugins, the support arrangements are described on each plugin's page in the *Atlassian Plugin Exchange* at `http://plugins.atlassian.com`.

Choose just the plugins you need. Evaluate each plugin carefully to make sure that it satisfies your requirements and that the core wiki functionality is not enough for your needs. Having too many plugins on the site can complicate your administrator's life when upgrading to a later version of the wiki software.

If you are planning to provide XML exports of the documentation to your customers, think about the impact on those customers if you use plugins and macros that are not part of the core wiki functionality. If your plugins affect the content of the pages, such as metadata, intelligent tables, or other formatting add-ons, then your customers will need to install those plugins onto their own Confluence site in order to use the imported documentation. Chapter 11 describes the reasons for providing XML exports.

Keep going back to the plugin exchange to see if new plugins have arrived or existing plugins have acquired just the feature you need to give you a killer wiki.

Get intimate with your wiki

Know the wiki inside out. Make full use of its best features and be ready to help other people become wiki huggers too. Technical communicators are in an excellent position to do this because they are accustomed to learning a new tool and getting down to the nitty-gritty of its capabilities. In many environments, technical writers use highly technical writing tools that other people in the organization never touch. With a wiki, everyone works on the same platform. People will need help getting started. This is a technical communicator's chance to shine!

Recognize and reward the community

Tap into the community, including work colleagues, customers, developers, and even the international community of technical communicators. I have found that people, and especially technical communicators, are very generous with their time and skills. Their passion for their role and for helping other people is boundless. It is good practice to acknowledge people's ideas and contributions. But more than that, it is rewarding and exciting to do so. The online world, and in particular social tools, thrive on a system of give and take.

Chapter 17 describes ways of engaging the community in the documentation, and Chapter 19 talks about managing contributions from external authors.

Help people to find the documentation

There is so much information zooming around the world, especially the online world. When searching for help about a product, people are likely to find a *Wikipedia* page or someone's blog post just as easily as the official documentation. There are ways, however, to make the documentation jump up and shout about itself. This will help people find the information they need. It is also good for us, the technical communicators. The documentation is our product. The more people who find it, the more our role is recognized and appreciated.

See Chapter 9 for guidelines on search engine optimization (SEO) and other aspects of searchability and findability. Good structure, as described in Chapter 7, is also important in helping people find what they need. Context-sensitive help links put people right where they need to be. Read all about wikis and online help in Chapter 12.

Answers to some common questions

One of the best things about working on a wiki is that people are always asking questions about it. This book is especially for the world's technical communicators, in recognition of their enthusiasm, ideas, and skill. The book is also a response to their questions, including those questions that have no good answer. Those are the ones that drive wiki developers to keep improving their product.

What is the learning curve for wiki authors and readers?

How easy is it for technical writers to learn how to use the wiki? What about readers and other contributors?

Readers who are accustomed to web pages will feel at home on a wiki page too. Wiki pages, built of HTML, CSS and JavaScript, make use of common web conventions such as hyperlinks and familiar text formats.

For people who will contribute to the content, the editing experience is paramount. The first thing that comes to most people's minds is wiki markup. Some wikis offer a rich text editor as an alternative to the wiki markup editor. Confluence 3.5 has both, and in Confluence 4 there is only a rich text editor (see Chapter 6). The Confluence 4 editor recognizes wiki markup in two ways:

- You can insert a block of markup into a dialog, and Confluence will immediately convert the markup to rich text format when you submit it to the editor.

- You can type wiki markup formatting characters directly into the editor, and Confluence will convert them as you type.

For business writers who dip into and out of the wiki rather than spending their entire day on it, the rich text editor is the thing. The functionality is similar enough to Microsoft Word for people to feel comfortable with it.

Most technical writers love learning new tools. Their experience with other tools puts them in good stead to understand quickly how a new tool works. They know what they want to do and what it should be possible to do. Finding the "how" is relatively easy. When using Confluence 3.5 or earlier, most technical writers choose wiki markup over the rich text editor for the sake of simplicity and consistency of formatting. They usually also learn the autoformatting and autocomplete techniques to speed up their content entry.

Technical writers also need to learn space administration, content import and export, and other tools required for document management. Again, people know what they want to do, and learning how is straightforward.

Can everyone update the pages?

Only if you let them. See Chapter 5 for guidelines on setting wiki permissions.

Can everyone add comments to the pages?

The wiki permissions determine who can add, edit, and remove comments.

Do people vandalize the documentation?

I have never seen an example of someone willfully spoiling an existing page. We do get spam on our documentation wiki in the form of comments added to the pages. We have locked down the permissions so that only known authors, both internal and community, can add and edit pages. But we do allow everyone to add comments.

Other spaces on the wiki have less restrictive permissions, allowing any wiki user to add pages or blog posts. This is where people come to hold discussions, and develop their own ideas. In those areas, we sometimes see spam added as a page or a blog post.

How can I manage updates to, and comments on, the documentation pages?

Chapter 14 describes the techniques for monitoring activity on the wiki. Chapter 18 describes how to invite and respond to feedback on the documentation. Chapter 19 has ideas on managing updates by community authors.

Can I protect the documentation from updates but still allow community contributions?

Some documentation is subject to strict regulation. It is therefore not possible to allow community updates. Or perhaps we need to write the documentation in some other medium but still want to publish it to the wiki for collaboration, feedback, and sharing. Chapter 19 describes how to set up different spaces with different collaboration models to cater for such scenarios.

Is it a good idea to keep internal procedures and customer-focused documentation on the same wiki?

It is certainly possible to use just one site, with different spaces containing the internal and external content. The wiki permissions allow such a division of content, with some spaces visible to staff members only and other spaces visible to the general public.

For reasons of security and simplicity, I recommend using two different sites. This will help prevent people from making sensitive information visible to the general public by mistake. Having two wikis will also reduce the amount of content on each wiki, making each of them simpler to use. The content on an intranet is different in nature to the content written for customers and community members. Staff members can read both wikis. Due to the different nature of the content on each wiki, they will find it easy to understand where to look for the information they need.

Will every piece of information on the wiki be correct?

No. But not everything in a published book is correct either. Here is a quick diversion into a horror story from my days working as a book indexer. In one of the non-fiction history books I indexed, to be published by a reputable publishing house, the same person died on one page and then died again 50 pages and 4 years later. The book had been through all the editorial and fact checking processes. The only thing left was to compile the index. Luckily, the index contained dates of death, and I picked up the anomaly. That error was caught, but it made me aware that others will probably slip through in any book.

Back to the matter at hand: at least with a wiki, there is a good chance that someone will happen along quickly and fix up the mistake someone else has made. Also big brother is watching, in the guise of technical writers wielding RSS feeds and wiki watches.

What about intellectual property?

People's rights need protection. Wiki pages should include a statement of copyright so readers know how they can use the documentation. If community authors are allowed to contribute to the documentation, they need a clear statement of their rights and duties, too. Atlassian has chosen to provide the documentation under a Creative Commons by Attribution (CC BY) license. Community authors sign a contributor license agreement. Chapter 19 has the details.

What is the difference between a wiki and a CMS?

A few years ago, it was easy to distinguish between a wiki and a content management system (CMS). A wiki was a fairly basic piece of software that put an edit button on a web page. A CMS assigns content development to a behind-the-scenes environment and provides a workflow capability for publishing the approved content to a website or another medium.

Recently the functionality of both wikis and CMSes has expanded and moved closer together. The primary distinction is that a wiki is designed for people to edit the live content, whereas CMSes are designed around a publication workflow. By "live content," I mean the version of the documentation that is visible to readers.

I think it is fair to say that a wiki is a content management system – one that focuses on simplicity and speed in getting the information to the reader. Other CMSes focus on the workflow aspect and on stricter control in the publication process. You can add workflow to a wiki, and you can simplify the workflow of a CMS so that there is minimal fuss between writing a page and showing it to readers. But that distinction is what determines the authors' and readers' experiences on the two types of platform.

A CMS is designed for use by a dedicated team of writers. A wiki is designed for collaboration by many authors, including technical writers and other people from the organization and the community.

What is the difference between Confluence and SharePoint?

The section titled "Confluence and SharePoint" (p. 310) is devoted to this topic.

Getting technical

These are common questions about the technical side of wiki writing.

Does the wiki include a spell checker?

Confluence does not have a built-in feature for checking your spelling. Instead, you can use your browser's spell checker. It will work within the Confluence editor and other areas of Confluence where you can enter text.

- Firefox[1] and Safari both have built-in spell checkers that support different languages and different varieties of English.

- Google Chrome also has a built-in spell checker[2] that supports multiple dictionaries.

- There are a few add-ons available for Microsoft Internet Explorer. Try ieSpell[3] or Speckie.[4]

How easy is it to add context to a page?

A wiki page is a web page. A person coming into the documentation for the first time may land on any page. They may come in via Google search, or by clicking a help link in the application that the documentation covers, or by clicking a link anywhere else on the web. How easy is it to add contextual information, to orient people and help them find their way around the documentation?

A table of contents is one of the best tools for giving people context. The table of contents should be constant, always present, and should expand and highlight the page that is currently open. Readers can then see which section of the manual they are in (user's guide, administrator's guide, installation guide) and all the pages in the vicinity of the current page. In addition, wikis often provide "bread crumbs," a trail of links across the top of the page that shows the parent pages of the current page.

Authors should include an introductory paragraph on each page and links to related topics at the top or bottom of the page, giving more contextual information to a reader who has just dropped in.

Chapter 7 is full of ideas on structuring a wiki page and a documentation space.

What about broken links on the documentation pages?

In Confluence, there are two types of links: internal and external. Internal links are links to other pages on the same wiki. Confluence "knows" all about those links. If someone changes a page name, Confluence will automatically update all the links pointing to that page name. If someone tries to delete a page that has links pointing to it, Confluence will warn them and show a list of the affected pages. Each page has an "Information" view that shows all incoming and outgoing links.

Nevertheless, it is quite possible to break a link. There is as yet no built-in way of finding broken links across the wiki, and no cross-reference tool that shows all links across the site.

But take a look at the *Busted Stuff Report* described in Chapter 23. This is a recently-developed tool for finding broken links, created by a technical writer and available for general use.

Can I add structure to the wiki?

Looking at *Wikipedia*, and some other wikis on the web, it seems that structure is anathema to the wiki philosophy. Isn't a wiki just a flat collection of pages, discoverable only by searching or clicking through a haphazard collection of links?

No. Technical writers can add that beloved table of contents, ensure consistency of page layout and content, and apply other techniques described in this book. See Chapter 7 for information about structure and style and Chapter 8 for information about templates and content reuse.

Is the page order always alphabetical or can I define a logical sequence?

In some wikis, page order is alphabetical only. There is no way to put pages into a logical order, such as the order of chapters in a manual, other than by numbering the chapters by hand. In Confluence (and maybe other wikis) you can put pages into any order you like. Chapter 7 describes the procedure in detail. In brief,

- To move pages around within the space, go to "**Browse**" > "**Pages**" > "**Tree**," and drag the pages into position.

- To move an individual page, choose "**Tools**" > "**Move**," then tick the "**Reorder**" check box and click "**Move**."

Can I push updates through a workflow before publishing them?

Yes. If you are using Confluence out of the box, then it takes a bit of manual management via page permissions and sharing. Alternatively, plugins provide sophisticated and customizable workflow. Chapter 10 tells you all about it.

How do I manage version control on a wiki?

Let's say you have written the documentation for version 1 of a product. What happens when version 2 is released – do you need a whole new wiki for the new set of documentation?

In Confluence, spaces are a good mechanism for release management. It is useful to have a separate documentation space for each major version of the product. The details are in Chapter 10.

Can I convert my existing documentation into wiki format?

This can be tricky. It depends on the format and volume of the existing documentation. Chapter 11 describes the options for importing content into Confluence from various formats.

What about people who need offline help?

Sometimes readers cannot come to a wiki on the web. They may be offline some of the time, or they may be stuck behind a corporate firewall. For these people, technical communicators need to provide downloadable and even printable forms of the documentation. See Chapter 11 for information about the export formats available.

Confluence and SharePoint

This is a question often asked: How would you compare SharePoint and Confluence?

I have done a comparative analysis of Confluence and SharePoint 2010, with an emphasis on using them for technical documentation. I examined the following aspects:

- **Designing and creating a documentation suite.** Developing a document. Providing structure via mechanisms like a table of contents, a left-hand navigation bar, and logical page ordering. Classifying documents via keywords and other metadata, for ease of browsing and search. Moving documents around.

- **Collaboration.** The features that enable collaborative content development and review. Comments. Concurrent editing. Check in and check out. Version control for individual documents and pages. Tracking of updates via page history, RSS, and notifications.

- **Workflow and permissions.** Using workflow features to create a document or page in draft status, have it reviewed, publish the final approved version, and when necessary, update the document or page after publication. Also includes controlling access to documents and pages via permissions.

- **Support for other formats.** Publishing to formats like PDF, DocBook XML, and HTML. Office integration. Use of the documentation site as destination for online help.

- **Managing attachments and legacy documentation.** Uploading and serving existing documents and attachments.

- **Overall usability and reader's experience.** Collaboration with readers and other authors. Engagement of the readers in the documentation. Efficient use of time in the authoring and publication workflow. Aesthetics and simplicity.

My conclusion was that the SharePoint 2010 wiki is not the same sort of tool as a standalone wiki platform. The SharePoint wiki is more like a set of web pages that are unrelated to each other. Their strength is that they form one of the components of the SharePoint platform.

In other words, SharePoint and Confluence are totally different things. Both have areas of strength and areas where they are less strong.

- SharePoint is an all-in-one portal development, and document management tool, with wiki pages tacked on.

- Confluence is an all-in-one document development, document publishing, and collaboration tool, with management of external documents tacked on.

In my opinion, SharePoint is the tool of choice if both the following are true:

- Your primary need is document management. There is a large set of existing documentation in various non-wiki formats, such as legacy Word and PDF documents, complex Visio diagrams and spreadsheet formulas. In addition, you have an existing, stable and tidy SharePoint installation with competent, full-time site administrators.

- You need a wide variety of discrete content types all on a single portal, such as discussion lists, calendars, task lists, non-wiki documents, and web pages.

Confluence is the right choice if both the following are true:

- Your primary need is document development and presentation. You want a single platform for designing, developing, and publishing your documentation.

- You want your documentation easily accessible to readers and authors, with a uniformity of interface that is unintrusive and predictable (in a good way). Content is king. Readers and authors collaborate on the page itself rather than in separate discussion lists.

Disclaimer: I am a self-confessed wiki hugger. I have provided the above notes in the hope that they will help people direct their inquiries and analysis when investigating the right tool for their environment.

A point of interest for people already using SharePoint: The *Confluence SharePoint Connector* is a plugin that offers integration between SharePoint and Confluence. Once the connector is installed, people can embed Confluence pages and blog posts into SharePoint web parts and in turn display SharePoint lists on Confluence pages.

Kay's chocolate cake: part 2

Part 1 of this recipe listed the ingredients and implements (see Chapter 5). Now let's make the cake:

Preparation:

1. Turn on the oven to warm up.

2. Grease the cake pans with margarine, then sprinkle with a fine layer of flour to prevent the mixture from sticking to the pans.

3. Sieve (sift) the flour, cocoa, baking powder, and salt together.

4. Cut one of the bottom corners out of the plastic bag to form a small hole about the size of a medium knitting needle. You will use this to pipe icing onto your cake.

Making the cake:

1. Beat the margarine and sugar together until creamy.

2. Add one egg and beat the mixture until smooth again.

3. Add the second and then the third egg in the same way.

4. Gradually pour in the sieved dry ingredients (flour, cocoa, baking powder, and salt), scraping the mixture off the sides of the bowl at regular intervals and mixing it in with the rest to ensure a consistent texture.

5. Gradually pour in some milk at the same time, so that the mixture is smooth, but not runny. It should slowly drip off a spoon, in dollops.

6. Add the vanilla essence.

7. Spoon the mixture into the cake pans. Make sure there is the same amount in each pan. Do not press the mixture up against the sides of the tin. If possible, it is a good thing to have a slight hump in the middle, sloping down towards the edges.

8. Put the pans in the oven and bake for approximately 20 minutes.

9. Check to see if the cakes are ready. The mixture should have pulled itself away from the sides of the tin just slightly and should have risen in the middle too. Press your finger lightly on the top of one of the cakes. It should not leave a dent, or just a very slight dent.

10. Take the cake pans out of the oven and rest them on a rack to cool for 5 minutes.

11. Turn the pans upside down onto the racks. The cakes should drop out. If not, tap the pans, or if necessary, help them with a knife.

12. Leave the cakes to cool for at least half an hour.

In the third part of the recipe you will make the icing, put it all together, and eat the cake! Can't wait? Jump ahead to Chapter 19.

Summary of this chapter

The best tip a wiki technical writer can give is this: Know thy wiki.

References

These are the tools, websites, blog posts, and other references relevant to the content of this chapter.

Useful Plugins

The *Confluence SharePoint Connector*, for integration between SharePoint and Confluence: `http://-www.atlassian.com/software/sharepoint-connector`

Endnotes

[1]   Guide to using the Firefox spell checker: `http://support.mozilla.com/en-US/kb/Using the spell checker`

[2]   Guide to using the Google Chrome spell checker: `http://support.google.com/chrome/-bin/answer.py?hl=en&answer=95604`

[3]   ieSpell, a spell checker add-on for Internet Explorer: `http://www.iespell.com/`

[4]   Speckie, a spell checker add-on for Internet Explorer: `http://www.speckie.com/home/`

Giving your wiki wings

A technical communicator's skill and passion. A wiki designed for collaboration and innovation. Together they make the documentation fly.

17

Engaging your readers

Thought leaders in the technical writing community are experimenting with ways of engaging readers in the documentation. Some techniques are already tried and true, others are innovative and adventurous. Is reader engagement a good thing, why should we bother, and how can we do it?

Getting involved in this chapter

Let's discuss reader engagement and the benefits it can bring to the readers, to the organization as a whole, and to the technical communication team in particular. This chapter has the following sections:

- **Why engage the readers?** Engaging readers is an excellent way of helping people and of helping them help each other.

- **The documentation as a hub of activity.** A documentation wiki, managed by technical communicators, becomes the central point for exchange of information and ideas.

- **Why use social media?** Social media tools are attracting plenty of attention in the world of technical communication. Let's take a look at what people are saying and the benefits we can enjoy from using social media in and around the documentation.

- **Engaging internal readers and teams.** It is important for a technical communicator to raise awareness inside the organization as well as outside. Try a doc sprint, a documentation blitz test, or a documentation blitz.

- **Engaging bloggers.** Bloggers are an excellent source of information and good people to have on your side. Perhaps the official documentation can link to relevant blog posts, offering readers an extended information base.

- **Engaging everyone.** Internal teams and bloggers are special cases. The rest of this book is about engaging everyone else as well.

- **Promoting the role of technical writer.** The role and value of technical writers is under scrutiny, both from technical writers themselves and from others. By getting to know our readers, and being out there in the open, we can promote the documentation and our role. Let's examine the debatably attractive idea of "the naked technical writer."

- **Features of the wiki that make people visible.** A big part of engaging readers is making sure that they can see and recognize each other. The social features of a wiki help to do just that.

- **Tearing yourself away from this chapter.** This chapter's summary gives some thoughts to take away and ponder over.

- **References.** The websites, blog posts, and other references relevant to the content of this chapter.

Why engage the readers?

Engaging readers means giving them a way to get involved in the documentation. More, it means giving them the opportunity to feel good about the documentation.

A proposition: Engaging readers is an excellent way of helping people and of helping them help each other.

- Engagement in the documentation is good for the readers, because it helps them find the information they need and remember where they found it.

- Reader engagement is good for the documentation, because it encourages readers to give feedback on the documentation and opens up new ways for that feedback to reach the technical writers.

- Reader engagement is good for the role and image of technical communicators. It brings documentation to the attention of the world's movers and shakers, inside and outside the organization.

In the greater world of customer engagement, businesses are finding that engaged customers buy more of their products, are more likely to be satisfied with them, and are more likely to help each other use the products. Customer engagement studies and reports are all over the World Wide Web. A good summary of the benefits of customer engagement is at *Voice of the Customer (VoC)* by *PeopleMetrics*.[1] *Wikipedia* has a good overview on its *Customer engagement* page.[2]

Following along the same lines, a thesis of this book is that engaged readers will keep coming back to the documentation to find the information they need, and are more likely to feel satisfied with the documentation. They will send each other links to specific pages and will help each other use the documentation as a source of information. They will even begin to feel a sense of ownership of and pride in the documentation.

The documentation as a hub of activity

The product documentation is where people go to find the information they need about obtaining, installing, and using the product. The intranet is where employees find the procedures and guidelines they need to be a valuable member of the organization. Information delivery is the accepted purpose of a documentation site.

What if we expand that purpose, so the site becomes a hub where people go to talk to each other, swap ideas, and offer workarounds for tricky configurations or glitches in the software? What if it becomes a repository of shared information, continuously updated with everyone's ideas? What if we succeed so well that people feel the need to subscribe to a news feed so they don't miss anything?

Today's readers are finding and building online social communities all over the place. These readers form a shifting, vociferous, generous body of smart people with ideas to share. Our documentation needs to become part of this trend. We need to offer a community and catch the waves of contributors as they ebb and flow. Otherwise we miss out.

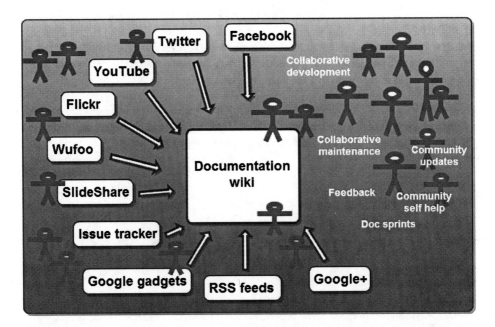

Figure 17.1. A documentation wiki as an interactive hub for information exchange

In Figure 17.1, the documentation wiki is the central hub supporting reader engagement via information exchange.

- Technical writers, product managers, and engineers collaborate to develop the documentation.

- Support engineers join in to keep the documentation up to date with the latest workarounds and configuration options.

- Community authors add their hints and tips.

- Readers provide feedback on the documentation.

- People ask for and offer help to each other.

- The technical writers organize doc sprints (see Chapter 22) on the wiki to focus community and organizational effort on a particular aspect of the documentation.

- People come to the documentation via links from social media, including Twitter, Google+, Facebook, and more.

- A page in the documentation displays a helpful YouTube video, a list of relevant feature requests from an issue tracker, and a presentation from SlideShare.

- Another page displays a live feed of updates from Twitter or FriendFeed.

- People subscribe to RSS feeds (see Chapter 14) from the wiki, to make sure they are informed when something changes.

- People display RSS feeds on a wiki page, to share news with their colleagues.

Many of the tools shown in Figure 17.1 are social media tools. In fact, a wiki is often classified as a social media tool.

This chapter and the rest of this book suggest ways to make that diagram a reality.

Why use social media?

There are other ways to bring people to the documentation site. Online help links are a good example. What specific benefits do social media tools bring to the mix?

Many businesses use social media as a way of communicating with customers. Marketing teams value such tools as a way of reaching people where those people are already talking to each other. When someone expresses a need for a particular tool or service, a savvy sales engineer can drop in and suggest the right product, exactly when that person needs it. Support teams can monitor social media and immediately help people with problems. In this way, companies can engage with customers by being there when needed, offering help at the opportune time, and not intruding when contact is irrelevant.

Peg Mulligan's post, *Social Business (also known as Enterprise 2.0),*[3] points out that it is not only the marketing teams that must learn this new way of reaching customers. Social media applies to all areas of the business that deal with customers. Even more, social media is relevant across the entire organization.

The topic of social media is gaining more and more attention in the technical communication field. In a blog post[4] about the Tekom 2010 conference, Jason Nichols advises technical communicators that we cannot ignore social media. Even if we do not currently participate in social media, we should learn about it, and think about how we can use social media tools to our advantage. The Spring 2010 edition of the ISTC's quarterly journal, *Communicator,* included a supplement sponsored by Adobe and titled *The role of social media in technical communication.* The supplement contains articles by David Farbey,

Gordon McLean, Noz Urbina and RJ Jacquez. Both David Farbey[5] and Gordon McLean[6] offer links to the supplement from their blogs. In a post titled *Social Support and Documentation Communities*,[7] Anne Gentle describes the intersection of classic user manuals, forums, and social media. David F. Carr gave a presentation to The Society for Technical Communication of the Palm Beaches in November 2011, called *How Social Media Is Changing Communications*.[8]

Social media offers new ways for getting readers involved in the documentation and for getting technical communicators involved with their readers. Never before has it been so easy to engage with so many people at once. Never before has it been so easy for customers and technical communicators to talk with each other, using the documentation as the focus of the conversation.

Below are some of the benefits we can enjoy from using social media in and around documentation.

- **Awareness that the documentation exists.** Use social media to let people know the documentation exists and where to find it. The documentation cannot help them if they do not know it is there. A suggestion to technical communicators: We have to stand on the rooftops and shout about the documentation! Let's find out where customers are holding relevant conversations and draw them into the documentation. Tweet about it, share links on Google+, add a Facebook page, and more.

- **Actionable feedback.** From a technical communicator's point of view, the documentation is our product. Just as marketers and designers find that they gain valuable insight into the way people use the product by talking to the people and watching them try out the product, we can do the same by watching people use the documentation and listening to them talk about it. If we give them a channel for providing feedback, they will tell us about gaps and mistakes in the documentation. We can respond instantly by plugging the gap and immediately find out how good the fix was.

- **Learning new ways of using the product and the documentation.** People are inventive. They experiment with the product and the documentation in ways that the developers never dreamed of. A technical writer may learn that customers are splitting the documentation into separate chunks and combining them in new ways to suit a different audience. Or we may learn about a new way of configuring the product that our manuals do not cover. If the readers are engaged in the documentation, they will let us and other readers know what they are doing. People love to help each other. We can help them do that.

- **Influencing people's opinions.** We can influence what people think and say about the documentation by solving the hard problems as soon as they come up and providing people with the means to help each other. Giving technical writers an online presence with a recognizable name and profile picture is an excellent way of engaging readers. The picture can be a photograph or a personal avatar. Readers are more likely to forgive gaps in the documentation if they feel like they know the technical writers who maintain it. It's even better if when a reader adds a comment pointing out a problem, the writers respond with an explanation of why the gap existed, a thank you to the person who pointed it out, and a note that the page is now fixed. A wiki encourages such personalized contact and provides the tools that make people visible. More about this later in the chapter.

- **Turning an existing phenomenon to our advantage.** People are talking about our products, companies, and documentation anyway, even if we do not see it happening. In many cases, they are doing it online and using social media. We need to listen in and help with their problems, and if appropriate, encourage them to use venues where it is easier for us to respond.

- **Reduced support and maintenance.** People will use the documentation as a tool to help each other. If permitted, they will even help us to update the documentation, adding small nuggets of information as comments or directly on the page. This reduces support and maintenance costs, and in some organizations may be the only scalable solution.

- **Fun.** In more traditional customer engagement models, only the outward bound departments in a company will have contact with customers. But given the opportunity, many other members of the company enjoy such contact, including technical communicators. It is fun and rewarding for technical writers and readers to have a way of talking to each other about their mutual interest, the documentation. Jokes and appreciation reach everyone in the organization, not just those in outward-focused roles. Customers also enjoy the contact with employees at all levels of the organization.

- **Documentation, the face of the company.** Engaging people with the documentation is one way of engaging them with the products. Documentation is the face of the company. It is often the first point of entry into the company's online presence. It is a good point of contact with customers on an ongoing basis. After buying a product, people are more likely to return to the documentation than to the corporate website.

Not everything is rosy in the world of social media and the enterprise. In *The Big Failure of Enterprise 2.0 Social Business*,[9] Laurie Buczek says that an important problem is the "lack of integration of social tools into the collaborative workflow." People use all sorts of tools, hopping from one network to another, picking and choosing where they get and share information. It is not easy to channel their activity into an organized workflow.

The result is that our readers are being bombarded with information from many different sources. They see Tweets, blog posts, Google+ posts, online videos and presentations, and more, all offering advice on the use of our products. If we want to add yet another tool, the documentation site, then we need to make sure it has sustained drawing power. When people find a central, authoritative documentation site that is at the same time open to discussion and collaboration, they will keep coming back to it. A wiki offers tools for aggregating information streams and providing them in a single, accessible, and interactive location.

Engaging internal readers and teams

Technical communicators have internal as well as external customers. It is to our advantage to ensure that the people who drive the organization are aware of the importance and excellence of the documentation. In many cases, the documentation is as useful to employees as it is to customers. This is true of internal procedures on an intranet. It is also true in software development companies, where both internal and external engineers use the API and other developer-focused guides.

How can the technical communication team encourage other areas of the organization to be aware of and get involved with the documentation?

Chapter 22 is all about organizing a **doc sprint**. Here is a technical writer's secret: only one of the aims of a doc sprint is to write the documentation. A major benefit is that a doc sprint gets people, both within the company and outside, working together with the technical writers. A wiki is the perfect platform for such an event.

Chapter 15 describes the **documentation blitz test.** This is a short period of time, usually an hour, in which people do concerted and focused testing of the documentation. The testers can include developers, technical writers, product managers, and QA engineers. This is a great way of testing the documentation and also fostering a sense of ownership and pride in the documentation.

Another idea is to hold a **documentation blitz**. This is similar to a blitz test, but instead of looking for problems, the idea of a documentation blitz is to fix a specific, known problem. Let's say the development team makes one of those small, dreaded changes to the product that necessitates sweeping changes to the documentation. Almost all pages and screenshots are affected. A small technical writing team will find it hard to apply the updates in the time allowed. On the other hand, the changes are fairly straightforward and follow an easily defined pattern. Just the job for an engineer! The technical writers define a plan that specifies the pages that need updating, the changes required, and the date, time, and location of the blitz. The plan also specifies the tools available: the wiki and a screenshot capture tool. The technical writers will be on call during the blitz to advise on wording and screenshot requirements.

Edwin Dawson has written a blog post about such a documentation blitz, held at Atlassian in August 2011: *Interviews from the JIRA Screenshot Doc Blitz*.[10] The story behind the blitz? Fairly late in the release cycle, the development team changed the administration user interface of JIRA, the issue tracking software that Atlassian develops. The UI changes were high level and significant, and they affected many pages in the JIRA administration guide, including all the screenshots. The engineers realized that this would put a heavy load on the technical writers, so they offered to help with the updates.

A number of good things resulted from the documentation blitz:

- People who took part in the blitz reported that they enjoyed it and felt a sense of achievement and an increased team spirit.

- The engineering team took ownership of the required documentation updates. The engineers saw at first hand how changes in the product require updates to the documentation. In future releases, they will know when to alert the technical writers of documentation-breaking changes.

- Engineers became more aware of the work that technical writers do. A couple of the doc blitz participants came up with the idea of automating the screenshot updates in future releases.

- The time-critical updates were done in time, without technical writer burnout.

Engaging bloggers

Bloggers often write targeted, information-rich posts about specific configurations of a product or workarounds for glitches or missing functionality. What if our documentation provided links to that information, so that customers could find it easily?

In the Atlassian documentation, *Tips of the Trade* pages link to "how to" articles written by bloggers.

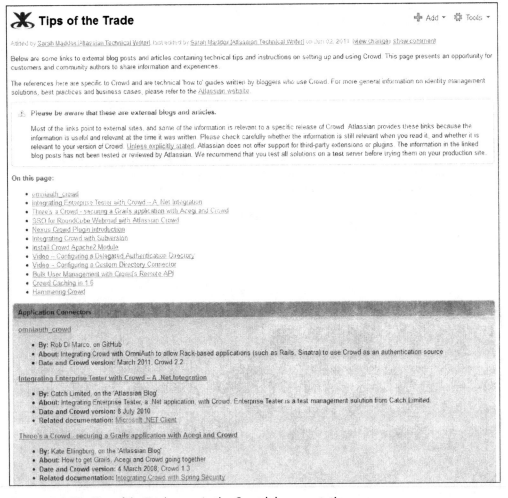

Figure 17.2. The *Tips of the Trade* page in the Crowd documentation

Figure 17.2 shows the *Tips of the Trade* page[11] for Atlassian Crowd, a user management application. Figure 17.2 shows the top of the page, including the following sections:

- Two introductory paragraphs, describing the purpose of the page.

- A warning block, colored yellow, explaining that the links on the page point to external sites and recommending that people apply their own judgment to what they read in the linked posts.

- A table of contents, listing the sections on the page.

- A categorized list of blog posts. For each post, the page gives the author, a short description, the date and relevant product version when known, and related documentation if relevant. The page also links directly to the blog post.

Bloggers can add a comment to the page asking the technical writers to add a link to their post. Some bloggers even have update rights to the documentation pages and can add the links themselves. (See Chapter 19 for tips on managing updates by external contributors.)

Before adding a link to the *Tips of the Trade* page, the technical writers confirm the following:

- It must be a "how to" article that contains good information and is well presented.

- The information must be correct. If necessary, an engineer should verify the information.

- It must be useful. The blog post may document an edge case, but not too much of an edge case.

After linking to someone's blog post, the technical writers add a comment on that blog post letting the blogger know that the product documentation now has a link to their post. This is a great way of engaging bloggers. They know that such links will drive more traffic to their blog. It is also a great way of providing additional information to readers.

Here are the benefits of linking to blog posts from the official documentation:

- **Provides useful edge-case information for readers.** It is not possible to develop and maintain official documentation about all possible use cases for a product. The official documentation tends to cover the most common scenarios – information that will be most useful to most customers. This keeps the documentation as simple and readable as possible. It also reduces the cost of maintaining the documentation through ongoing product releases. On the other hand, blog posts can offer useful, detailed instructions about edge cases that the official documentation cannot hope to cover.

- **Requires little maintenance.** Readers are aware that blog posts are point-in-time publications. They do not expect the information to be updated for each new product release. They are willing to draw as much information from the post as possible and extrapolate to suit their own environment. Provided we warn our readers – explaining that we are linking to an external article and that readers should check that it is relevant to their situation – there is no need to ensure that the blog post matches the current product release. We do need to check the links periodically and cull those that have become obsolete, but that is a small cost for a high return.

- **Expands our own knowledge.** When researching the blog posts, we learn new ways of configuring and using the product. We may even learn new ways of documenting it.

- **Fosters a good relationship with bloggers.** Bloggers are good people to have on your side. They are enthusiastic, knowledgeable, and resourceful. They tend to do a lot of research and experimentation before writing a post, and their opinion is valued in their field. Many of them are highly influential. They appreciate it when we link to their posts, and they will usually reciprocate with links back into our documentation. They will often respond with hints and tips about the product and documentation.

Engaging everyone

Internal teams and bloggers are special cases, so I have covered them in this chapter. The book is also about engaging everyone else! Chapters 18 to 22 are full of ideas and techniques for doing that.

Promoting the role of technical writer

Is our role as technical writers obscure? Do people understand what we do? At recent conferences, and online, I have seen technical writers express the concern that the work we do is not recognized, or at least is not given its full due. In a blog post about the tekom/tcworld 2011 conference,[12] Sarah O'Keefe wrote a section titled "Fear, uncertainty, and doubt" in which she discussed this concern. Conference attendees felt that they do not have the power to effect change because management does not hear their concerns. When Sarah made suggestions about possible strategies for approaching management, many people said, "They don't listen. They don't care. I can't get them to pay attention to this."

A related discussion started up on *Technical Writing World* in November 2011. The post mentions the results of a survey conducted by WebWorks in which 95 percent of respondents said that technical communication needs to evolve. The comments on the post are still coming in as I write this chapter.[13] Roger Hart puts it strongly, and perhaps even ruthlessly, in a post titled *Pandas: below "C-level," and drowning*: "You're sleepwalking into obsolescence…"[14] He is talking about the overlap in the roles of content strategists and technical authors, and the fact that technical authors do not advertise their skills and products as well as content strategists do.

If we find ourselves in a situation where our contribution is not recognized, how can we help our colleagues understand our role in the product development process? How can we make sure they recognize the value of our work, and therefore of us? How can we work towards a change in the attitude towards technical communicators and a strategic change in technical communication itself?

Jason Nichols wrote a blog post called *The role of TechComm in reaching the Holy Grail*[15] where he discusses how his technical communication team has helped increase sales. He lists an impressive number of deliverables, including documentation, training material, videos, guides to best practices, and more. The final sentence in the post is a call to action: "The next time you think about what a technical writer does, or should or can do, consider the bigger picture and the Holy Grail."

Let's take it even further: how can we foster in other people a sense of pride in, ownership of, and responsibility for the documentation?

It is essential that we write good documentation in the first place. We can contribute to a project in many ways, over and above the documentation itself. We can ensure that the user interface text is useful and concise, report bugs when we find them, let the developers know about the user experience of the product, and more. Yet, when it comes down to it, as technical writers the documentation is our product, and that is what people judge us by.

Therefore, we need to make sure people find and appreciate the documentation.

Recognizing the value of documentation

Smart companies are coming to realize that documentation promotes products, strengthens the corporate brand, and draws customers to the company website. Technical documentation is, by its very nature, good at attracting search engines. The content is concise and on topic, makes good use of key terms, includes similes, has good structure, and includes contextual information at the top of the page that the search engine can incorporate into its summary. (See Chapter 9 for guidelines on search engine optimization.)

If the organization recognizes the value of the documentation, that has to be a good thing for technical communicators.

In a post called *Trust is Essential to Creating Great Docs*,[16] Mark Baker says that success in technical communication is all about relationships. The organization must recognize the value of the documentation and trust the technical communication team. His post examines many aspects of the problem.

MindTouch, makers of the MindTouch collaboration platform, are very good at researching and publicizing the value of technical documentation as a way of drawing people to the company and its products. In a recent example, Mike Puterbaugh, then VP of marketing at MindTouch, wrote an article on *Mashable* titled *5 Reasons Your Product Documentation Is a Marketing Asset.*[17] He starts the article with these words:

As a CMO, it's important to understand what clever technology developers and open source leaders have known for years: Great product documentation isn't loathsome – it's marketing, and darn good marketing at that.

Quite apart from being a marketing asset, good documentation is what customers need. Smart product owners already know that. Overheard: words from Daniel Franz, product manager for developer relations at Atlassian, in June 2011:

If we don't get the documentation right they [the community developers who form Daniel's customer base] will hate us, no matter what else we give them.

The naked technical writer

The beauty of a well-crafted document is obvious to everyone, right? Certainly technical writers recognize good documentation, and so do other people, at times. For example, take a look at the inspiring presentation called *Documentation is freaking awesome,*[18] by Kyle Neath, Director of Design at GitHub.

Yet in the main, good documentation is one of those things that customers do not notice until it is missing. Sometimes people do not even know the documentation exists. They just battle along without it. How can we make sure our documentation site is one of those that people do find and remember? It comes down to making ourselves seen and heard. A wiki is a very good platform for doing that.

With tongue in cheek, this section is titled "the naked technical writer" because the word "naked" hints at something out in the open for all to see. That is true of people writing documentation on a wiki.

Readers add comments to the documentation pages and get to know the technical writers who reply. While the documentation itself remains devoid of personal style, the comments on the pages often reflect the character of the technical writer. It is rewarding to converse with readers. Sometimes it is just plain fun. In June 2011, a whimsical comment came in on an Atlassian documentation page:[19]

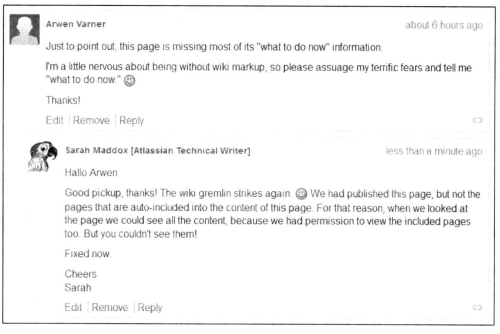

Figure 17.3. Comments on the Confluence 4.0 change management guide

In the comment (see Figure 17.3) Arwen Varner pointed out a problem in the documentation. I investigated, fixed the problem, and thanked Arwen for her good pickup. Later came a happy reply from Arwen. Note that she added a profile picture in the interim:

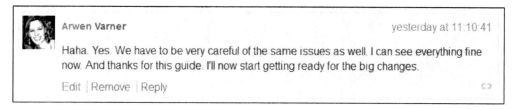

Figure 17.4. A further comment

People who get to know you as the technical writer on the wiki may sometimes email you directly. They prefer to contact someone they know, rather than an anonymous help desk or issue tracker. A lively correspondence often results where both parties learn new techniques.

There are a couple of things on the downside of having such a visible profile. It can be time consuming to respond to comments on the documentation pages and answer email messages personally. It can even be a bit disconcerting to receive direct email messages. But most people enjoy the experience. It is up to each person to decide how much involvement is appropriate and how personalized his or her presence on the wiki and the web will be.

Features of the wiki that make people visible

A big part of engaging readers is making sure they can see and recognize each other.

Let me tell you a story. A couple of days before writing this paragraph, I tweeted a link to one of the pages on our documentation wiki. My husband, Peter, clicked the link and looked at the page with moderate interest. He scrolled down and found some comments from readers, which he found much more interesting. Then he clicked through to the wiki dashboard and saw the activity stream, which shows a list of the most recent updates on the wiki along with the author's names, comments, and profile pictures. Peter was spell bound. He recognized some of the names and asked why only some people had personalized their profile pictures while others were using the standard Confluence images. He examined what each person had been doing and wondered why my name did not appear in the activity stream. (I was working on a different wiki at the time.) Peter does not use any of the products or documentation mentioned on that wiki, yet he still found it interesting. This illustrates the power of a personalized, online presence in engaging readers.

Figure 3.1 in Chapter 3 gives an overview of the content elements in Confluence wiki, but the diagram leaves out the most important part: the people! Let's rectify that now (see Figure 17.5).

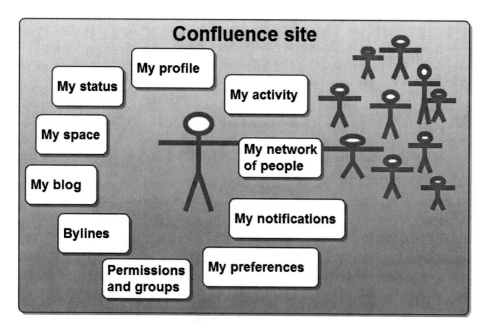

Figure 17.5. Confluence elements that make people visible

Wikis encourage collaboration and reader engagement by making people visible to each other.

- **User profiles and avatars.** In Confluence, people have profiles where they can tell other people something about themselves. Take a look at the screenshot of Ganache's profile in Figure 17.6. She has added a picture of herself to her user profile. Confluence will display this picture on comments that Ganache adds to the wiki and in various other places that show Ganache's activity on the wiki. People can also come to Ganache's profile to see a short description of her and her position in the organization and a link to her website (see Figure 17.6).

- **Personal spaces.** Each person can have a personal space. A personal space is very similar to a standard wiki space, which Confluence calls a "global" space. People use personal spaces to keep their notes, ideas, and drafts. It can become a place to express your personality. You can change the look and feel of your space by adding a theme that is different from the rest of the wiki.

- **Personal blog.** A blog post is a news item. It has a date stamp and is typically a topical piece of information that will not necessarily be kept up to date after publication. People can add blog posts in Confluence spaces, including their personal spaces.

- **Status updates.** Status updates are short messages telling everyone on the wiki what you are doing. They are similar to Twitter messages. When people spend much of their day on the wiki, status updates can offer a welcome moment of diversion. See what other people are up to and tell them what you think. Perhaps counter-intuitively, many people find that such a moment of personal contact makes it easier to focus their attention back on their work after the short break.

- **Activity streams.** Confluence shows an activity stream on the wiki dashboard and on each person's profile page. The activity stream shows a short summary of each update made to the wiki along with the name and profile picture of the person who made the change. This offers a great way of getting to know the wiki dwellers and seeing which pages are currently being updated. Chapter 14 has more about activity streams.

- **Profile hover.** If you hover your cursor over someone's name on the activity stream, on a comment, or on a page, Confluence will display a popup window with that person's profile picture and give you the option to follow that person or view his or her profile.

- **Network.** People can "follow" each other on the wiki. This is a way of keeping track of the updates and comments made by people who are important to you. It is also interesting to see the list of people who are following you!

- **Bylines.** Every Confluence page displays a byline under the page title, saying who created and who last updated the page. This promotes a sense of ownership of the content. It also means that readers get to know the regular authors.

A wiki gives people a fair amount of control over their presence on the wiki.

- **Notifications and feeds.** People can specify the pages or spaces that they are interested in, so they receive notifications of activity on those areas of the wiki. Chapter 14 describes wiki watches and RSS feeds, two ways of monitoring wiki activity.

- **Preferences.** Chapter 14 also describes how readers can set notification and email options. Other configurable preferences include time zone and language. Each person can select a site home page that Confluence will open first when that person visits the wiki. There are also some editor preferences such as enabling or disabling the autocomplete and autoformatting functionality.

- **Permissions and groups.** Each person is the space administrator of his or her own personal space. This means that each person can manage the visibility of content and the update rights that other people have within that space.

Figure 17.6. Ganache's user profile on the Confluence, Tech Comm, Chocolate wiki

Tearing yourself away from this chapter

Thought leaders and movers and shakers aside, what do you think? Is all this talk about engaging readers just a fuss about nothing, is it the way forward for technical documentation, or does the answer lie somewhere between? The debate continues! It is probably happening right now, on the *Confluence, Tech Comm, Chocolate* wiki at `http://wikitechcomm.onconfluence.com`.

We want people to find the documentation. Once they have found it, we want them to keep coming back to it when they need help. There are a number of ways to engage readers, both in and around the documentation. A wiki makes things easier, but inventive technical writers can do things with any platform. They will mix and match ideas and use what works in their environment with their tools and for their audience.

Engagement is all about the people and the ideas they exchange. The technical communicators are the designers and curators of the platform on which the information exchange happens. In a later chapter we will see how technical communicators can work with wiki developers to extend the platform itself or even build their own add-ons to customize the platform to their exact needs.

Before that, let's look at some ways of getting feedback on the documentation and statistics about comments received. After all, one of the best ways to engage readers is to ask for their opinion. Onward to Chapter 18!

References

These are the tools, websites, blog posts, and other references relevant to the content of this chapter.

Further reading

Scott Nesbitt wrote a post called *Change your writing style to make documentation more usable and user friendly*, in which he discussed the advantages of adopting a more casual style than is traditional in technical documentation. I agree with Scott that there is a place for a lighter touch in much of the online documentation we write. It is a delicate balance. On the one hand, it is important that the writing style does not annoy or offend the reader and does not detract from the content. We also need to be aware of people whose first language is not the one we are writing in. On the other hand, the occasional touch of humor or personality can focus the reader's attention onto the page. Scott's neat, concise post gives good advice: *Communications from DMN*, October 2009: `http://www.dmncommu-nications.com/weblog/?p=1459`

Endnotes

[1] *Voice of the Customer (VoC)* by *PeopleMetrics*: `http://www.peoplemetrics.com/voice-of-customer.htm`

[2] A page about *Customer engagement* on *Wikipedia*: `http://en.wikipedia.org/wiki/Customer_engagement`

[3] *Social Business (also known as Enterprise 2.0)*, a blog post by Peg Mulligan on *Content for a Convergent World*, February 2010: `http://pegmulligan.com/2010/02/09/social-business-also-known-as-enterprise-2-0/`

[4] *Social media and trends in software user assistance: Twitter and rich media*, a blog post by Jason Nichols on *TechComm for Life*, November 2010: `http://www.jasonanichols.net/?p=426`

[5] *Social Media in Technical Communication*, a blog post by David Farbey on *David Farbey's Marginal Notes*, March 2010: `http://www.farbey.co.uk/index.php/2010/03/social-media-in-technical-communication/`

[6] *Role of Social Media*, a blog post by Gordon McLean on *one man writes*, March 2010: `http://www.onemanwrites.co.uk/2010/03/21/role-of-social-media/`

[7] *Social Support and Documentation Communities*, a blog post by Anne Gentle on *Just Write Click*, July 2011: `http://justwriteclick.com/2011/07/14/social-support-and-documentation-communities/`

[8] *How Social Media Is Changing Communications*, a presentation given to The Society for Technical Communication of the Palm Beaches by David F. Carr, on *Carr Communications Inc*, November 2011: `http://www.carrcommunications.com/wp-content/uploads/2011/11/How-Social-Media-Is-Changing-Communications.pdf`

[9] *The Big Failure of Enterprise 2.0 Social Business*, a blog post by Laurie Buczek on *Beyond the Cube*, August 2011: `http://www.lauriebuczek.com/2011/08/23/the-big-failure-of-enterprise-2-0-social-business/`

[10] *Interviews from the JIRA Screenshot Doc Blitz*, a blog post by Edwin Dawson on *Atlassian blogs*, September 2011: `http://blogs.atlassian.com/developer/2011/09/interviews_from_the_jira_screenshot_doc_blitz.html`

[11] The *Tips of the Trade* page in the Atlassian Crowd documentation: `http://confluence.atlassian.com/display/CROWD/Tips+of+the+Trade`

[12] *Tekom thoughts*, a blog post by Sarah O'Keefe on the *Scriptorium Publishing Blog*, October 2011: `http://www.scriptorium.com/2011/10/tekom-thoughts/`

[13] *The True Objective of a Technical Writer: Does strategy even matter?!* a discussion on *Technical Writing World* started in November 2011: `http://technicalwritingworld.com/forum/topics/the-true-objective-of-a-technical-writer-does-strategy-even`

[14] *Pandas: below "C-level," and drowning*, a blog post by Roger Hart on *simple-talk*, September 2011: `http://www.simple-talk.com/community/blogs/roger/archive/2011/09/26/103488.aspx`

[15] *The role of TechComm in reaching the Holy Grail*, a blog post by Jason Nichols on *TechComm for Life*, September 2011: `http://www.jasonanichols.net/?p=735`

[16] *Trust is Essential to Creating Great Docs*, a blog post by Mark Baker on *Every Page is Page One*, October 2011: `http://everypageispageone.com/2011/10/31/trust-essential-to-creating-great-docs/`

[17] *5 Reasons Your Product Documentation Is a Marketing Asset*, an article by Mike Puterbaugh on *Mashable*, August 2011: `http://mashable.com/2011/08/12/product-documentation-marketing/`

[18] *Documentation is freaking awesome*, a presentation by Kyle Neath on *Warpspire*: `http://-warpspire.com/talks/documentation/`

[19] A page about wiki markup in the Confluence 4.0 change management guide: `http://conflu-ence.atlassian.com/display/DOC/Confluence+4.0+Editor+-+What%27s+Changed+for+Wiki+Markup+Users`. At time of writing, the page has the comments from Arwen Varner as described in this chapter.

18

Gathering feedback on the documentation

Gone are the days when a writer wrote a manual, had it printed, bound, and packaged, then sent it out to unseen readers. Equally gone are the days when customers had no contact with each other. Most significantly, gone are the days when readers had little opportunity to give feedback on the manual itself and writers had no chance to respond to feedback until the next edition of the manual.

This chapter was fun to write. It is a blend of known techniques and innovative ideas. At this very moment, technical writers all over the world are experimenting with new tools in search of the perfect solution for inviting and managing readers' feedback. If you find any of the tools in this chapter useful, or have discovered another good idea, it would be great to hear about it. Come and chat on the *Confluence, Tech Comm, Chocolate* wiki at `https://wikitechcomm.onconfluence.com`.

Let's talk about our readers talking to us

This chapter discusses ways of getting feedback about the documentation. The chapter has the following sections:

- **Gathering feedback – why and how.** It is amazing how much we learn from our readers, and also how much they have to say. A pilot project helps to gauge how much feedback we can expect.

- **Inviting comments on the documentation.** Real-life examples illustrate the type of feedback given in comments and the number of comments received. Managing comments involves deciding who can add a comment, monitoring the comments, responding to them, and minimizing spam.

- **Adding feedback forms via plugins or other customizations.** Forms are useful for inviting structured feedback. Smart and innovative people design new form builders all the time and make them available as plugins or add-ons to the wiki.

- **External tools for gathering feedback.** Online form builders such as Wufoo and Google Docs offer a great way of designing a sophisticated form. Grab the HTML and copy it into a wiki page, or use a built-in widget to display the form, and readers can complete it right there on the page.

- **Using Twitter for feedback on your docs.** A teaser. Can we use Twitter to gather feedback?

- **The talk of the chapter.** A short summary.

- **References.** The websites, blog posts, and other references mentioned in this chapter.

Gathering feedback – why and how

Readers' feedback enriches both the documentation and our understanding of our customers. It shows us how people use the product and what they expect from the documentation. Sometimes it is amazing to see how inventive people are, putting the product to uses the developers never thought of. We have the opportunity to improve the documentation via direct contact with our readers. This is a luxury that technical writers in the past could only dream of. Chapter 18 is all about engaging readers in the documentation.

If we offer people a way of communicating with us, they will use it. Often enthusiastically and multi-fariously. We need to set up ways of managing and responding to the feedback. If the number of readers is large, it is a good idea to run a pilot project to measure the amount of feedback that we can expect and to judge how much time we will need to spend responding to it. A pilot project runs for a short period of time, such as two or three months. You can also consider limiting the scope of the feedback during the pilot project. For example, allow feedback on only a section of the documentation, or from only a segment of the customer base.

Here is an idea for making your pilot project even more useful. You can use it as an opportunity for publicizing the documentation and engaging readers. Write a blog post or send email messages letting people know that the pilot is happening. At the end of the pilot period, publish the results and let customers know the results of your pilot. They will appreciate and enjoy being "in the know."

The tools we use depend on our audience and our primary reason for asking for feedback. Some tools, such as feedback forms, offer direct communication between the reader and the document curators. Other tools, such as comments on the pages, allow all readers to join in the conversation.

The documentation platform also plays a part in determining the feedback tools. Some of the techniques discussed in this chapter rely on features in Confluence, but some can be used with other platforms, too.

Inviting comments on the documentation

Comments or discussion on the pages are a characteristic feature of wikis. Some wikis, like MediaWiki, put discussion on a separate tab next to the relevant page. Other wikis, like Confluence, put the comments at the bottom of the page. Confluence allows threaded comments, meaning that people can reply to specific comments and see a cascading flow of comments and replies.

A big advantage of comments over feedback forms is that readers can see and respond to each other's comments. People can instantly benefit from advice from other readers. They can hold a conversation and help each other solve problems independently. This takes the pressure off an overloaded technical communication team. Ideally, we would update the documentation as soon as a reader points out a problem. But if the reader's comment is there for everyone to see and provides useful information, that gives us leeway to tackle other urgent updates first.

An advantage that comments have over discussion forums is that the comments are right there on the relevant page. People do not need to go looking for them on a separate discussion site. The information in the comments complements what is on the page.

Examples of real comments on documentation pages

Let's take a look at some comments that people have added to the Atlassian documentation wiki.

- **A quick fix to a publishing glitch.** Chapter 17 gives a good example of a reader helping the documentation team pick up a publishing error. Arwen Varner added a comment to a page, saying that she could not see any information in most of the sections on the page. (See Figure 17.3 and Figure 17.4.) The cause turned out to be a glitch in the publishing process. We had published the main page, but not the pages that contain the reused chunks of text. We did not notice the problem, because we had permission to see all the pages. Arwen's comments are amusing and useful, a great example of a good interaction between technical communicators and readers. To see screenshots of the comments, look at the section titled "The naked technical writer" (p. 330).

- **Positive feedback telling us what readers like.** Matt Doar wrote a comment on a page in the Atlassian JIRA documentation:[1]

Figure 18.1. Comment giving positive feedback on a document

It lifts a technical writer's spirits to receive such a comment. It also tells us what readers like. That is useful input into our documentation design.

- **A small change in the product that passed us by.** An anonymous reader (someone who had not logged in to the wiki) let us know that an option had moved from the "Administration" section to the "Configuration" section of the user interface:[2]

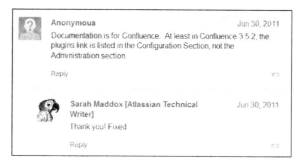

Figure 18.2. Comment about a change required in the documentation

- **Additional information for us and other readers.** On the installation guide for the Crowd Apache Connector on Ubuntu Linux, Bertrand Paquet added information about two Debian packages that he had created and made publicly available. An Atlassian Crowd developer, Adrian Hempel, responded with thanks:[3]

Figure 18.3. Comment giving readers additional information

Managing comments

The first thing to decide is who should be able to add and remove comments on the documentation pages. Using the Confluence space permissions, you can assign the right to add comments, which includes the right to edit existing comments. You can also assign the right to remove comments. And you can assign these permissions to individual users, groups of users, and anonymous users (people who have not logged in). The details about Confluence permissions are in Chapter 5.

What part does chocolate play in your life as technical writer?

From Harold Moses on Technical Writing World:[4] "We usually have chocolate in the IT department at the University of Arkansas @ Little Rock spread to be shared with the programmers and myself (the sole tek.writer on staff). Maybe it has some cognitive powers….idk…ijs."

Note: "idk" and "ijs" are abbreviations commonly used in Internet chat and short message services. They mean "I don't know" and "I'm just saying."

In the Atlassian product documentation we allow everyone to add and respond to comments, even people who have not logged in to the wiki. If the person has logged in, his or her username and profile picture will appear next to the comment, otherwise the comment is marked as anonymous.

Only Atlassian employees can remove comments. (If people want to remove a comment but do not have permission to do so, they can simply edit the comment and remove all its content. This is a little-known fact that does occur to some people, resulting in a few empty comments here and there. Sometimes people replace the content with the words, "Please delete this comment," or something more quirky. Often good for a laugh.)

We have also enabled public signup on the wiki. That means that anyone can sign up for a username and log in immediately. They need to enter an email address with any name and username they like.

When the comments start rolling in, you will need to monitor and respond to them. Confluence does not currently offer a way of managing all the comments in a space. The solution is to monitor incoming comments via RSS feeds or wiki watches, as described in Chapter 14. While writing this section, I raised a feature request on the Confluence issue tracker for better comment management, which may generate some interesting discussion.[5]

How long does it take to respond to comments? It depends on the size of the documentation set, the number of readers, and the complexity of the product or procedure. My experience is that responding to comments could take up all your time if you let it. Not all of the comments on your pages will be about the documentation. In fact, it is likely that most of them will be calls for help with using the product or suggestions for improvement and new features. (See the next two sections for an analysis of comments received.) It is a good idea to arrange for ongoing assistance from the product management and support teams in responding to such comments.

A good practice is to allocate a set period of time each day for comment management. In that period (half an hour or, if necessary, an hour), the first task is to triage the comments. Decide which ones you can handle. Pass the others on to the support or development teams by, for example, raising issues in their issue tracker. Then respond to the comments that are your responsibility. If the volume of comments is still too high, raise the problem with your team leader and ask for more help.

Chocolate and stamina

Both Casanova and Sherlock Holmes downed mugs of cocoa for breakfast. So, we must deduce, chocolate builds stamina in the brain cells as well as in the other bits.

Here are some guidelines for responding to comments:

- **Be direct and positive.** Comments benefit from the concise and clear language that characterizes other technical writing too.

- **Be professional and polite.** The tone of the comments should be professional and polite, especially when responding to someone who may be frustrated because some feature is not working or they cannot find the help they need.

- **Be light-hearted.** Where appropriate, a light-hearted tone has its place, especially when replying to someone who is already a known or frequent contributor, or when the comments are themselves light-hearted.

- **Be yourself.** Let the reader know that you are a real person. Personalize your wiki profile, by uploading a photograph or other avatar and adding a short description of yourself. Even just your role (technical writer at *Choc Ink*) is enough.

- **Be appreciative.** If someone has contributed useful information or helped someone else, thank them. Let them know if you copy their text into the main body of the documentation.

Once the question has been resolved, let the comment and response remain on the page for a given period. At Atlassian, we aim for a week. Then remove both comment and response. This will give the readers time to see the exchange of information, and at the same time it will keep the pages clean. Too many comments make a page difficult to read.

Now, if you look at some of the Atlassian documentation, you will see a long stream of comments on certain pages. This is often because the technical writers do not have time to clean them up. We have discussed running an automated task to clear the comments after a given time, but we are afraid of losing the valuable information in those comments.

Comment statistics

Confluence does not supply a tool for analyzing the number and type of comments received. The solution is to build a Confluence RSS feed that retrieves the comments over the time period you are interested in, then use another tool to crunch the numbers.

Let's take a look at the number and type of comments received on the Atlassian documentation wiki between the 7th and 14th of January 2011. First, here are some statistics about the overall traffic that the documentation site received.

Across the entire wiki:

- 99,329 people visited the site, in a total of 157,312 visits.
- 45,549 pages were viewed a total of 540,488 times.
- Number of comments on all spaces: 98

In the Confluence documentation (DOC space) only:

- 1,793 pages were viewed a total of 99,310 times.
- Number of unique page views: 78,037
- Number of comments on this space: 39

Let's analyze the 39 comments in the Confluence documentation (DOC space) in that week.

Who made the comments:

- 18 comments were anonymous – added by someone who was not logged in to the wiki.
- 21 comments were from logged-in users.

What the comments were about:

- 9 offered information, hints, or tips.
- 2 suggested an addition/correction to the documentation.
- 1 was an unhelpful comment, possibly meant to be humorous.
- 9 requested help on functionality ("how do I…" or "is it possible to…").
- 13 requested support (when something did not work as expected).
- 5 suggested a feature or improvement in the product.

My blog post has more details about how we handle comments on the Atlassian documentation and more examples of real comments from readers.[6]

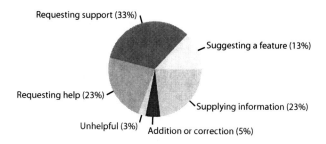

Figure 18.4. Comments on the DOC space

Rackspace's experiences with comments

Anne Gentle gave me some good information about handling comments on the Rackspace document-ation site. When they were about to turn on the commenting facility, Anne gave these guidelines to the support group who were preparing to moderate the comments:

- Aim to respond to a comment within 24 hours, or perhaps 48 hours, for at least the first six months. After that period, assess the situation to see if readers have started answering each other's questions. The aim is to make sure that people can expect a response.

- Do a bit of Internet research, before responding, to discover something about the reader's technical background and motivation for asking the question.

Anne also categorized about six months' worth of comments appearing in the OpenStack document-ation. These are the patterns that she found:

- People add a number of comments within a short period of time, to gain an understanding in their first few days of using the product.
- They point out typographical errors.
- They request specific examples, especially of code.
- They want details about supported features and platforms.
- They ask when features will be available.

Minimizing spam

Confluence uses CAPTCHA for spam prevention. CAPTCHA is a common technique used to test whether input is coming from a real person or an automated spam generator (spambot). The term is an acronym of the rather unwieldy "Completely Automated Public Turing Test to Tell Computers and Humans Apart." The behavior of Confluence's CAPTCHA is configurable.[7] On the Atlassian documentation wiki, CAPTCHA is enabled for anonymous users. Whenever an anonymous user adds a comment, CAPTCHA prompts the user to read some text from an image and type it into a form.

Even with CAPTCHA enabled, you may still get the occasional spam comment from real people. These may be people trying to sell their product or to entice people to come to their site. You will need to remove these spam comments when you see them.

Adding feedback forms via plugins or other customizations

Comments are good, but unstructured. People are free to write whatever they want, in whatever style and layout they choose. This is good for many purposes, but forms are useful when we need to request a specific type of feedback, and they help us extract reports and statistics from the structured data in the responses.

Feedback forms come in many shapes and sizes. Talented people are continuously designing new solutions. For example, take another look at the attractive form on the *ExactTarget Documentation Wiki* mentioned in Chapter 2 and shown in Figure 2.3.

The following sections describe some Confluence plugins that offer feedback forms as an integral part of a page. By the time this book is published, more plugins may be available. You can find them on the *Atlassian Plugin Exchange* at `https://plugins.atlassian.com`.

Content Survey and Reporting plugin

Jeremy Largman, support engineer at Atlassian, spearheaded the design and development of the *Content Survey and Reporting plugin*.[8] It was originally intended for Confluence spaces that host a knowledge base created by support teams to help customers troubleshoot and find work arounds for known problems. People quickly noticed that the plugin is useful in technical documentation spaces, too. You can see the plugin in action on the *Confluence Knowledge Base* wiki space.[9] The feedback form appears at the bottom of every page in the space.

Help us improve!

Is this article helpful?	○ Yes ○ No
Is the content complete?	○ Yes ○ No
Is it well written?	○ Yes ○ No

Figure 18.5. Feedback form generated by the Content Survey and Reporting plugin

Space administrators can specify the questions to appear on the feedback form. The plugin accepts simple yes/no answers only, a typical example being "Was this article helpful?" To add the feedback form to a page, use the KB survey macro. In wiki markup, it is simply {kbsurvey} – no parameters or other input required. In most cases, you would want the form to appear at the bottom of every page. The simplest way to do that is to use the Confluence Documentation theme and add the KB survey macro to the page footer. (See Chapter 7 for guidelines on setting up the Documentation theme.) The plugin's documentation has a number of useful tips on configuring the plugin and using additional macros.[10]

The magic happens in the analysis and reporting tools that the plugin offers. A detailed report shows the following information for every page that has received feedback:

- The total number of feedback ballots received.

- A composite score for the page, based on the last 32 votes cast and using a configurable algorithm.

- The percentage of "Yes" answers for each of the questions.

There is also a list of pages that have not yet received feedback and a rollup report showing statistics for the whole space. The information is designed to help the document curators identify the most useful pages and take action on areas in the documentation that need improvement.

Another useful feature of the plugin is the enhanced search functionality. Provided that you are using the Confluence Documentation theme, you can replace the standard search with one that prioritizes search results based on the popularity of the pages. It uses the composite scores of the pages, described above, to determine their relative popularity.

Interesting to note: You can configure the plugin's enhanced search to cover more than one space, and you can specify the spaces individually. The standard Confluence search covers all spaces or just one space at a time.

A custom feedback plugin under development

The *Atlassian Developers* site at `https://developer.atlassian.com` is hosted on a Confluence wiki. At first glance the site does not look like a wiki at all. The site is under rapid development as I'm writing this chapter, and one of the features being crafted is a feedback form.

Figure 18.6 shows the feedback panel, which is at the bottom of each page in the documentation section of the site.

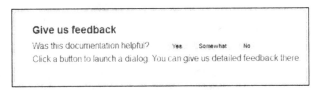

Give us feedback

Was this documentation helpful?   Yes   Somewhat   No
Click a button to launch a dialog. You can give us detailed feedback there.

Figure 18.6. Feedback panel on the Atlassian Developers site

The panel asks, "Was this documentation helpful?" and offers a choice of "Yes," "Somewhat," or "No." When the reader chooses one of those options, the feedback form opens up. The questions on the form differ depending on the reader's choice. Figure 18.7 shows the set of questions for "Somewhat."

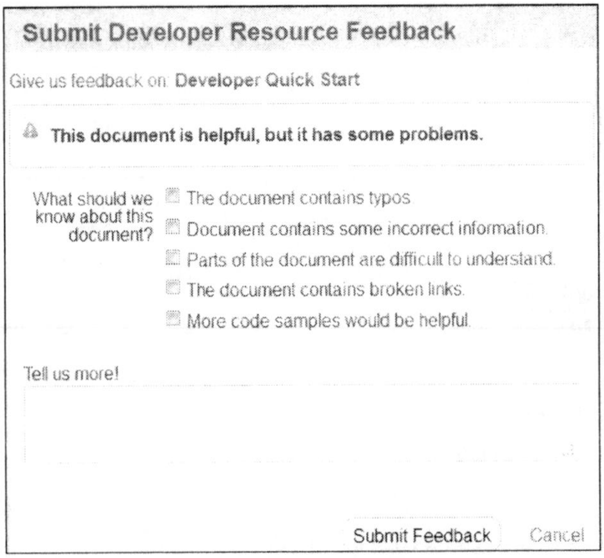

Submit Developer Resource Feedback

Give us feedback on: Developer Quick Start

⚠ This document is helpful, but it has some problems.

What should we
know about this
document?
☐ The document contains typos

☐ Document contains some incorrect information.

☐ Parts of the document are difficult to understand.

☐ The document contains broken links.

☐ More code samples would be helpful.

Tell us more!

Submit Feedback   Cancel

Figure 18.7. Feedback form on the Atlassian Developers site

The form submits the response to an issue tracking system, which the developers and technical writers monitor. This solution is still under development and the plugin is not currently available for general use. If it does become available, it will be listed on the *Atlassian Plugin Exchange*.

Survey plugin

The *Survey plugin* by Adaptavist[11] offers fully-fledged survey forms with configurable sections and multiple field types (text, single choice, dropdown selections, and more). You can limit the number of responses or set an end date for the survey, and you can manage surveys across the Confluence site.

Figure 18.8 is a view of the survey designer showing just two field types, text and single choice:

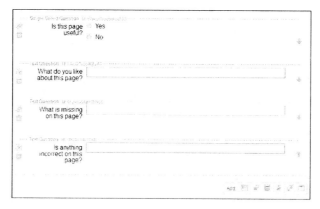

Figure 18.8. Form designer provided by the Survey plugin

You can view the survey results inside Confluence or export them to a .csv file for crunching in Microsoft Excel or other software. The plugin documentation has all the details.[12]

Other voting and ranking plugins

Two more plugins provide the ability to collect votes on and rank Confluence pages according to popularity and usefulness:

- The *Rate Macro plugin*.[13]
- The *Content Voting plugin*.[14]

External tools for gathering feedback

External tools provide a website where you can design a feedback form that can then be embedded in your documentation. In most cases, the website gives you a chunk of HTML that you can insert into your documentation page. Such forms can be used on any platform that lets you add HTML to the page, including Confluence.

Wufoo

Wufoo (http://wufoo.com) is an online web service that you can use to create an HTML form and collect information from people over the Internet. You can also choose to build a form on Wufoo and then grab the HTML to embed into your page. When readers fill in the form, the responses are stored in the Wufoo database, and you can extract the results from there.

For documentation on Confluence, Wufoo is a good option because Confluence's widget macro[15] provides a secure and easy way to embed a Wufoo form onto a page. There is no need to create and embed raw HTML. In many public-facing wiki sites, raw HTML is not allowed because it can make the wiki vulnerable to security attacks. The macros in Confluence are protected against such attacks.

Ganache, our technical writer extraordinaire, has decided to add a Wufoo form at the bottom of every page in her Twitter guide.[16] She wants a very simple form:

- Form title: "**Is this documentation useful?**"

- A field offering a choice of three options: "**Very,**" "**Somewhat,**" and "**Not at all.**"

- A text box where people can add more feedback if they want to: "**Tell us more…**"

The first step is to design and build the form. Ganache goes to the Wufoo website at http://wufoo.com and signs up for a user account. Then she follows the instructions to add the form. Wufoo has an impressive gallery of sample forms and a flexible theme designer. The form designer offers a number of different field types, including single and multi-line text boxes, multiple choice, number, check boxes, name, address, email address, and more.

The next step is to add the form to a Confluence site. Ganache uses Confluence's widget macro. All she needs is the URL of the Wufoo form. If you want the form on just a single page, simply add the widget macro to the page. Ganache, however, wants the form to appear at the bottom of every page in her Twitter guide. Since she is using the Documentation theme, she can easily add the widget macro to the page footer. In Chapter 7 we saw how Ganache added a copyright statement to the page footer. Now she adds the Wufoo form above the copyright statement. As mentioned in Chapter 7 ("wiki markup is even less dead than we thought") the configuration sections of the Documentation theme accept wiki markup in both Confluence 3.5 and Confluence 4. This is the markup that Ganache uses to add the Wufoo form, limit its height to 220 pixels, and wrap it in a panel:

```
{panel}
{widget:url=http://myexample.wufoo.com/forms/is-this-documentation-useful/|height=220px}
{panel}
```

The screenshot in Figure 18.9 shows the resulting form at the bottom of a page.

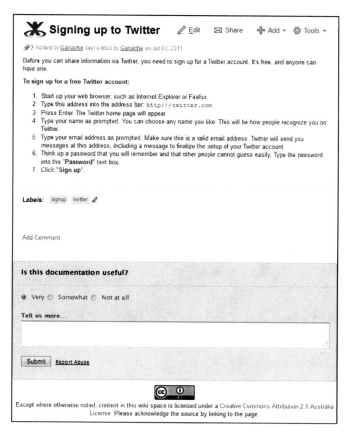

Figure 18.9. A Wufoo form in a Confluence page footer

What happens when someone completes the form and submits their feedback? You can choose to receive notifications via email or via a text message to a mobile phone. In addition, Wufoo provides an RSS feed for each form. Wufoo also integrates with a number of other services, including Twitter, Salesforce, MailChimp, and more. For example, if you hook your form up to a Twitter account, Wufoo will send a Tweet to that Twitter feed when someone submits a form entry. The Wufoo documentation has all the details about the integration partners.[17]

A live example of a Wufoo form in Confluence

Although not a feedback form, the Atlassian Contributor License Agreement is a live example of a Wufoo form on a Confluence site. People who want to contribute to the Atlassian documentation can submit an application via this form.[18]

Wufoo provides a detailed view of all entries submitted for the form. To view the aggregated results, use the Wufoo report builder or export the results to Microsoft Excel, a .csv file, or a tab delimited text file.

For more advanced uses, it would be good if the form contained information about who submitted the form and the URL of the page being read. You can include this information by adding parameters to the form's URL that cause the data to be sent to the Wufoo form each time someone opens

it. A Wufoo blog post[19] and the Wufoo documentation on URL modifications[20] explain how to do that. You will need to embed the HTML for the Wufoo form into Confluence using the HTML macro instead of the widget macro. The use of the HTML macro is described below, where our gallant Ganache experiments with Google forms.

Google Forms

Google forms offer another simple way of gathering feedback. A form is one of the document types available in Google Docs, at `http://docs.google.com`. You can build a form and then email it to people or embed it into a web page. Google automatically creates a spreadsheet for the form with the same title as the form. When someone submits a form entry, the information is stored in the spreadsheet. The Google Docs documentation gives all the details.[21]

Ganache makes a quick foray into the world of Google forms. Just as in the Wufoo example above, she keeps the Google form simple:

- Form title: "**Is this documentation useful?**"

- A field offering a choice of three options: "**Very,**" "**Somewhat,**" and "**Not at all.**"

- A text box where people can add more feedback if they want to: "**Tell us more**"

Ganache goes to Google Docs at `http://docs.google.com` and designs a form. Then she chooses the option to embed the form in a web page. Google gives her HTML code that looks like this:

```
<iframe src="https://docs.google.com/spreadsheet/embeddedform?formkey=xxxxxx"
width="760" height="638" frameborder="0" marginheight="0"
marginwidth="0">Loading...</iframe>
```

Instead of all the Xs in '`formkey=xxxxxx`', there is a long string of text identifying the specific online form.

Now to add the form to Confluence. At time of writing, Confluence does not provide a built-in macro for embedding Google forms onto a wiki page. The widget macro does not support Google forms. Instead, Ganache needs to embed the raw HTML that she has copied from the Google form. By default, Confluence does not permit people to embed raw HTML into its pages, but it does provide the HTML macro for this purpose. The HTML macro is disabled when you first install Confluence. A Confluence administrator can enable the macro.[22]

Note: Allowing people to include raw HTML on a page can be risky. It can open your Confluence site to cross-site scripting (XSS) attacks, where an attacker embeds their own JavaScript into a web page.[23] Do not enable the HTML macro unless you trust all the people who have permission to add or update any content on the wiki, including comments. Also, the macro needs to remain enabled in order for the embedded HTML to work. You cannot enable it, add the HTML, and then disable the macro again.

To enable the HTML macro:

1. Choose "**Browse**" > "**Confluence Admin**" > "**Plugins**."

2. View the "**Managing Existing**" tab and click "**Show System Plugins**."

3. Find "**Confluence HTML Macros**" in the list of system plugins, and click it to see the plugin details.

4. Click "**Manage Plugin Modules**."

5. Enable the "**html**" module.

Ganache considers the pros and cons of enabling the macro, then goes ahead and enables it. The next step is simple. Ganache is using the Documentation theme, so all she needs to do is add the HTML macro in the theme configuration for the page footer and put the Google form code inside the macro. She also adjusts the width and height of the `iframe` in the Google code to suit her form.

Example 18.1 is the markup to add the Google form and wrap it in a panel in the page footer:

```
{panel}
 
{html}
<iframe src="https://docs.google.com/spreadsheet/embeddedform?formkey=xxxxxx"
width="600" height="400" frameborder="0" marginheight="0"
marginwidth="0">Loading...</iframe>
{html}
 
{panel}
```

Example 18.1 – Google form wrapped in a panel in the page footer

Figure 18.10 shows the resulting form at the bottom of a page.

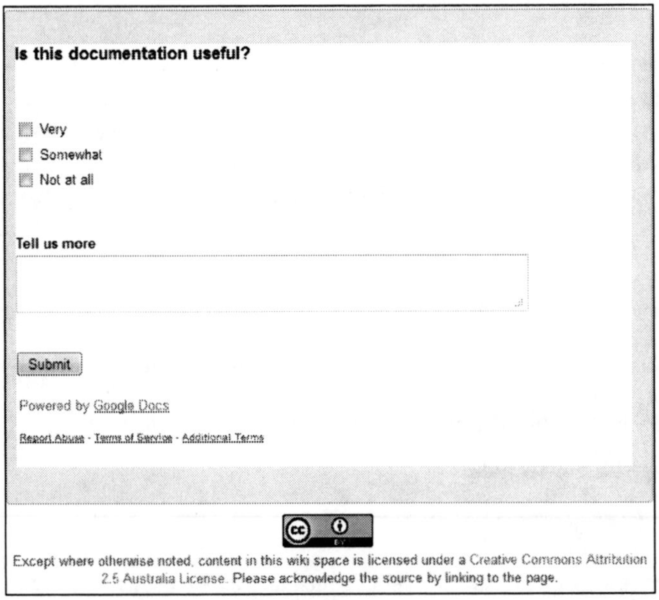

Figure 18.10. A Google Docs form in a Confluence page footer

Advanced styling and processing of Google forms

Google offers a set of themes for styling the form, with choices of font, color and background image. For more sophisticated styling, you can grab the form submission HTML from the Google form view and tweak it to suit your needs. This is also useful if you want to change what happens after a person has submitted a response. By default, the form will display a message and offer the person the chance to go to the online form on the Google site or to create another form. As the form designer, you can tailor the words of the message but not the links to the online form. However, to prevent a break in user experience on your Confluence page, you may want to remove those links and change the behavior to suit your site.

The first step is to grab the form submission code from Google. Go to the live Google form (as if you were about to complete the form online) and view the source of the web page. Copy the HTML element that starts with something like this:

```
<form action="https://docs.google.com/spreadsheet/formResponse?hl=en_GB&amp;
formkey=xxxxxx;ifq" method="POST" id="ss-form">
```

The element is a few lines long and ends with `</form>`.

Now you can add your own JavaScript, CSS, and HTML to make a really shiny form. When you have finished, just wrap all your code in the HTML macro and add it to the footer in the Documentation theme. Magic!

Thank you to Paul Watson for the idea of using Google forms to gather feedback. Thanks also to Mark Halvorson for the hint about grabbing and customizing the form submission code.

Other online form builders

These are just a few of the online form builders available. A web search will reveal many more. In all the cases shown below, you can either ask people to visit the form at the online survey site or grab the code to display the form on your own site. All these services offer data reporting and export functionality. It is worth examining the licensing arrangements and specific features of each to choose the one that suits your requirements.

- FormSite: `http://www.formsite.com/`

- JotForm: `http://www.jotform.com/`

- KISSinsights: `http://kissinsights.com/`

- Rational Survey: `http://www.rationalsurvey.com/`

Using Twitter for feedback on your docs

This is a new idea. I have not heard of anyone doing it. How about using Twitter as a mechanism for gathering feedback from readers? Take a sneak peek at Chapter 21, where Ganache tries it out.

The talk of the chapter

Comments, forms supplied by plugins and add-ons, external forms designed in online form builders... Possibly a Tweet or two. There are a number of useful and even exciting tools for inviting readers' feedback. Give people a place to talk, and they will come. Perhaps in great numbers. Before adding feedback mechanisms to the documentation, it may be wise to run a pilot project. This will help to manage expectations, both within the organization and among the readers. It will also help you decide on the people and tools needed to monitor the feedback and respond to people's questions.

When I started working on a wiki a few years ago, it was the first time I had experienced near-instant feedback from readers. Even now, a few years later, being able to ask for and respond to people's feedback still feels like a luxury and a privilege. And some comments brighten my day.

References

These are the tools, websites, blog posts, and other references mentioned in this chapter.

Endnotes

[1] A page about nested groups in the Atlassian JIRA documentation: `http://confluence.atlassian.com/display/JIRA/Managing+Nested+Groups`. At time of writing, the page has the comment from Matt Doar as described in this chapter.

[2] A page in the Atlassian documentation for the Universal Plugin Manager: `http://confluence.atlassian.com/display/UPM/Navigating+to+the+Universal+Plugin+Manager`. At time of writing, the page has the comment from an anonymous reader as described in this chapter.

[3] The Atlassian Crowd installation guide for the Crowd Apache Connector on Ubuntu Linux: `http://confluence.atlassian.com/display/CROWD/Installing+the+Crowd+Apache+Con-`

nector+on+Ubuntu+Linux. At time of writing, the page has the comments from Bertrand Paquet and Adrian Hempel as described in this chapter.

[4] *What part does chocolate play in your life as technical writer*, a discussion on *Technical Writing World*, started in May 2011: http://technicalwritingworld.com/forum/topics/what-part-does-chocolate-play

[5] A request for better comment management in Confluence wiki: https://jira.atlassian.com/-browse/CONF-23442

[6] *Experiences with readers' comments on the Atlassian documentation wiki*, a blog post by Sarah Maddox on *ffeathers*, January 2011: http://ffeathers.wordpress.com/2011/01/29/ex-periences-with-readers-comments-on-the-atlassian-documentation-wiki/

[7] *Configuring Captcha for Spam Prevention*, a page in the *Confluence Administrator's Guide*: http://confluence.atlassian.com/display/DOC/Configuring+Captcha+for+Spam+Pre-vention

[8] *Content Survey and Reporting plugin* by Atlassian: https://plugins.atlassian.com/plu-gin/details/24440

[9] The *Confluence Knowledge Base* space on the Atlassian documentation wiki: http://conflu-ence.atlassian.com/display/CONFKB

[10] Documentation for the *Content Survey and Reporting plugin*: https://studio.plugins.at-lassian.com/wiki/display/CKBSP/Home

[11] *Survey plugin* by Adaptavist: https://plugins.atlassian.com/plugin/details/311029

[12] Documentation for the *Survey plugin* by Adaptavist: http://www.adaptavist.com/display/-Plugins/Survey

[13] *Rate Macro plugin* by Adaptavist: https://plugins.atlassian.com/plugin/details/181

[14] *Content Voting plugin* by Adaptavist: https://plugins.atlassian.com/plugin/details/-310964

[15] The widget macro described in the *Confluence User's Guide*: http://confluence.atlassi-an.com/display/DOC/Widget+Connector+Macro

[16] *Using Twitter*, a guide to Twitter for technical communicators on the *Confluence, Tech Comm, Chocolate* wiki: https://wikitechcomm.onconfluence.com/display/DOC/

[17] Wufoo integration settings, in the Wufoo documentation: http://www.wufoo.com/docs/in-tegrations/

[18] The Atlassian Contributor License Agreement, an example of a Wufoo form on a Confluence site: http://confluence.atlassian.com/display/ALLDOC/ACLA+v2.0

[19] *How to Track Users with Wufoo*, a blog post by Ryan Campbell on *The Wufoo Blog*: http://-www.wufoo.com/2008/03/25/how-to-track-users-with-wufoo/

[20] Wufoo documentation on URL modifications: `http://wufoo.com/docs/url-modifications/`

[21] Google Docs documentation on Google forms: {{`http://docs.google.com/support/bin/-topic.py?topic=1360904`

[22] Page about the HTML macro in the *Confluence User's Guide*: `http://confluence.atlassian.com/display/DOC/HTML+Macro`

[23] Cross-site scripting, or XSS, vulnerabilities on a website may make it possible for an attacker to embed their own JavaScript into a web page. *The Web Application Security Consortium* has a good description and examples: `http://projects.webappsec.org/w/page/13246920/-Cross%20Site%20Scripting`

19

Updates by everyman

Our hero Ganache works at *Choc Ink*, purveyors of chocolate-focused software. We first encountered Caraque, a customer of *Choc Ink*, in Chapter 13. Now let's peek into an email conversation between Caraque and Ganache.

Caraque:

> Hey Ganache, Thank you for that document I requested. How on earth did you get it up so quickly? Max chocolate kudos to you!

Ganache:

> I wrote the first draft yesterday. The product manager and another technical writer reviewed it and made some good changes. I published it this morning. Since then, a support engineer has added a note. And a community author added a section about a specific configuration I missed. Oh wait, that was you! Thanks. :-)

In this chapter

Collaboration. It is where a wiki flies. How about allowing community updates – what are the advantages, what are the concerns, and how can we make it all work? This chapter has the following sections:

- **A model for collaboration.** There is life after publication. People work together to keep a page alive and up to date. Let's look at some examples of collaboration before and after publication and get an idea of the number of community updates we can expect.

- **Is it safe to allow readers to update the documentation?** What are the concerns and how can we address them?

- **Managing updates by everyman.** Let's do a quick brush up on wiki permissions, RSS feeds, and watches. Then we will examine how style guides, templates, and structure affect would-be contributors. And what about deleting pages that are redundant or obsolete?

- **Documentation that cannot be updated.** Sometimes it is not practical to allow community updates, but it may still be useful to publish the documentation on a wiki and allow community feedback.

- **Safeguarding copyright and intellectual property.** A contributor license agreement is useful, and a copyright license is essential.

- **How to encourage contributions.** People will not know that they can contribute unless we tell them.

- **Kay's chocolate cake: part 3.** Yum.

- **The Perfect Technical Document.** A summary of this chapter.

- **References.** The websites, blog posts, and other references relevant to the content of this chapter.

A model for collaboration

Technical writing teams are made of skilled, passionate and knowledgeable people. That is a large part of what this book is about. Such a team represents a core of brilliance in the organization and in the larger group of people who depend on the documentation. But there is brilliance around the edges too. For every skilled and passionate member of the team, there are a few hundred outside the team. Some of them are inside the organization and some are in the wider community. Smart organizations and smart teams know this. They find ways of harnessing that brilliance.

Many technical communication teams are moving towards a new model, where the team focuses on its core competencies and on ways of encouraging external people to contribute their skills and knowledge. The team's focus may be on organizing and curating the information. Or perhaps the team's skill lies in a particular type of information, such as the user guides, rather than some other type, such as developer guides. In a fast-growing organization, this tactic of combined specialization and outreach is the only way a small team can keep its head above water. It is also a good way of ensuring that readers have access to the best and most up-to-date information possible.

Of course, people can review or give feedback on a document without updating its content. In Chapter 18 we saw the role comments can play in giving readers a place to point out areas for improvement and contribute additional information. But, consider how much faster the content management cycle becomes if contributors can update the content themselves. All the technical writer needs to do is verify the change and tweak it if necessary.

Chapter 10 is all about the documentation workflow. Figure 10.1 depicts a simple workflow from document creation to publication. Now let's take a look at Figure 19.1. The life of a page does not stop at publication. If asked, a page would probably say that life only starts there.

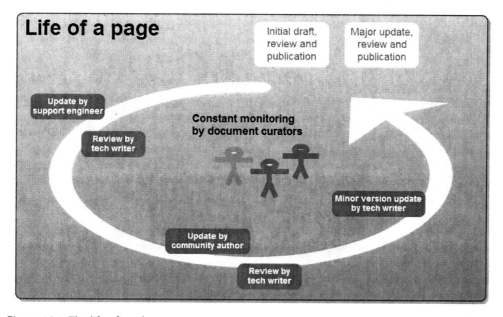

Figure 19.1. The life of a wiki page

Chapter 10 describes the mechanics of using Confluence to manage the workflow from draft to publication. In particular, the section on organizing a review describes the use of page restrictions to allow reviewers to update the page. It also gives guidelines on responding to and incorporating the reviewers' feedback. What happens if we extend the model beyond the initial publishing workflow? Consider ongoing updates by a widening funnel of authors, including technical writers, other employees, community authors, customers, the whole wide world. The principles are the same. The mechanics do not change. It is primarily the role of the technical communication team that evolves. As well as authors, we become document curators. We manage the channel of communication between our organization and its customers, and among the customers themselves.

A quick clarification of terminology: In this book I use the term "document curator" a few times, but I have steered away from the term "content curator." By "curator," I mean someone who takes care of the content and guards it from harm. In contrast, the accepted meaning of the term "content curator" is someone whose primary role is to collect information from various sources and present it in a central location. Bill Ives has written a good post about the role of content curator titled, *Good Advice for Content Curators*.[1] He gives some useful tips for content curators and provides a clear and current definition of the role. A content curator is someone who sifts through the plethora of information in the world today, gathering the bits that are relevant to a particular community or interest group, and publishing them in an accessible form. That is certainly akin to what a technical communicator does. The roles overlap. But in this book, let's talk about "document curators" when we mean someone who looks after the content on a technical documentation site.

The scenarios in this chapter focus on product documentation, but the concepts and ideas are applicable to internal procedures, too. Collaboration is equally valuable when developing procedures and guidelines for the members of an organization on either an intranet or a publicly available wiki for an open organization. Take a look at the ideas in this chapter for managing updates by contributors, safeguarding intellectual property, and encouraging contributions.

A scenario for collaboration while drafting a document

Let's say that Ganache is putting together the release notes for a major product release. In fact, it is the upcoming release of *ChocTactics 2.0*, one of *Choc Ink*'s flagship products.

She adds the page to the wiki. At this point the page is visible only to *Choc Ink* employees. Ganache designs the structure of the page to highlight the main features of the release and writes a draft of the introductory paragraph. She adds a placeholder for a "what's new" video and lets the marketing team know where to put the video when they are ready. Then she invites Charlie, the product manager, to add his material. You can see it all happening in Chapter 13.

Release notes are an excellent candidate for collaboration during the initial development of the document. In many organizations, release notes are as much a marketing tool as a technical document. There are a number of stakeholders, including product management, the marketing team, the engineering team, and the technical communicators. Given a framework, these people can work productively and quickly to make the release notes a shiny communication tool.

An example of collaboration before and after publication

In mid 2011, Atlassian released Confluence 4.0, a major update to the wiki software. The highlight of this release was a smart new all-in-one editor. It replaced the two editors available in earlier versions of the wiki. Now, as you might guess, the editor is the heart of the wiki. When the editor changes, the world changes. Well, that is true at least for people who spend a significant part of their day on the wiki! Atlassian needed to make sure people were aware of the impact of the upcoming release and give them some material to use in preparing for it.

The Confluence 4.0 documentation includes a special set of change management guides headed by the page titled *Planning for Confluence 4.0*.[2] The guides are designed primarily for managers and team leaders who will need to train their staff and update their in-house procedural documentation. When an organization decides to upgrade its Confluence site to Confluence 4.0, the change management team can use these guides to let the wiki users know what is about to happen. They can prepare training material and quick reference guides based on the material in the guides.

Part of the guide is a page of frequently asked questions and answers (FAQ) about the new editor.[3] The team published this page very early in the release cycle, long before the actual release date, to give people a place to ask questions and a page to watch for news and updates. This page immediately became very busy. A number of customers posted their concerns, questions, and suggestions. They asked for information, reassurance, and clarification, and let the developers know their needs. The product managers, developers, and quality assurance engineers were kept busy responding and updating the page.

From the very start, the change management guides were a highly collaborative effort, and collaboration has continued well after publication. The most often updated page is the one titled *Confluence 4.0 Editor – What's Changed for Wiki Markup Users*.[4] At the time of writing this chapter, the page history shows 75 updates between January and September 2011. The authors are:

- Four technical writers.

- A product manager.

- A CEO (one of the company's chief executive officers).

A note about restricted editing rights: An interesting point about the FAQ page in the change management guide is that the page restrictions allow only the technical writers, development team lead, and product managers to edit the page. This is not something that Atlassian does often. Almost all of the documentation pages are open to editing by all employees and community authors. In this case, the product managers felt it was necessary because of the importance customers give to the information on this page and the level of passion and interest it generated.

Examples of community updates

Up to this point we have primarily discussed updates by people within the organization. What about other people? A number of community authors have update rights on the Atlassian documentation wiki. Later in this chapter we will take a look at the agreement that people sign if they want to contribute to the documentation. First, let's see some examples of updates that community authors made in the six months from May to September 2011.

- **Update to the JIRA upgrade notes.** JIRA is an issue tracking software application. The *JIRA 4.4 Upgrade Notes*[5] contain information that people need when upgrading to version 4.4 from an earlier version of JIRA. The technical writers and development team wrote the document in preparation for the software release and continued adding notes to it as new issues and workarounds came to light around the time of the release. A community author, Adam Saint-Prix, noticed a comment that another reader had added to the page. The reader had spotted a typographical error in an XML element quoted on the page: `<vaidation-query>` instead of `<validation-query>`. Adam fixed the error. A small fix, but a useful one indeed.

- **Update to a page about Confluence logging.** A page in the Confluence configuration FAQ describes how to enable user access logging on a Confluence site.[6] One of the code blocks on the page contained some superfluous backslashes intended to escape the braces and square brackets used in the code. Instead, the backslashes were showing up as part of the code. This probably happened when someone converted the code block from one format to another. Sim Hua Soon, one of the community authors, noticed and fixed the error.

- **New page about disabling a feedback button.** Matt Doar, an active community author and developer, recently wrote a new page in the JIRA documentation: *How to disable or enable the GOT FEEDBACK button.*[7] The "got feedback" link is a relatively new feature on the JIRA user interface. People can click the link to provide feedback directly to Atlassian. As Matt points out, some JIRA administrators prefer to manage feedback in other ways and will therefore be grateful for the guidelines on removing the link. Matt created the page, and the technical writers saw it pop up in their RSS feeds and reviewed it.

These are just a few examples of the contributions that community authors make to the documentation. Speaking as one of the Atlassian technical writers, I am very grateful for the value these dedicated people add to the documentation in our care.

How many updates are we talking about?

In practice, how many community authors will update the documentation? There is a lot of material on the Internet analyzing how many people join and actively participate in online communities. The 1% rule states that in every 100 members of an online community, one person will create content while the other 99 will view the content without contributing. Then there is the 90-9-1 principle, which holds that one percent of people create content, nine percent edit or modify the content, and 90 percent just view it. For more research on this topic, read *The 90-9-1 Rule in Reality*, a blog post by Dr. Michael Wu.[8]

A technical documentation wiki is not the same thing as an online community. The above rules, and the research from which they originated, are interesting but not entirely relevant. For one thing, the distinction between creating content and updating existing content is not very useful when talking about community contributions to documentation that is managed by a group of technical writers. The division seems to imply that newly-created content is more valuable than updates to existing content. But this is not true in a documentation wiki. Indeed, fixes and additions to existing content may be considered the most useful, especially when the documentation has a carefully-designed structure and a consistent depth of coverage. A community author can easily fall into the trap of adding a page that already exists somewhere else or adding too much detail to a section and thus causing an imbalance in the documentation set.

Let's take a look at the number of updates made by community authors to the Atlassian documentation in the six months from May to October 2011:

- Number of community authors signed up to date: 36

- Number of community authors who updated pages in the 6 month period: 11 (31% of the number of people signed up)

- Number of updates made: 32

- Largest number of updates by one person: 14 (44% of the number of updates)

Of course, all documentation wikis are different and so are the communities they serve. If you need to estimate the number of updates that your documentation may attract, a pilot project will be useful. See the section called "Gathering feedback – why and how" in Chapter 18 for a discussion of pilot projects.

Is it safe to allow readers to update the documentation?

The answer is, "It depends." It depends on the type of documentation, the nature of the organization publishing that documentation, the pool of contributors, and the requirements of the readers.

There are a number concerns common to all types of documentation:

- **Technology.** The documentation platform must be able to handle version control, concurrent updates, permissions, and monitoring of updates.

- **Trust.** Should we allow the whole wide world to update the documentation or just people we know and trust? How can we check that people have not abused our trust, and how can we make sure people know they can trust us and the content?

- **Quality.** We want the highest quality possible, including a consistent style and accurate information.

- **Intellectual property.** We need to guard the rights of all: the organization publishing the documents, the contributors, and even third parties who own copyrighted content elsewhere.

Let's cover the simplest first, the technical aspects of version control and concurrent updating. Confluence keeps a history of all the updates made to a page. You can compare the versions of the page, see who did what when, and even revert to a previous version. The wiki also manages concurrent editing. If a number of people edit the same page at the same time, Confluence will do its best to work out if any of the changes conflict. If they do, then the last person to save the page will see a message describing the conflict and asking for a resolution.

The rest of this chapter addresses ways of preserving trust, quality, and intellectual property rights.

Managing updates by everyman

Here are some practical guidelines on how to manage a documentation wiki where community updates are allowed. The primary aim is to preserve the trust of readers and the quality of the documentation by ensuring that only authorized people can update the content and that all updates are monitored and corrected where necessary. It is also important to preserve the trust of the contributors by communicating with them if we need to remove the content they have added.

Managing update rights

In Confluence, you can grant users and groups permission to update the pages in your documentation space. The most useful practice is to define a group in Confluence for each group of people who will use the wiki, then grant appropriate permissions to each group. Chapter 4 has guidelines on planning your user management and permissions. Chapter 5 describes how to add users and groups to the wiki and how to set the permissions in your documentation space so that only trusted people can update the content.

Here is an example that works well. It is the permissions model used in the Atlassian documentation.

Example groups:

- **Anonymous users.** This is not really a group, but a permission type defined by Confluence itself. Anyone who is not logged in to the wiki is considered an anonymous user, even if that person has a username and could log in.

- **Confluence users.** This is a special group, `confluence-users`, predefined by Confluence. Everyone who has a username and can use Confluence is a member of this group. If an administrator moves a username out of this group, then that person still has a username but can no longer log in.

- **Staff members.** You may find it useful to add this group to your Confluence site. For example, Ganache's documentation wiki might have a group called `chocink-employees`.

- **Technical writers.** This is a useful group to add to your Confluence site. It is likely that the technical writing team will need administrative rights on the documentation spaces. If all members of the team are members of a group in Confluence, it is easy to assign them permissions when you add new spaces.

- **Community authors.** This group could contain people external to the organization who have some update rights on the documentation.

Table 19.1 shows one possible way to assign permissions to group

Table 19.1 – Example of permission assignment to groups

Permission	Groups	Remarks
View pages	Anonymous users Confluence users	In this example, everyone has permission to view the pages, including anonymous users and everyone who has logged in.
Add and update pages	Staff members Community authors	Technical writers are part of the "Staff members" group.
Remove pages	Staff members	
Add and update comments	Anonymous users Confluence users	In this example, everyone has permission to add comments to the pages, whether logged in or not.
Remove comments	Staff members	
Add attachments	Staff members Community authors	
Remove attachments	Staff members	
Administer the space	Technical writers	

Monitoring the updates

When the updates start rolling in, check them for accuracy and consistency across the documentation suite. Chapter 14 describes tools for monitoring updates. Here are two ways to use them:

- **Request notification of every change to a given page or space.** Wiki "watches" are a good tool for monitoring activity on a specific area of the wiki. If you add yourself as a watcher of a page or a space, Confluence will send you email whenever anyone updates that page or space.

- **Request a list of updates over a given period of time in a given area of the wiki.** An RSS feed gives you targeted information on wiki activity. Upon request, Confluence will generate an XML file containing the most recent updates matching a set of selection criteria. You can read the result in an RSS feed reader, such as Sage or Google Reader, or display the feed on a Confluence page.

It can also be useful to see how many people are using the wiki and where they are spending their time. Web analytics tools, such as Google Analytics and Webtrends, are useful for this purpose and are discussed in Chapter 14.

Deleting redundant and obsolete pages

Sometimes a contributor may add a page that is unnecessary or incorrect. They may have missed the fact that the information already exists somewhere in the documentation set, or they may have made a mistake and added the page to the wrong documentation set.

Funnily enough, it is more likely to be a staff member, rather than an external contributor, who adds such pages. For example, the support engineers may need to add some information quickly because they have a customer breathing down their necks. They know that the technical writers are watching and will fix the documentation. Sometimes they even let us know what they have done and ask us to move the information into the right spot. External contributors tend to be more conservative when it comes to adding pages.

Whoever the contributor is, the technical writer's task is the same. We need to remove the page if it is completely irrelevant or move the information into the right area. First, make a copy of the information on the page and email it to the contributor explaining your actions. If some of the information is valid, copy it and add it to the relevant page. Thank the contributor for that information and explain where you have put it. Then remove the page.

When reviewing the documentation at each major release, you will find pages that have become obsolete. This may be because they document a feature that has been removed or because they describe a workaround for a problem that no longer exists. In such a case, you can simply remove the page as part of the normal documentation release procedures. If you provide archived versions of the documentation for previous releases (see Chapter 10), then the page will remain available in the archive.

Providing style guides, templates and structure

Practically speaking, it is often the job of the technical writers to apply a consistent style to the documentation. When someone updates a page, the technical writer will perform a review and tweak the wording and layout.

Style guides that are very restrictive may discourage people from contributing to the content. Even so, both internal and external authors find a style guide useful. This kind of guide is very different from detailed references like *The Chicago Manual of Style*[9] or the *Microsoft Manual of Style for Technical Publications*.[10] Community authors often have little time to read a long document. Instead, consider publishing a light-weight style guide that is a summary of information about the best way to do things. A mantra for such a guide might be: "Don't scare off the readers." Briefly describe the style and philosophy of the documentation suite. Give useful, concise information that most contributors will need to know.[11] Refer to the heavy-weight style manual and dictionary, which lurk behind the scenes ready to settle disputes.

Templates offer a good way of helping people write documentation in the preferred style and layout. Chapter 8 describes the features and plugins available for creating templates in Confluence. As with the style guide, the use of templates should be optional and should be seen as a guideline, rather than a mandate, when it comes to community contributions.

A well-structured documentation set makes it easy for community authors find existing pages to update. It is therefore less likely that they will unnecessarily add a new page when the content already exists elsewhere. Good structure encourages contributions and helps to make those contributions more valuable. See Chapter 7 for guidelines on structuring the content in the wiki.

Documentation that cannot be updated

In some cases it is just not practical to allow community updates.

- Some technical documentation is subject to strict regulation. It is not feasible or even desirable to allow community updates. But perhaps we do want to allow people to give feedback about the published content.

- Single source publishing is another interesting case. How can we use the wiki if we write our documentation in some other medium and want to publish it to the wiki for collaboration, feedback, and sharing?

Here is a solution: Set up different spaces, with different collaboration models, on the same wiki.

For want of better terminology, let's talk about the "official documentation" and the "community documentation." The official documentation is the content published by the organization, where for one reason or another we cannot allow community updates. The community documentation is a collection of content that community authors have permission to update. They may have created the content themselves, or perhaps the organization has published the initial framework and documents, and the community authors contribute as well.

How does it work? Here is one model:

- Create a space for the official documentation. Set the wiki permissions so that only the technical writers (for example) can add and update the pages. Allow other people to add comments.

- Create another space for the community documentation. Set the wiki permissions on this space so that anyone (or approved groups) can add and update content as well as add comments.

In the above model, the core documents are untouchable, but everyone can give their views and add related content, hints and tips, fixes that need applying, and more. The example suggests a single space for the official documentation and another for the community documentation, but you can have a number of spaces for each.

When you update the official documentation and publish the updates to the wiki, the comments attached to the updated pages will remain unaffected. The content in the community documentation space will also be unaffected. Only the core documents will be updated to the latest version.

Safeguarding copyright and intellectual property

When other people have the right to update the documentation, a number of new and interesting requirements arise.

- We need to protect the rights of the organization, the contributors, and any third parties whose work may be intentionally or unintentionally included in the documentation.

- Contributors need to know their rights and duties as defined by the copyright license under which their contributions will be made available to the public.

- Readers and customers need to know how they can use the documentation.

To satisfy the above goals, at the very least we need a clear statement of copyright and a clear statement of the rights and duties of contributors. It may be a good idea to ask contributors to sign a declaration acknowledging that they understand their rights and duties.

The following section describes a contributor license agreement and a Creative Commons copyright license, two mechanisms that address the above needs.

Putting a contributor license agreement in place

A contributor license agreement is a document that explains the rights and duties of a person who contributes to a repository, such as a source repository or a documentation wiki. The *Atlassian Contributor License Agreement*[12] is an example. Putting this agreement in place was a very interesting, exciting and rewarding experience. I have described it here in the hope that it will be useful to other organizations too. Anyone is free to copy the text of the agreement and adapt it to develop a similar agreement for their own purposes.

I was part of the team of people who wrote the agreement. We started in late 2008. As a first step, we looked around for similar agreements in use by other organizations. The closest we could find was the Apache Software Foundation's *Individual Contributor License Agreement*, which is intended for source code rather than documentation.[13] Based on that document, we drafted an agreement that would work for documentation. We then consulted an intellectual property lawyer, who shaped and perfected the document to protect everyone's rights.

Then we had to make the agreement available for people to sign. We wanted contributors to acknowledge their understanding of the agreement, so some sort of signup was necessary. But we wanted it to be as simple as possible, both for the contributors to sign and for us to process and administer.

In its first incarnation, the agreement took the form of a PDF document that people could download from a wiki page.

- The would-be contributor downloaded the PDF form, signed it, and faxed or emailed it to us.

- The fax or email triggered the creation of an issue in our issue tracker.

- A technical writer verified the form, granted the person update rights on the wiki, and emailed the person to let them know they were now authorized to update the documentation.

That process was not too bad, but it was a bit clunky from the contributor's point of view. Downloading, signing, and emailing a PDF form? In this day and age?

Then we cottoned on to the fact that Confluence supports Wufoo forms. This support came in Confluence 3.0, after the initial publication of the contributor license agreement. We redesigned the agreement to use a Wufoo form instead of PDF. Now all the would-be contributor has to do is fill in the information and submit the form online. Wufoo creates an issue in the issue tracker, and the backend process is the same as before. Chapter 18 describes the use of Wufoo forms in Confluence.

Finding a copyright license that fits the bill

A copyright license lets readers and customers know what they are permitted to do with the documentation. Can they copy it, adapt it, and even sell their adapted version? A copyright license also tells contributors how other people may use the material they have contributed.

There are many flavors of copyright license, and the legal situation differs in each country. For example, Australia does not require any registration of copyright. The author has automatic copyright upon publication.[14] It is essential to research the options available and find the one that suits your requirements. For people based in the United States, the United States Copyright Office publishes information at http://www.copyright.gov/.

The Creative Commons organization offers a useful array of licenses and tools. The first part of their mission statement reads, "Creative Commons helps you share your knowledge and creativity with the world."[15] Atlassian has chosen a Creative Commons Attribution license, also known as CC BY. Specifically, the documentation uses Creative Commons Attribution 2.5 Australia.[16] Under this license, anyone can use the documentation for any purpose, provided they acknowledge the source of the material. A number of customers have let us know that they use and adapt our documentation as a basis for their internal procedures.

How to encourage contributions

It is safe to say that document curators will not be flooded with updates from community authors. Reading about the 90-9-1 principle, the 1% rule, and the numbers quoted from experience above, it seems the opposite is true. We will need to encourage people to contribute. How?

- Advertise and promote the documentation and the fact that it is open to community updates. Tell it on Twitter, blogs, forums, anywhere potential contributors spend their time.

- Try some more advanced tricks with Twitter. This is another teaser for Chapter 21!

- Add style guides and information about how to edit the documentation. Draw people in.

- Hold doc sprints. (See Chapter 22.)

- Employ all of the techniques discussed in Chapter 17 on how to engage your readers.

We need to let community authors know that we want and appreciate their updates. We do the baking, they put the icing on the cake.

Kay's chocolate cake: part 3

In part 2 of this recipe you baked the cake. (See Chapter 16.) Now let's make the icing and put the two halves of the cake together.

Making the icing:

1. Put the cocoa into a mug and add a very small amount of boiling hot water.

2. Stir and add more hot water, until the cocoa forms a paste.

3. Pour about half of the icing sugar into a bowl.

4. Add the cocoa paste and mix with a metal spoon. The mixture will turn surprisingly runny. Keep mixing, adding water as needed. Add more water as needed. When you have finished mixing, the icing should be dark brown in color and smooth but not runny. It should slowly drip off a spoon. (If you add too much water by mistake, top up with icing sugar.)

5. Mix approximately 3 ounces (100 g) of the icing sugar with water, so that you end up with a small amount of white icing too.

Filling and icing the cake:

1. Put one half of the cake on a plate, with the smooth side down.

2. Spread a layer of raspberry jam on the cake. Not too thick, just enough to give a nice sharp taste to counter the sweetness of the cake.

3. Put the other half of the cake on top, with the smooth side up.

4. Pour boiling water into a deep bowl and put the knife into it, to warm up the knife.

5. Spread the chocolate icing onto the cake, using the hot knife to smooth it down.

6. Put the knife back into the hot water.

7. Use a small spoon to put the white icing into the plastic bag, being careful not to let it all drip out of the hole just yet.

8. Decide whether you want straight lines or concentric circles on your cake. Straight lines are easier.

9. Squeeze the white icing onto the chocolate icing layer.

- If you decide on straight lines, start at one side of the cake and draw 5 or 6 lines parallel to each other across the cake.

- If you decide on circles, start with a small circle in the middle of the cake, draw a larger one around it, and another until you reach the edge of the cake. Four circles is a good number.

10. Drag the hot knife lightly over the lines, catching the white icing to form a feathery pattern. Make sure to drag the knife in opposite directions each time.

- If you have straight lines, drag the knife at 90° angles to the lines, from left to right then from right to left, from left to right again and so on.

- If you have circles, drag the knife from the middle of the cake towards the edge, then from the edge towards the middle, and so on.

There is a photograph of the cake on the *Confluence, Tech Comm, Chocolate* wiki.[17] Join us there to swap stories about baking and eating Kay's chocolate cake, and to share more scrumptious recipes.

Happy chocfest!

The Perfect Technical Document

Some things just do not exist and never will. One is the Great American Novel. (Substitute your own nationality at will.) Another is the Perfect Technical Document. The reason? Things change. The minute the author has finished writing the GAN or the PTD, something happens to make it out of date. People are doing things to make the work obsolete even as the author is developing it.

The only way to solve this problem is to get those people involved in keeping the work current, alive, and valuable.

This chapter presented a model for collaboration with both internal and external contributors. Examples showed the model working in real life. We did a quick brush up on the use of wiki permissions to control who can do what and where, and we touched again on RSS feeds and wiki watches, both useful for keeping track of what people are up to in the documentation. We discussed light-weight style guides as a useful tool for contributors as well as technical writers. Then we examined intellectual property rights: A contributor license agreement helps to guard the rights of all, and a clear statement of copyright lets everyone know what's what.

Here's the takeaway from this chapter: encouraging people to contribute to the documentation is as important and rewarding as the frosting on a chocolate cake.

References

These are the tools, websites, blog posts, and other references relevant to the content of this chapter.

Further reading

Rahul Mehrotra has published a paper titled *Printed Books and Online Help Using a Wiki* about his team's move from traditional desktop publishing to a distributed, wiki-based collaborative authoring and publishing model. The paper describes the problems encountered with the original content development and publishing process, the testing of different solutions, and the eventual migration to a wiki. It includes a report of the rollout and training and the results seen in collaborative authoring and simplified processes. Rahul Mehrotra is the technical writing Lead for Agilent EEsof's documentation group. The paper is published on the *WritersUA* site: `http://www.writersua.com/articles/-WIKI/index.html`

Endnotes

[1] *Good Advice for Content Curators*, a blog post by Bill Ives on *Portals and KM*, September 2011: `http://billives.typepad.com/portals_and_km/2011/09/good-advice-for-content-curators.html`

[2] *Planning for Confluence 4.0*, the Confluence 4.0 change management guides: `http://confluence.atlassian.com/display/DOC/Planning+for+Confluence+4.0`

[3] *Confluence 4.0 Editor FAQ*, a page in the *Confluence User's Guide*: `http://confluence.atlassian.com/display/DOC/Confluence+4.0+Editor+FAQ`

[4] *Confluence 4.0 Editor - What's Changed for Wiki Markup Users*, a page in the *Confluence User's Guide*: `http://confluence.atlassian.com/display/DOC/Confluence+4.0+Editor+-+What%27s+Changed+for+Wiki+Markup+Users`

[5] *JIRA 4.4 Upgrade Notes*: `http://confluence.atlassian.com/display/JIRA/JIRA+4.4+Upgrade+Notes`

[6] *How to audit Confluence - enabling user access logging*, a page in the Confluence documentation: `http://confluence.atlassian.com/display/DOC/How+to+audit+Confluence+-+enabling+user+access+logging`

[7] *How to disable or enable the GOT FEEDBACK button*, a page in the JIRA administrator's FAQ: https://confluence.atlassian.com/display/JIRA/How+to+disable+or+en-able+the+GOT+FEEDBACK+button

[8] *The 90-9-1 Rule in Reality*, a blog post by Dr. Michael Wu on *Lithosphere Community*, March 2010: http://lithosphere.lithium.com/t5/Building-Community-the-Platform/The-90-9-1-Rule-in-Reality/ba-p/5463

[9] *The Chicago Manual of Style:* http://www.chicagomanualofstyle.org/home.html

[10] *Microsoft Manual of Style for Technical Publications, Third Edition* http://www.microsoft.com/-learning/en/us/book.aspx?ID=6074&locale=en-us

[11] *Style Guide for the Atlassian Developer Documentation*, an example of a style guide designed for internal and external contributors to the documentation: https://developer.atlassian.com/-display/DOCS/Style+Guide+for+the+Atlassian+Developer+Documentation

[12] *Atlassian Contributor License Agreement*: http://confluence.atlassian.com/display/-ALLDOC/Atlassian+Contributor+License+Agreement

[13] Apache Software Foundation's *Individual Contributor License Agreement*: http://-www.apache.org/licenses/icla.txt

[14] *How you get copyright*, a document by the Australian Copyright Council: http://archive.copy-right.org.au/information/cit020/wp0019

[15] The Creative Commons organization: http://creativecommons.org/

[16] Creative Commons Attribution 2.5 Australia: http://creativecommons.org/licenses/by/-2.5/au/

[17] A photograph of Kay's chocolate cake, on the *Confluence, Tech Comm, Chocolate* wiki: https://-wikitechcomm.onconfluence.com/display/CHAT/Kay%27s+chocolate+cake

20

Pages that walk and talk

Gadgets and widgets and streams, oh my! Visual, dynamic, and real-time content offers a great way to convert a page from documentation to communication.

Strolling into this chapter

This chapter examines the use of dynamic, visual, and real-time information in the documentation. The chapter has the following sections:

- **Lively, shiny, and useful too.** The reasons why dynamic, visual, and real-time information is a Good Thing.

- **Risks, and how to manage them.** A look at the not-so-rosy side of things, and how to manage the risks.

- **Video and audio.** Displaying multimedia files hosted on the Confluence site and embedding videos from YouTube, Dailymotion, Episodic, Vimeo, and more.

- **Photographs and other images.** Screenshots attached to the wiki page, diagrams supplied by plugins, and slide shows from Flickr.

- **Presentations and other documents.** PowerPoint presentations, Microsoft Office documents, PDF files, and presentations hosted on SlideShare, SlideRocket, Scribd, and Google Docs.

- **Gadgets.** A way of displaying information from external sources on a wiki page, and a way of displaying Confluence information on another site.

- **Issue trackers.** Drawing a filtered stream of items from the issue tracker to enrich the release notes or keep readers informed about the development of a new feature.

- **Other integration points.** Activity streams, news and blog posts, micro blogging, calendars, and our imagination.

- **Talking about shininess.** A summary of this chapter.

- **References.** The plugins, tools, and references relevant to this chapter.

Lively, shiny, and useful too

Integration with other web services and tools is an area where wikis shine. They are likely to do so more and more, as the world of online information expands and becomes more and more interrelated. Confluence provides macros for embedding live content drawn from other sites as well as multimedia files that are hosted on the Confluence site.

These are some of the reasons why dynamic, visual, and real-time information is a Good Thing:

- Attractive and interesting pages attract people. By engaging our readers' interest, we are making it easier for them to focus on and understand the information on the page. People are more likely to return to the documentation rather than other forms of help out there on the web. This means that we can be sure they are getting the best information possible: ours!

- Most of the tools discussed here are essentially visual in nature. Many people find it easier to learn and remember something that is presented in a visual way.

- The technical communication team can make efficient use of information that already exists. Many people make videos and presentations: the marketing team, community authors, bloggers, customers. We can tap into resources that are already out there to save time and resources, while at the same time giving our readers good information.

- Live feeds from web services provide information that is up to date at the very moment when the reader sees it.

Risks, and how to manage them

Not all is rosy in the world of dynamic and real-time information.

- If we are pulling information from another web service in real time, there is a chance that the other site is unavailable when the reader views the page.

- Information coming from another site is out of our control. It may go out of date. It may even change without our knowledge.

- Dynamic information does not work well in printed or static documentation such as PDF. This is true for multimedia files that are hosted on the wiki as well as a stream of information that is pulled from another site.

There are good ways of managing the risks. When adding a widget to a page, test the result in a PDF export and a printed page. If it does not look reasonable, do not use the widget. Raise a request with the wiki developers to get the print format fixed. By "look reasonable," I mean that the printed format should show some sort of icon or text indicating that there is a non-printable object on the page. When pulling information from a website that is outside the control of the technical documentation team, add a note to the page explaining that the information is drawn from an external site, is unsupported, and is made available because at the time of adding it, it gave readers useful supplementary information.

As a mitigation factor, in many cases we can assume readers understand the online environment. If the dynamic list of issues in the release notes is empty and displays a message saying that the issue tracker is currently unavailable, the reader will recognize this familiar scenario and come back later.

Video and audio

Sometimes an instructional video is the best way to give people a quick understanding of a concept or task. A podcast or other audio recording may add valuable information to a written topic. Perhaps the multimedia file is available to the author and can be hosted on the Confluence site. In other cases the file may be hosted on a public site such as YouTube.

Multimedia files hosted on the Confluence site

Confluence offers the multimedia macro for displaying content that is hosted on the Confluence site. First upload the video or audio file as an attachment to a Confluence page, then add the multimedia macro to display it. Chapter 6 is all about becoming a macro magician. Here is a quick refresher.

To add the multimedia macro using the macro browser:

1. Edit the page and place your cursor in the right spot.

2. Click the **"Insert/Edit Macro"** icon (in Confluence 3.5) or **"Insert"** > **"Other Macros"** (in Confluence 4) in the editor toolbar.

3. A list of macros will appear. Start typing "multimedia" in the search box. The list of macros will shrink until you can see the one you need.

4. Click the macro block and supply values for the parameters as prompted.

5. Click **"Insert"** to add the macro to the page.

6. Save the page.

The *Confluence User's Guide* lists all the supported file types.[1] At time of writing, they are:

- Adobe Flash (.swf)

- Apple QuickTime (.mov)

- Windows Media (.wma, .wmv)

- Real Media (.rm, .ram)

- MP3 and MP4 files (.mp3, .mp4)

- MPEG files (.mpeg, .mpg)

- AVI files (.avi) You may need to enable an avi decoder within your browser.

Videos displayed directly from external sites

What if the video is hosted on an external site? There are a couple of options. The first is to use Confluence's widget macro.[2] It supports videos from YouTube, MySpace Video, Yahoo Video, Dailymotion, Episodic, Vimeo, Metacafe, blip.tv, and Viddler. Add the widget macro to your Confluence page and supply the URL of the video.

The second option is to use raw HTML to display the video. Most video hosting sites offer code that you can grab to embed the video into your own page. For this to work, the Confluence site must allow you to add raw HTML to a wiki page. Take a look at the section called "Google Forms" in Chapter 18, where our dauntless Ganache enabled the HTML macro and then embedded the code for a form into her page. The procedure is the same for embedding a video.

For an example where an externally hosted video is useful, take a look at Figure 20.1. It shows some release notes[3] with a video in a prominent position at the top of the page. The video makes the page more appealing to the eye and provides the reader with useful information. The marketing team made the video to give an overview of the release, and posted it on the website and in a blog post. The technical communication team used it to add value to the documentation too.

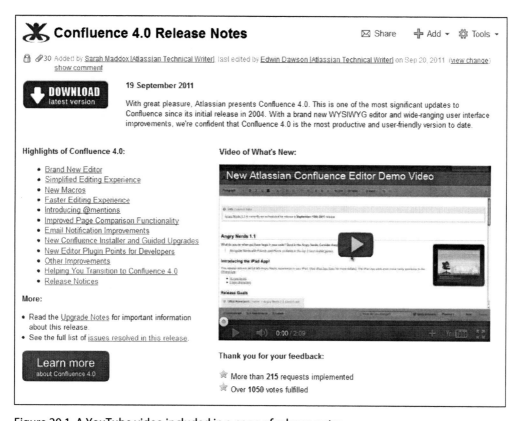

Figure 20.1. A YouTube video included in a page of release notes

Photographs and other images

Chapter 6 describes how to add screenshots and other images to a Confluence page by attaching the image to the page and then displaying it. That chapter also describes the use of plugins such as *Gliffy*, *Balsamiq*, and *Creately* for drawing and displaying diagrams and mockups. Now let's examine the use of photographs and images that are hosted on external sites. Once again, the widget macro is the star. It displays images from Flickr and Skitch, both online services that host images. All you need is the URL of the image or set of images. Let's look at Flickr in detail.

Flickr, at `http://www.flickr.com/`, is an online service where people can upload pictures and share them with colleagues, friends, and the world. Flickr also offers an API (application programming interface) which gives other web services access to the images on Flickr. The Confluence widget macro is built upon the Flickr API. Confluence authors can use the widget macro to embed a Flickr image or a set of images onto a wiki page by specifying the URL of the Flickr image or set. If you embed a set of images, Confluence displays them as a slide show.

Where would such images be useful? One example is to show the usage scenarios for a product. Another is to illustrate the unpacking and setup procedures. Because Flickr is a public web service, customers and community authors can upload their own photographs and tag them. You can use this social and collaborative aspect of Flickr to enrich the documentation with images from the community.

Another example is to show the participants and activities in a doc sprint. You could publish a "hall of fame"[4] to record the event and to encourage people to join in next time. Chapter 22 describes this idea in more detail.

Presentations and other documents

A presentation, PDF document, or spreadsheet may add information to a documentation page. Confluence can display documents that are attached to wiki pages and those that are hosted on external sites.

You can attach a Microsoft Office document or a PDF file to a Confluence page and display its content in Confluence without the need to convert the content to wiki format. The view-file macro[5] is the tool here, offering options to display Microsoft Word, Excel, PowerPoint, and PDF files. This is a good way to use existing documents or information that is best compiled in a format other than HTML.

For example, use the power of an Excel spreadsheet to crunch the numbers, then attach the spreadsheet to a wiki page. Readers can see the information without having Office installed on their computers.

What about presentations hosted on external sites? Let's say your organization has recently held a conference at which a number of information-rich presentations were given. After the conference, the slide shows were published online. Many of them complement the topics in the documentation. Instead of laboriously transcribing each one, you can simply embed the presentation into the page. The widget macro displays presentations hosted on SlideShare, SlideRocket, Scribd, and Google Docs.

Gadgets

A gadget is a small object (a piece of functionality) offering dynamic content that you can put onto a dashboard, a wiki page, or some other web page. When displayed, a gadget most often appears within a frame, dividing the content from the rest of the page. As a well known example, take a look at the Google gadgets that people use to build their iGoogle home pages or enliven their Gmail inboxes. (Want to try it out? To build your own iGoogle home page, go to http://www.google.com/ig and follow the prompts.)

Chocolate robots

Artist, engineer and tinkerer PenfoldPlant has posted an endearing recipe for chocolate robots. The recipe looks yummy. The post on *instructables* and the resulting comments are fun to read. Who would have thought the First Law of Robotics had anything to do with eating chocolate?[6]

In Confluence, gadgets provide a way of displaying information from external sources on a wiki page. They also provide a way of displaying Confluence information on another site, be it a Confluence site or a different application. Confluence gadgets are based on the OpenSocial[7] gadget specification, which is in turn based on Google's gadget technology. The aim of OpenSocial is to help websites share their social information across the web. To this end, OpenSocial defines a common interface for web applications to use. The gadgets in Confluence and other Atlassian applications are built upon the data exchange part of the OpenSocial specification.

At this time there are some limitations and complexities in the gadget implementation in Confluence. Still, gadgets are worth playing with and represent an area of functionality that Atlassian may decide to expand and improve upon.

Gadgets that bring external content onto a Confluence page

A gadget can display information drawn from another Confluence site, another Atlassian application or another website. Here are some examples of where a gadget may be useful in technical documentation:

- If the organization has two Confluence installations, a gadget on one site can offer a search for related information on the other site. For example, people reading the technical documentation can search for information in the knowledge base.

- For organizations using more than one Atlassian product, gadget technology is one of the mechanisms that Atlassian provides for integrating the individual applications to form a cohesive developer toolset. In developer-focused documentation, a gadget can present a graphical view of activity on the development issue tracker (JIRA[8]), build server (Bamboo[9]), and source repository (Subversion, Git, Mercurial and others, as viewed through FishEye.[10]) The big list of Atlassian gadgets[11] shows all the gadgets available (a nice round fifty at time of writing) to display information from each of these applications with instructions on how to do it.

- On a page of frequently asked questions or troubleshooting tips, a gadget can display the related entries from an external discussion forum.

- The world's your oyster! There are plenty of Google gadgets available. For the more technically inclined, Google gadgets are fairly easy to develop, and the Google documentation[12] is excellent.

Gadgets that display information from external sources may introduce a security risk. For that reason, the Confluence administrator must register the gadget before anyone can add it to a page.

To make an external gadget available in a Confluence site:

1. Choose "**Browse**" > "**Confluence Admin**" > "**External Gadgets.**"

2. Paste the URL of the gadget into the field labeled "**Gadget Specification URL.**" The URL should point to the XML file that describes the gadget. This file is called the gadget specification. The URL looks something like this: `http://example.com/my-gadget-location/my-gadget.xml`

Now wiki authors can add the gadget to their pages by simply adding a macro. The gadget is available as a macro with a name corresponding to the gadget name.

Some gadgets require extra authorization to allow the gadget server to send information to the server where the gadget is displayed. This is when things become a little complex. There are two parts to the authorization, one performed by the administrator and one by the person adding the gadget to or viewing the gadget on the page. The administrator needs to set up a trust relationship between the two servers by configuring an "application link."[13] After that, everyone who views the page will need to authorize the display of the information under the authority of their username. The authorization lasts for a given period (a week, by default) unless the person revokes it. The gadget will prompt the reader to log in and approve the display of the information.

Gadgets that display Confluence content on another site

Imagine if readers could search Confluence from within their email program or see a Confluence page on their iGoogle home page or on their JIRA dashboard…

These are the gadgets that Confluence supplies:

- The page gadget displays content from a Confluence page. If so configured, it will provide links that people can click to view and/or edit the page on the Confluence site.

- The activity stream gadget lists the most recently changed content on a Confluence site.

- The quicknav gadget provides a quick navigation text box, allowing people to search the Confluence site that publishes the gadget. For example, they can use their iGoogle or JIRA dashboard to manage their work. When an item comes up relating to the documentation, they can immediately search for the page without having to open up a browser window and find their Confluence site.

A Confluence gadget should work on iGoogle, Gmail, and other web applications that support OpenSocial gadgets. The documentation points out that the use of gadgets on external websites is still experimental. People do encounter problems with them. From the Atlassian documentation:[14]

> Please do have fun playing around with them, but be aware that Atlassian does not support the use of Atlassian gadgets on iGoogle, Gmail or other external web sites.

External gadgets via the widget macro

By now you will not be surprised to learn that the widget macro can display gadgets, too. It supports Google Gadgets and Widgetbox.

Issue trackers

It is useful to display a list of resolved issues in the product release notes. Instead of the technical writers having to constantly return to the page and add items during the release cycle, how about including a macro that draws the information directly from the issue tracker when the reader opens the page?

There are two macros in Confluence that do just that: the JIRA issues macro[15] and the JIRA macro.[16] Using the JIRA issues macro, you can display a list of items based on a filter that you have previously set up in the issue tracker. Using the JIRA macro, you can define the list of items when you add the macro to the page (see Figure 20.2).[17]

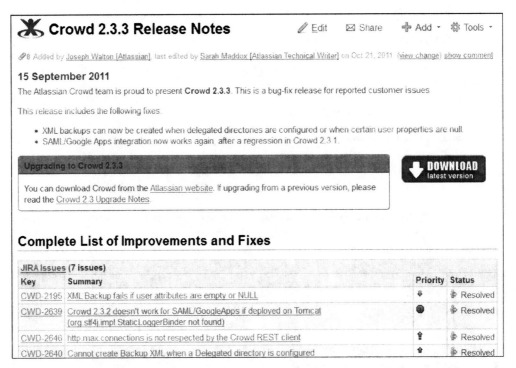

Figure 20.2. A list of fixes displayed in the Crowd 2.3.3 release notes via a JIRA issues macro

Aside from release notes, what are the other scenarios where it is useful to show up-to-date information from an issue tracker? Let's consider the documentation of a new or experimental feature. Perhaps the feature is still in beta release. Readers, customers, and support teams will benefit from a list of known issues and requests for improvement. People can click an item in the list to hop over to the issue tracker and see the current status of the fixes and any responses from the developers.

Other integration points

Here are some more points of integration between a wiki and other applications and ideas about where they are useful in technical documentation.

- **Activity streams.** We have mentioned these a few times in the book already, primarily in Chapter 14. Activity streams show people what is happening on the wiki. They are useful for building a community around the documentation. Activity streams can also show what people are doing outside the wiki. Developers of other applications can inject their own events into an activity stream. Confluence does not yet support this feature, but the possibilities are just starting to open up, and we can expect much more development in this area in the future.

- **News and blog posts.** A wiki page can display a list of external news items via an RSS feed using the RSS macro (see Chapter 14) or a list of blog posts from within the wiki using the blog posts macro. This is handy for listing technical alerts that have been posted recently.

- **Micro blogging.** The widget macro displays status updates from Twitter, FriendFeed, and BackType. Twitter? Yes! The next chapter is dedicated to the story of Twitter, wikis, and documentation.

- **Calendars.** An inventive use of online calendars might be to display a release schedule for the product, or a list of upcoming user group meetings. The widget macro supports Google Calendar. The *Team Calendars plugin* adds embeddable, sharable calendars to Confluence.

There are others. Many of them are mentioned in this book, and more are under development all the time!

Talking about shininess

This chapter was all about bringing the documentation to life by adding dynamic and visual information such as videos, audio files, slide shows, presentations, gadgets, and filtered lists from issue trackers. By integrating the documentation with web services and tools we can ensure that it is up to date, to the minute. No need for manual or even automated updates to keep it that way. When someone opens the page, the information is loaded and ready for absorption.

Let's move on to perhaps the shiniest integration of all: Twitter plus wiki.

References

These are the tools, websites, blog posts, and other references relevant to the content of this chapter.

Useful plugins

Team Calendars for Confluence by Atlassian, a plugin for embedding and sharing calendars: `https://-plugins.atlassian.com/plugin/details/72307`

Endnotes

[1]   A page about embedding multimedia content, in the *Confluence User's Guide*: `http://confluence.atlassian.com/display/DOC/Embedding+Multimedia+Content`

[2]   The documentation for the widget macro, in the *Confluence User's Guide*: `http://confluence.atlassian.com/display/DOC/Widget+Connector+Macro`

[3]   *Confluence 4.0 Release Notes*, an example of a video used on a wiki page: `http://confluence.atlassian.com/display/DOC/Confluence+4.0+Release+Notes`

[4]   The doc sprint hall of fame on the Atlassian documentation wiki: `http://confluence.atlassian.com/display/DOCSPRINT/Doc+Sprint+Hall+of+Fame`

[5]   A page about the view file macro, in the *Confluence User's Guide*: `http://confluence.atlassian.com/display/DOC/View+File+Macro`

[6]   *Chocolate robot recipe* by PenfoldPlant on *instructables*: `http://www.instructables.com/id/Chocolate-Instructables-Robots/`

[7]   The OpenSocial website: `http://www.opensocial.org/`

[8]   JIRA is an issue tracker developed by Atlassian: `http://www.atlassian.com/software/jira/`

[9]   Bamboo is a continuous integration server developed by Atlassian: `http://www.atlassian.com/software/bamboo/`

[10]   FishEye, developed by Atlassian, gives developers access to their source repositories in the browser: `http://www.atlassian.com/software/fisheye/`

[11]   The big list of Atlassian gadgets: `http://confluence.atlassian.com/display/GADGETS/All+Atlassian+Gadgets`

[12]   *iGoogle Developer's Guide*: `http://code.google.com/apis/igoogle/docs/igoogle-devguide.html`

[13]   The documentation on configuring application links in Confluence: `http://confluence.atlassian.com/display/DOC/Configuring+Application+Links`

[14] Documentation on adding Confluence Gadgets to external websites: `http://confluence.at-lassian.com/display/DOC/Configuring+Confluence+Gadgets+for+Use+in+Other+Ap-plications`

[15] Confluence documentation on the JIRA issues macro: `http://confluence.atlassian.com/-display/DOC/JIRA+Issues+Macro`

[16] Confluence documentation on the JIRA macro: `http://confluence.atlassian.com/dis-play/DOC/Inserting+JIRA+Issues`

[17] The Crowd 2.3.3 release notes: `http://confluence.atlassian.com/display/CROWD/-Crowd+2.3.3+Release+Notes`

21

Twitter, wiki and games

Let's take a look at Twitter, communication tool extraordinaire. Tweets are punchy and engaging. People read them, exchange them, play a global game of tag with them. With Twitter and a wiki, we can find people and draw them into the documentation. We can put other people's words onto our pages, dynamically, as the people write them. We can even have a document suggest a tweet that the reader may like to send. Our readers can use Twitter to communicate with each other while reading the documentation. That could be the start of something special. Anyone for a game?

All atwitter about this chapter

The first few sections of this chapter are about the Twitter service and how to use it:

- **Introducing Twitter.** A bit about the web service called Twitter, and the small chunks of information called Tweets.

- **Getting started with Twitter.** How to send your first Tweet, gain some followers, and find a good Twitter client.

- **Useful Twitter conventions.** The special language that people have devised for Twitter: @Replies and @Mentions, reTweeting, shortened URLs and hash tags.

- **How to embed Tweets onto a web page.** The widgets that Twitter provides for displaying Twitter information on a page.

- **How to embed Tweets onto a wiki page.** Our old friend, the widget macro in Confluence.

- **How to suggest the words for a Tweet.** Adding information to a Twitter URL to populate a Tweet with words.

The next sections discuss the use of Twitter in technical communication, including the special case of a configuration guide turned into a game with Twitter for interactivity:

- **Inviting and publishing tips via Twitter.** Encouraging customers to Tweet their hints and tips about a product, then publishing the Tweets in the documentation.

- **Twitter as a medium for release notes.** Tweeting the highlights of a release and using Twitter's hash tags to display the related Tweets in a continuous stream.

- **Twitter for gathering reader feedback.** A new idea for technical writers to try. If you dare.

- **Saving Tweets.** Using RSS feeds to retrieve and store Tweets.

- **Interactivity and games in documentation.** Positing the idea that technical documentation can benefit from a laugh and a game.

- **Case study: the Dragon Slayer docs.** A configuration guide tricked out as a game, with Twitter as the tool that the participants use to communicate.

- **Lessons learned.** The generosity of people. Strategies for coping with the number of people on Twitter. Making the most of your Tweets. Observing some etiquette and safeguarding people's rights.

Wrapping up:

- **A little bird told you.** A short summary of this chapter.

- **References.** The tools, websites, blog posts, and other references relevant to the content of this chapter.

Introducing Twitter

Twitter is a web service that people use to send short messages to anyone who is interested in reading them. For us as technical communicators, Twitter offers a way of publishing and sharing information. It is a medium quite different from others that we use. It provides unique opportunities. And like any other tool, it imposes its own requirements on information design.

A message sent via Twitter is called a Tweet. Tweets are very short, a maximum of 140 characters. There are advantages to publishing such small chunks of information:

- **Easy to digest.** A person can read a Tweet at a glance. Twitter client programs (more about these below) display Tweets in a stream, ordered by date and time, classified by category or person. People run their eyes down the stream, focusing here and there on a message that interests them.

- **Shareable.** Twitter is a social medium, engineered to encourage people to copy and promote the Tweets sent by their friends. A Tweet can "go viral" within a few hours, reTweeted around the world in ever-increasing numbers.

- **Conversational.** Using @Replies and @Mentions, people hold conversations across the globe and its time zones, drawing other people in just by mentioning their Twitter names in the Tweet.

- **Interactive.** Tweets can include clickable links to web pages, photographs, and videos.

- **Cross-platform and mobile.** One of the most magical aspects of Twitter is that people do not have to go to *Twitter.com* to use the service. People read and send Tweets via gadgets on other websites, via desktop widgets and applications, and on their phones and tablets. This means that we can use Twitter to reach people wherever and whenever they are online.

Figure 21.1. A Tweet

Twitter is not necessarily associated with wikis. People use Twitter and wikis completely independently. So why does Twitter deserve a whole chapter in this book? Because of the unique opportunities it offers for technical communication, and especially, because of the ways that you can integrate information from Twitter into the wiki documentation.

Getting started with Twitter

Sign up for a free account at `http://twitter.com`. Type a message where Twitter asks "What's happening?" and click "Tweet." That's it, you have Tweeted. The next step is to "follow" some people who Tweet about interesting things. Twitter will send you their Tweets.

Investigate some Twitter clients to set yourself free from the Twitter website. There are a number of clients that add useful functionality to your desktop and make it possible to Tweet from your phone, too.

- TweetDeck for iPhone, Android and desktop: `http://tweetdeck.com/`

- Echofon for iPhone: `http://www.echofon.com/twitter/iphone/`

- Seesmic: `https://seesmic.com/seesmic-social/desktop`, supports a number of other social networks as well as Twitter

- HootSuite: `http://hootsuite.com//`, another tool for managing multiple social networks

- TwitPic: `http://twitpic.com/`, for uploading pictures and other media and Tweeting at the same time

There are many more clients available, as a quick search of the Internet will reveal.

Useful Twitter conventions

People love holding conversations, and Twitter is a social tool. Each Tweet can be a maximum of 140 characters in length, including spaces. That is not very long. People have devised a number of conventions to make the best use of those characters and of the conversational aspects of Twitter. First we take a look at the conventions themselves. Later in the chapter we will see how to take advantage of them in technical communication.

@Replies and @Mentions

When you want to reply or refer to someone by name, include an "@" and the person's Twitter user-name. For example, "@sarahmaddox." Most Twitter clients provide a "reply" option that does this for you. Clients have a separate column or tab showing replies and mentions directed to you so you can easily find them.

ReTweet (RT)

If you like what someone said and want to spread the word while acknowledging the original Tweeter, you can "reTweet" their message. Add the letters "RT" and an @Mention of the person's Twitter name at the start of the message, copy the original text, and Tweet away. Most Twitter clients offer a reTweet option and will compile the Tweet for you.

Shortened URLs

A URL can be very long, consuming far too many of those valuable 140 characters. Luckily, there are services on the web that will shorten a URL for you. The service gives you a short string that you can include in your Tweet (or in an email, or a blog post, and so on). When other people click the short string, they are bounced through the service's URL and directly to your original URL. Examples of such services are *TinyURL.com* and *bitly*. Most Twitter clients will automatically shorten URLs, and Twitter itself does the same.

Some of the URL shortening sites, such as *bitly*, offer a professional service that you can sign up for. Such services provide statistics of the number of people who have clicked through to each shortened URL, their geographical location, and more.

Hash tags and search

Another very useful convention is the hash tag, or # tag, which is used to collect related Tweets. Let's say you are sending a message about a particular subject, and you think other people would like to read and share Tweets about that subject on an ongoing basis. This could be for fun or for a more serious purpose. Think up a keyword, prefix it with a "#," and include it in your message.

Most Twitter clients automatically provide a link on any hash tag that appears within a Tweet. Clicking the link brings up the Twitter search and shows all recent Tweets that contain the same term.

How to embed Tweets onto a web page

In December 2011 Twitter announced a new feature. You can now embed a single Tweet into a web page. Go to the page that shows the details for the Tweet by clicking the date/time stamp at the bottom of the Tweet. Then click 'Embed this Tweet', near the bottom of the message. A dialog will pop up, offering the HTML, which you can grab and insert onto your web page.

Twitter also supplies widgets that you can add to a web page to display a live stream of Tweets. The widgets are bits of JavaScript and HTML made to your specification. To build a widget, go to the Twitter widget builder at `http://twitter.com/goodies/widgets`. There are a few different types of widget on offer, such as a search widget, a profile widget, a favorites widget, and so on. Pick the widget you want, then supply the information as prompted. For example, if you pick a profile widget. then a form will prompt you for the name of the Twitter user whose profile you wish to display, a title, and a caption. Tweak the colors and other settings, then grab the code that appears on the screen and paste it into your document.

How to embed Tweets onto a wiki page

The above methods for inserting an embeddable Tweet or a Twitter widget will work on a Confluence page, too, provided that you can insert raw HTML onto the page. Take a look at the section titled "Google Forms" (p. 354), where our hero Ganache used the HTML macro to embed the code for a form into her page. Embedding a Tweet or a Twitter widget works in the same way.

For security reasons, your Confluence system administrator may disable the HTML macro. This will prevent authors from using raw HTML on a page. Instead, you will need to use a Confluence-specific macro.

At time of writing this chapter, there is no Confluence macro that embeds a single Tweet into a page. (I am sure that feature will come along soon now that Twitter has released the feature.) But the wiki does supply a widget that wraps the Twitter widget in a safe container: the widget macro.[1] This is the same macro described in the sections about videos, other multimedia, and forms for gathering reader feedback. Let's see how to use it to embed a stream of Tweets.

In Chapter 5 we saw Ganache build an attractive home page for her Twitter guide (Figure 21.2):

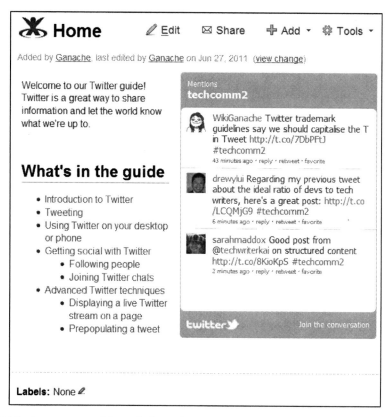

Figure 21.2. The home page of Ganache's guide to using Twitter

How did Ganache add the Twitter stream to the page? All she needed was the widget macro and the URL of the Twitter search to pick up all Tweets that use the keyword "techcomm2:" `http://-search.twitter.com/search?q=techcomm2`. (Why "techcomm2" and not "techcomm"? For her screenshots, Ganache decided to narrow down the search to just a few messages from known Tweeters. So she asked Andrew Lui and me to Tweet with the #techcomm2 hash tag.)

Using wiki markup, this is what the widget macro looks like when wrapped around the Twitter URL. The height and width of the widget are set to 300 pixels:

```
{widget:url=http://search.twitter.com/search?q=techcomm2|width=300|height=300}
```

Instead of wiki markup, you can insert the widget macro via the tool bar in the Confluence editor, and supply the Twitter URL in the dialog that appears. Chapter 7 describes how to add a macro to a page.

How to suggest the words for a Tweet

Chocolate at LavaCon 2011

LavaCon[2] is an annual conference about digital media and content strategies run by Jack Molisani. If the Tweets sent by attendees are anything to go by, chocolate plays no small part. From Kirsty Taylor (@kirstyt) on 16 November 2011:[3]

"Idea: let's swap chocolate at the stitch'n'bitch in the Omni bar tonight. #Lavacon. Don't make me take TimTams home!! #Lavacon"

It can be fun and useful to give your readers something to say and encourage them to say it in a public forum like Twitter. This gives them the opportunity to involve their community (their Twitter followers) in what they are doing with your product or documentation.

To suggest the words for a Tweet, you set up a link for people to click. That will open Twitter in their web browser and put some words into their Twitter message. If they have not yet logged in, Twitter will prompt them to log in. They can choose to edit the words, or just leave them as they are, and then send the Tweet as usual.

How? Add an HTML link on your page that points to the reader's Twitter home page. In the URL, specify a "status" parameter that contains your suggested text. Here is an example of the HTML required:

```
<a href="http://twitter.com/home?status=Hello world">
Say hello to the world on Twitter
</a>
```

In Confluence, just enter the text, "Say hello to the world on Twitter." Then select the text and insert a link: `http://twitter.com/home?status=Hello world`.

If the message includes funny characters, such as a # sign, then you will need to URL-encode the message. For example, use this link to suggest the Tweet, "Hello world #testing":

```
{{http://twitter.com/home?status=Hello+world+%23testing}}
```

There are websites that will URL-encode your text for you. One example is *The URLEncode and URLDecode Page* from Albion Research Ltd.[4]

Now that we have covered how to use Twitter, let's look at some ways in which these techniques are useful in technical documentation.

Inviting and publishing tips via Twitter

Many organizations are lucky enough to have an enthusiastic and active group of customers or community developers who discover new ways of using the product or clever techniques for configuring it. Those people are keen to share their discoveries with other people. Twitter offers a quick and easy way to do that. There is no need to find the relevant page in the documentation or discussion forum and then submit a comment. Instead, people can pull up their favorite Twitter client, bang out a Tweet with a hash tag and a link for more information, and move on. Other customers or prospective customers will subscribe to the list of Tweets to see what the experts are saying.

This is how a technical communicator can set it up:

- Publish a hash tag that will identify the Tweets for a given product. For example, if the product is *ChocTactics* then the hash tag could be "#ChocTacticsTips."

- Encourage customers, developers, and others to Tweet their hints and tips using the given hash tag. Write blog posts, send out Tweets, and add a page to the documentation to let people know that they can share their hints in this way.

- Publish the stream of Tweets on a "Hints and tips" page in the documentation, on the corporate website, and in blog posts. Not only can people see the hints and tips from everyone via Twitter, they can also see their own Twitter names appearing in the documentation.

The Confluence *Tips via Twitter* page[5] is an example. Figure 21.3 shows a screenshot of the Twitter widget on that page.

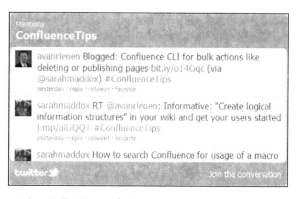

Figure 21.3. Confluence tips via Twitter widget

Twitter as a medium for release notes

Tip: Hold a Tweet-a-Choc

Writing Tweets can be hard work. Everyone knows that the shorter the document, the more difficult it is to write. Now imagine a document that is about 110 characters long, including spaces! That is the length of a Tweet once you have factored in the hash tag and the URL.

A good way of composing an alluring set of Tweets is to gather the technical writers, product managers, and marketers in a room, supply them with hot chocolate, and brainstorm the release notes. We call such a gathering a Tweet-a-Choc.

Let's say your company has released a major version of a product, and you have written the release notes. Now you want people to come and read them. The marketing team will blog about the release, send out newsletters, and add a link from the website. Twitter is another channel for getting the news out and drawing people to the documentation.

Many organizations already send out a Tweet or two on release date. But how about taking it further by writing a Tweet about each of the highlights in the release notes and sending those Tweets out over a few days?

That is a good way of reaching people who may be offline for a day or two, and of giving people useful information about exactly what is in the release and how it is relevant to them.

These are the aims when using Twitter as a medium for release notes:

- Send a Tweet for each major point in the release notes.

- Include a link to the release notes, using a shortened URL, so that people can read the full details if they want to.

- Use a hash tag to tie the Tweets together. For example, if the release is *ChocTactics 2.0*, then the hash tag could be "#ChocTactics20." This provides a way for Tweet consumers to see a collection of related Tweets. And, of course, a collection of release highlights is in itself the release notes. In that way, the release is fully, if briefly, documented on Twitter.

If the Tweets are compelling and rich in information, people will share them and respond to them. Is there a chance people will add unfavorable information to the stream? Yes. Someone could write something nasty and add the hash tag to their Tweet. But that is all part of using Twitter and other social media. It is up to us to reply and to keep the conversation going.

An additional idea for the release notes: We could add a live stream of Tweets to the release notes themselves. This is something I have not yet tried. It should work well. The mechanism is the same as for tips via Twitter, discussed in the section above. One disadvantage is that the stream of Tweets for a given release would dry up pretty quickly, so we would need to remove the widget a couple of weeks after the release date.

Twitter for gathering reader feedback

This is another new idea that I have not yet tried, nor have I heard of anyone else doing it. In Chapter 18 we looked at a number of tried and true methods for gathering feedback, including comments, forms generated by plugins, and forms built on external web services. Instead, could we use Twitter as a mechanism for gathering feedback from readers?

The method would be similar to gathering hints and tips. Designate a hash tag and encourage people to Tweet comments about the documentation. Put suggested wording for the Tweets in the footer of each page.

Let's say Ganache wants to gather feedback on the *ChocTactics* documentation. She adds the following suggested Tweets, which people can use or change as they like, and asks them to supply the text for "X:"

- This documentation needs X #ChocTacticsDocs

- Found a useful page about X #ChocTacticsDocs

- I couldn't find any information about X #ChocTacticsDocs

Ganache can collate the feedback by gathering all Tweets containing the key word "#ChocTacticsDocs." Anyone else can gather that feedback too. Ganache can even display the stream on a page for others to see. Ganache would need to be fairly confident about the quality of her documentation. But what a great way to build community spirit!

Saving Tweets

In the sections above we used a Twitter widget to publish a live stream of Tweets. The widget uses the Twitter search API to find the relevant Tweets. However, there is one problem: a Twitter search only retrieves the results for the last few days. Twitter needs to limit the number of Tweets retrieved because of the huge number of Tweets posted. The exact period of time fluctuates, but it is rarely more than a week.

Here is an idea that may solve the problem: subscribe to an RSS feed for a given hash tag, save the results somewhere, and publish them as required. For example:

- This URL generates an RSS feed containing the Twitter search results for the term "techcomm:" `http://search.twitter.com/search.rss?q=techcomm`

- This URL includes the hash tag in the search term, "#techcomm:" `http://search.twitter.com/-search.rss?q=%23techcomm`

You can use an RSS reader to gather the updates from the RSS feed. (Chapter 14 has more about RSS feeds and readers.) A good way to gather and save the updates is using an email client such as Mozilla Thunderbird. When you pull an RSS feed into an email client, the feed items look and behave like email messages. You can therefore save them in an email folder.[6]

Interactivity and games in documentation

Technical writers are talking about using games and fun in user assistance. In a post called *Business intelligence, intelligent content and devices, games, and noise*[7] Lisa Dyer remarks, "You know this intuitively: games are a great way to learn even the most arduous stuff." She takes it even further, saying that people who have an interest in gaming also have the mindset for creating good user assistance. She makes the interesting point that games lack "an artificial separation of function and information."

Bill Kerschbaum, writing about humor and interactivity in documentation, suggests that documentation can be a "service that interacts with customers and gives them an experience."[8] Ellis Pratt says that many companies are putting effort into a customer experience strategy, which includes ways to increase brand loyalty and to engage customers with the company on an emotional level. Ellis suggests that organizations should include user documentation in their engagement budget and plans and that technical authors should devise a customer experience management strategy of their own.[9]

Twitter offers a way of adding interactivity and reader engagement to technical documentation. Let's take a look at a set of technical documentation that combines a bit of a game and a lot of interactivity with a hard core configuration guide. There are some pretty pictures in there too.

Case study: the Dragon Slayer docs

The Atlassian *Dragon Slayer* documentation[10] leads people, or "dragon slayers," through the process of installing and integrating a suite of Atlassian applications. It is styled as a quest. The dragon slayers acquire better armor and more strength as they pass each of the stages. When they reach the end, they can claim a T-shirt. A real one.

The documentation was first published in October 2009 to achieve a very specific set of goals. Atlassian provides a number of products that can work together to provide an integrated development environment: bug tracking, agile project management, wiki, gadgets, source repository navigation, continuous integration, and build management. But installation and configuration are complex. The applications were developed at different times by different teams and were not originally designed to work together. They use newly emerging technologies, such as OpenSocial gadgets, which are themselves under rapid development.

Documentation is a good way of fixing a painful procedure in the short term. So the Dragon Slayer documents have a four-fold purpose:

- Lead people through the installation and configuration steps.

- Demonstrate the integration possibilities by showing an example of each type.

- Highlight the steps required and the pain points that Atlassian should address first. Help the product managers to track the improvements by seeing the steps reduce in number and complexity. Let customers know that the company is aware of the problems.

- Convert the pain to fun.

Dragon Slayer: the game

The documentation consists of a number of pages, each page representing a stage in the quest. The number of stages has changed in the years since first publication as we have added new products to the suite and simultaneously removed some of the complexity from the procedures. At time of writing, there are eight stages.

Pictures at the top and bottom of each page show the hero, Charlie, in various states of attire (see Figure 21.4). At the beginning of the quest he has very little clothing or other accoutrements. He holds a feeble-looking staff in one hand. As the dragon slayer works his way through the quest, the images of Charlie become more adorned. By the end, he has a big strong sword, a shield, and a suit of armor to protect him when he finally meets the dragon. The dragon turns out to be a softy, and all's well that ends well.

The people doing the quest feel part of a community of dragon slayers. They develop a sense of camaraderie with the support team and technical writers. They get the difficult task done by following the step-by-step, top-down instructions. They feel a sense of achievement, having conquered the dragon and told everyone else about it too. They also have fun and win a T-shirt. Everyone loves a T-shirt.

Figure 21.4. The top part of a page in the Dragon Slayer documentation

Dragon Slayer interactivity via Twitter and more

The interactivity in the *Dragon Slayer* documentation is provided primarily by Twitter. People can Tweet their status directly from the wiki pages. The suggested tweets are funny, in context, and represent a call to arms. The first page displays a stream of Tweets showing the dragon slayers who are currently doing the quest and the stage they have reached. Sometimes other people add Tweets to the stream, just out of interest. Each Tweet includes a link back to the documentation so people can see what it's all about.

In addition to the Twitter activity, each page has a link to a forum where people can meet other dragon slayers and help each other. The Atlassian support team monitors the forum and helps with problems encountered and dragons not yet tamed.

Who is Charlie?

Charlie is the hero of the dragon quest. Charlie is also the nickname of the figure in the Atlassian logo. In the original logo, he is dark blue with a round head, arms raised high and feet placed solidly apart. The logo was restyled in October 2011 and is now less obviously humanoid. But those in the know can still see Charlie in there. By the way, Charlie the dragon slayer is quite distinct from Charlie the product manager, another character in this book.

Is this really technical documentation?

Yes, it is. The quest, Tweets, and pretty pictures are at the top and bottom of each page. The central section consists of solid, well tested, and carefully designed technical instruction.

Lessons learned

These are the lessons our technical communication team has learned from experimenting with social tools, specifically Twitter, combined with wiki documentation.

People are generous

When using a social tool, people are the stars of the show.

- **People love a game.** Offer them the chance to enjoy something, and they will do so.

- **People are social beings.** Both our team members and our customers benefit from communicating directly with each other.

- **People like to help other people.** It makes everyone feel good.

- **People like it when others acknowledge their hard work and expertise.** They also like being given the chance to pat someone else on the back.

- **The technical communication team learns a lot.** We learn about how people use the products and about the products themselves.

Numbers can be overwhelming

More chocolate at LavaCon 2011

From Alyssa Fox (@afox98) on 16 November 2011:[11] "Great chocolate exchange and chatting at #lavacon tonight. In early for some work stuff and reading."

There are many, many people out there. It is easy to be flooded with responses and hard to manage them all. The organization needs to develop a strategy for social media and put teams in place to implement that strategy. The technical communication team can then decide its own priorities. It will always be necessary to work with other teams in the organization because it is unlikely that the tech comm team will be big enough to manage the flood of responses. Instead, work with the support and marketing teams to parcel out the work according to each team's priorities and core capabilities.

Make the most of your Tweets

Use Twitter in a way that suits the medium. Here are some suggestions:

- **Set up a communication plan that considers timing and time zones.** In general, people read only the last few minutes' worth of tweets. Twitter is not like email, where people make an effort to at least scan their inbox when they have been out of touch for a few hours or days. For example, when Tweeting the release notes, spread the Tweets over a period of three or four days. Tweet at least twice a day, with a gap of a few hours, to catch people in different time zones. Do not repeat a Tweet. Use a different one each time. Some people will catch them all.

- **Be real.** Spend time on the content of each Tweet, making sure that it is fresh and engaging and that it speaks from the heart. For example, when Tweeting the release notes, add a use case that makes the content relevant to a particular group of people.

- **Be topical.** Do not Tweet old news.

- **Stay on topic.** People follow a specific Twitter account because they are interested in the main topic advertised by that account.

- **Use colloquial language.** Anything else sounds odd and uses up too much space.

- **Link to the relevant documentation page.** For us as technical communicators, a primary aim is to let people know the documentation exists.

Etiquette matters

Even in something as fleeting as a Tweet, etiquette matters.

- Do not send the same Tweet more than once.
- Use a corporate Twitter account rather than a personal account so people know what they will get when they follow the account.
- Do not spam people. Make sure that the Tweets contain real, useful, and targeted information.
- Do not automate replies.
- Treat other people well. Do not flame or belittle anyone.

Safeguard other people's rights

Be aware of possible concerns with intellectual property and privacy when publishing or saving Tweets. People own the content of their Tweets, just as they own the content of their other publications. If you are planning to publish other people's Tweets in some way other than via a Twitter widget, you will need to get the permission of the person who Tweeted. The Twitter terms of service[12] and the Twitter Rules[13] are often updated as Twitter attempts to keep them up to date with the fast-changing environment of online media. It is worth checking the rules before saving and republishing any Tweets. Even after checking the rules, be aware that they may change (they changed a couple of times while I was writing this book) and apply your general knowledge of copyright rules and respect for other people's privacy and intellectual property.

A little bird told you

Who would ever have guessed there were so many ways to use Twitter in technical documentation? Ask people to share tips about your company's products and publish those tips in the documentation. Tweet about the release notes and use hash tags to display the release highlights within Twitter itself. Take the bull by the horns and ask people to Tweet their feedback about the documentation. Turn a configuration guide into an interactive game where people communicate across Twitter, boasting about how far they have progressed in the configuration quest.

Twitter and other social media are expanding in functionality and in numbers of participants. Twitter is a mature technology. Google+ is the new kid on the block, and people are already playing with it as a technical communication tool. The web is all about communication, and technical communication is no small part of it. The web is also becoming increasingly social and interactive. Technical communication can take advantage of that too.

References

These are the tools, websites, blog posts, and other references relevant to the content of this chapter.

Further reading

- Anne Gentle wrote a summary of how Twitter is useful to technical writers in which she explains that we need to monitor and listen first, decide on our role in the conversation, be generous in the information we give, and measure the results. The article is *Focus on Twitter for Technical Documentation*, published on *twitip* in March 2010: `http://www.twitip.com/focus-on-twitter-for-technical-documentation/`

- For more in-depth integration of Twitter on a website, take a look at the page about Twitter's @Anywhere platform: `http://dev.twitter.com/anywhere/begin/`. I have not tried it, but it looks easy enough and quite a lot of fun.

- In a post on *GigaOM*, Ryan Kim discusses gamification, the idea of using techniques from the game industry to encourage people to do specific tasks. The article discusses Badgeville's launch of a new behavior management platform: *Badgeville looks beyond gamification, launches a behavior platform*, November 2011: `http://gigaom.com/2011/11/07/badgeville-looks-beyond-gamification-launches-a-behavior-platform/`.

- Twitter offers a lot more to technical communicators than is covered in this chapter. It opens up opportunities for conversing with other technical writers, for learning, and for broadening the scope of our careers. Funnily enough, the loquacious Ganache has started writing a technical communicator's guide to using Twitter. If you like, you can visit the guide, and contribute to it, on the *Confluence, Tech Comm, Chocolate* wiki at `http://wikitechcomm.onconfluence.com`.

Endnotes

[1] The widget macro described in the *Confluence User's Guide*: `http://confluence.atlassian.com/display/DOC/Widget+Connector+Macro`

[2] You can find out more about LavaCon at `http://lavacon.org/`

[3] Tweet about LavaCon and chocolate from Kirsty Taylor, 16 November 2011: `http://twitter.com/#!/kirstyt/status/136466252666372096`

[4] *The URLEncode and URLDecode Page* from Albion Research Ltd `http://www.albionresearch.com/misc/urlencode.php`

[5] The Confluence *Tips via Twitter* page: `http://confluence.atlassian.com/display/DOC/-Tips+via+Twitter`

[6] *RSS basics - Thunderbird*, in the *mozillaZine* knowledge base: `http://kb.mozillazine.org/-Thunderbird_:_FAQs_:_RSS_Basics`

[7] *Business intelligence, intelligent content and devices, games, and noise*, a blog post by Lisa Dyer on *Just enough code to be dangerous*, January 2010: `http://jectbd.com/?p=1191`

[8] *Did You Hear the One About the User Guide with a Sense of Humor?* a blog post by Bill Kerschbaum on *Intext Writing*, February 2010: `http://intextwriting.com/2010/02/25/did-you-hear-the-one-about-the-user-guide-with-a-sense-of-humor/`

[9] *Turning technical documentation into an emotional experience (for the customer)*, a blog post by Ellis Pratt on *Cherryleaf Technical Authors Blog*, February 2010: `http://www.cherryleaf.com/-blog/2010/02/turning-technical-documentation-into-an-emotional-experience-for-the-customer/`

[10] The Atlassian *Dragon Slayer* documentation, first published in October 2009: `http://confluence.atlassian.com/display/ATLAS/Here+Be+Dragons`

[11] Tweet about LavaCon and chocolate from Alyssa Fox, 16 November 2011: `http://twitter.com/-#!/afox98/status/136638597238489090`

[12] *Twitter Terms of Service*: `http://twitter.com/tos`

[13] *The Twitter Rules*: `http://support.twitter.com/articles/18311-the-twitter-rules`

22

Doc sprints

We put twenty developers in a room for three days with their computers. We didn't know whether to expect shiny new tutorials or the complete works of Shakespeare. Read on to find out what happened.

Sprinting into this chapter

This chapter is about holding a doc sprint to improve your documentation and build community awareness. It is also about a wiki as the ideal platform for hosting such an event. In this chapter:

- **What is a doc sprint?** A description of the event and its aims.

- **Some examples of doc sprints.** The story of three Atlassian doc sprints and a list of sprints held by other organizations.

- **Planning a doc sprint.** A detailed guide to organizing a doc sprint and using a wiki as the platform for the sprint.

- **Fun.** How to add some light relief.

- **Demo and retrospective.** A session where sprinters show what they have achieved and give feedback on the sprint itself.

- **From an organizer's point of view.** Some tips about what it is like to organize a doc sprint.

- **Aftermath.** The things that need doing when the sprint is over.

- **A doc sprint hall of fame.** A good way of acknowledging the contributions of the sprinters.

- **Unusual kinds of doc sprint.** A reminder about innovation sprints, documentation blitz tests, and documentation blitzes.

- **A retrospective on this chapter.** Wrapping up the chapter.

- **References.** The websites, blog posts, and other references relevant to the content of this chapter.

What is a doc sprint?

A doc sprint is an event in which a group of people collaborate to write a specific set of documents. It usually spans two or three days. Doc sprints are similar to the book sprints popular in the open source community. *BookSprints.net* (`http://www.booksprints.net/`) gives a good introduction to book sprints.

A doc sprint focuses on technical documentation and on developing a set of documents rather than a book. The primary aim is to produce tutorials or guides of high quality. We do this by getting the experts to write the documentation and by getting them all together at the same time so they can consult each other and the technical writers.

Another, and very important, aim of a doc sprint is to build community spirit and encourage people to continue contributing to the documentation after the sprint. A buzz builds up around the sprint, both within and outside the organization. Sprinters tell their colleagues what they are doing. They Tweet and blog about it. As a result, more people know about the documentation and are therefore more likely to come to the documentation when they have questions. In many cases, the sprinters are developers and other subject matter experts as well as technical writers. Working in a doc sprint helps people feel that they own the documentation, that it is a good resource, and that they can make a valuable contribution to the organization by working on the documentation. They become familiar with the documentation platform and tools and with the content and its structure. They also become accustomed to working with the technical writers and seeing how the skills of the technical writers and other sprinters complement each other.

Some examples of doc sprints

Atlassian has held three doc sprints to date and intends to organize more. Two of the sprints focused on tutorials for developers. The other focused on quick-start guides for people using the products.

First doc sprint: tutorials for plugin developers

The first Atlassian doc sprint took place in February 2010, lasting three days. The aim was to develop some plugin and gadget tutorials for developers who want to build add-ons for the Atlassian products. Twenty-three developers took part in the sprint. Four of them were community developers and the rest were Atlassians. Most sprinters worked in the Sydney and San Francisco offices, while a few worked remotely.

The result was nineteen tutorials on how to develop gadgets and plugins. The tutorials include documents and source code that people can use to build a plugin or gadget.

Rather than jumping into a new document immediately, a few developers decided to spend the first day of the sprint reorganizing the existing reference documentation. Some of them even decided that this was the most valuable exercise of the entire sprint. One person remarked, "The biggest sign of victory is how many pages I managed to delete."

The sprint had a momentum of its own, especially from the second day when everyone was feeling comfortable about what they were doing. I did wonder if some of the sprinters might get bored or less than enthusiastic about spending so much time on documentation. After all, most of them were developers. But quite the opposite happened. The sprinters became absorbed in developing the best possible tutorials. A number of them said we should hold doc sprints regularly. Now that we have held two more sprints, this has become a common theme. Once involved, developers and subject matter experts are impressed with the usefulness of what they are doing. They also remark that it gives them a different view of the company's products, since they are seeing and using them from the customers' or external developers' point of view.

Overheard

On the last day of the sprint we held a session where people presented the work they had done. I overheard this affectionately ironic exchange between two developers walking into the room:

"This is the doc sprint presentations, right?"

"Yes, mate. We're going to talk about documentation for the next two hours. It's going to be awesome!"

Second doc sprint: quick-start guides for users

The focus of the second Atlassian doc sprint was user documentation, specifically quick-start guides. Giles Gaskell was the organizer of this event, in November 2010. Thirty sprinters took part, of whom fourteen were from outside the company. We had people from all over the world, including Russia and Israel, working remotely as well as in the Sydney and San Francisco offices. They were developers, technical writers, support engineers, business analysts, and other people who use our products and were generous and interested enough to take part in the doc sprint.

The sprinters produced twenty-three guides, some of which consist of many pages and were worked on by multiple authors. One of the guides is *Developing Technical Documentation on Confluence Wiki*,[1] which is now part of the Confluence product documentation. A number of people collaborated on writing this guide. Two are Atlassians and the others are from other organizations:

- Jodie Miners
- Ekaterina Stepalina
- Eliane Pohl
- Matt Doar
- Doug Morrison
- Matt Hodges
- Sarah Maddox

It is inspiring to see people's ideas for the documentation when working with them in the intense focus of a sprint.

Third doc sprint: tutorials for plugin developers

The third Atlassian doc sprint happened in November 2011, over a period of two days. It involved internal developers only. We did not invite community developers to this one, chiefly because there was no time. We decided to hold the sprint only a month before the documentation was required. That is not enough time to get everything organized and to expect other people to be able to make time in their schedules.

Overheard

From a developer: "I enjoyed the doc sprint so much I am doing it again today."

The target of this sprint was very specific. We needed some tutorials on the Java APIs for an upcoming release of JIRA. In addition, feedback from plugin developers indicated that the existing tutorials were either very simple or very complex. We needed to develop something to fill the gap between those two extremes.

What was the reason for holding a doc sprint instead of assigning developers and technical writers to write the tutorials individually? A doc sprint is a good way of ensuring the developers can focus on building the plugins and associated tutorials. We needed to move people away from their desks and their day jobs. It is also a good way of getting people together to help each other.

Fifteen developers and two technical writers took part in the sprint, and the other technical writers were on call to help with wording, chocolate, and anything else that came to mind. The outcome was eleven tutorials positioned nicely between "hello world" and "build a battleship."

We discovered that two days is not a long time. Some people wanted to develop a multipart tutorial, but found that there was only time to do the first part. A couple of people found that they were diving too deep into the code and decided to rethink and simplify their tutorials on the morning of the second day. Most people were happy to finish and polish their work in the few days following the concentrated forty-eight hours of the sprint.

Other doc sprints

A number of organizations hold doc sprints and book sprints, some regularly. Here are just a few.

- Adam Hyde writes about the Ogg Theora book sprint held in Berlin in September 2009. This is one of many book sprints organized by *FLOSS Manuals*.[2]

- Andy Oram describes the challenges of a book sprint and how taking part in one made him reconsider the conventions used in computer manuals.[3] This is a result of his attending a *Floss Manuals* book sprint in March 2009.

- Julie Stickler blogs about holding a doc sprint in an agile environment as an efficient way of getting a manual written.[4]

- OpenStack has a page on its wiki devoted to the topic of doc sprints: `http://wiki.open-stack.org/DocSprint`. Anne Gentle reports on one such sprint held in November 2010 as part of an OpenStack Design Summit. She tells how people approached her throughout the summit expressing interest in contributing to the documentation.[5]

- Mozilla is another keen host of doc sprints. Sprinters develop documentation on the MDN wiki for JavaScript, CSS, HTML, and other open web technologies. Janet Swisher is chief organizer. A *Mozilla Hacks* page[6] shows a list of recent sprints. Janet describes a sprint held in August 2011 where there was no central location. Instead, sprinters from around the globe worked remotely, with a few local groups getting together here and there.[7]

- Andy Oram writes about a set of four book sprints held by FLOSS Manuals at the Google Summer of Code summit in October 2011.[8]

Planning a doc sprint

Good advice is to start planning early, two or three months ahead, especially when organizing your first doc sprint. If sprinters will need to travel long distances, or if the sprint is longer than a day, they will need plenty of time to make arrangements.

A doc sprint takes a lot of planning and hard work up front, then review and tidying up afterwards. But it is worth it because the result is high quality documentation written by experts in the field. Good preparation means that the sprinters get off to a flying start when the day of the sprint dawns.

This is a summary of the tasks involved in planning a doc sprint:

- Define a focus for the sprint, and a wish list of the tutorials or other documents that you would like to develop.

- Configure a documentation development platform that can handle multiple authors. If external sprinters will be taking part, then you may need to make special arrangements to give them access to the platform. A wiki is useful here.

- Invite the people. It is all about the people!

- Set up a communication plan.

- Create a template for each type of document required.

- Write a style guide and technical guidelines on how to use the documentation platform. If the sprinters will write code as well as documentation, they will need technical guidelines for the coding part too.

- Set up the development environment including: any software development kits required, a source repository, and a build server.

- Set up and give people access to an online chat room, video conferencing facilities, webinars (web-based meetings), and an email group.

- Organize the facilities at the sprint venue: a room, network hub and cables, extra power points, and notices to tell people where to go.

- Organize the catering. It saves time if you have lunch delivered. Chocolate is inevitable.

- Publish the schedule and details of the online facilities and the venue.

The sections below give more detail about these tasks.

A focus and a wish list

A doc sprint needs a well-defined focus. This helps to choose the people whom you want to take part as well as the documents that will be written. In most cases, the need for the documents is what drives the decision to hold the sprint. If there are a number of documentation sets that need writing, or if the requirement is still rather vague or all-encompassing, then the technical writers and product managers should narrow the focus for the sprint.

A focus is akin to a subject area and document type, such as plugin development tutorials using a given set of APIs or quick-start guides for a given set of products. When refining the focus for a doc sprint, consider the following points:

- Target the document sets that are most urgently required.

- Target the document sets that will benefit most from being written by subject matter experts like developers or system administrators, rather than by technical writers.

- Make sure that the subject matter experts you need will be available during the sprint.

- Check that you will be able to set up the technical environment needed to develop the tutorials.

Once the focus is decided, it is a good idea to define a wish list of the specific documents required. This will help to encourage people to sign up for the sprint because they know what they will be doing and they can judge whether it is within their skill set.

If possible, it is also good to let people add their own items to the wish list. Customers and community developers, in particular, will have a first-hand knowledge of the gaps in the documentation.

A wiki or other documentation platform

During the sprint itself, people will need to work together on the same platform. They will probably even want to collaborate on the same document. A wiki is a good choice because it is designed with collaboration in mind (see Chapter 19). Some wikis, including Confluence, offer a rich text editor that makes it easy for people to develop content even if they have not used the platform before.

As part of the preparation, the technical writers will set up the framework for the documentation to be developed during the doc sprint. This may mean creating a space especially for the sprint and adding the basic structure of the manual required. (See Chapter 5 for information on adding and structuring a space.)

All sprinters will need access to the wiki, including any external sprinters. The easiest is to let them sign up themselves by enabling public access on the wiki. If your organization's security protocols do not allow that, then you will need to add the usernames yourself. Then grant all sprinters update rights in the appropriate spaces. (See Chapter 5 for guidelines on setting wiki permissions.)

Even before the sprint starts, the wiki comes into play. It is very useful to publish details of the doc sprint to outsiders, and even more useful to allow all comers to update that information. For example, set the permissions of the doc sprint space so that all logged-in users can edit the pages. Would-be attendees can sign up as users of the wiki, then add themselves to the list and put their names down next to the tutorials they want to develop. They can even add suggestions to the tutorial wish list. This helps people to feel involved immediately, long before the sprint starts.

Let's summarize the information that the wiki will contain, then we will go into more detail about some of the points mentioned. Firstly, in the planning phase the wiki contains at least the following information:

- The list of attendees, where people can add their names to join the sprint.

- The wish list of documents to be developed.

- A schedule showing the activities for each day of the sprint, split by time zone. The activities include webinars, catered meals, demonstrations, and retrospectives.

- Signup details for the online facilities: chat room, email group, and webinars.

- Templates, style guide, and technical guidelines.

During the sprint, the wiki is the central location for document development and collaboration:

- The documents under development.

- The fun activities: crossword puzzles, haiku competitions, and whatever else you can think of.

- The feedback gathered during the retrospective.

When the sprint is over, the technical writing team will want to publish the wrap up:

- The results of the sprint.

- A hall of fame showing the activities and the people.

The sprinters

The people are the best part of a doc sprint. They are also the ones who do all the work! Before getting deep into the planning, and especially before writing the invitations or announcements, it is useful to consider why anyone would like to contribute their time and knowledge. Andy Oram conducted a survey on O'Reilly's web site and wrote up the results in an article called *Why do people write free documentation?*[9]

A doc sprint that focuses on documenting a commercial product is not the same as free or open source documentation, but many of the same motivations apply.

- People find value in learning from the other experts taking part in the sprint.

- External developers and authors enjoy the contact with your organization, and employees enjoy and learn from the external developers who are using your tools.

- People like helping other people.

- People get a sense of satisfaction from fixing documentation that is out of date.

- People feel good about developing a new tutorial that is as near perfect as their skills can make it.

Think about the people you would like to invite. Don't be shy. Invite everyone from within the company: developers, support engineers, product marketing, business analysts, and CEOs. As technical writers, we know that everyone has an interest in the documentation. But do they know that?

Consider inviting people from outside the company too. Community developers, community authors, partners of the organization – all will have useful information to contribute and will in turn benefit from taking part.

Send the invitations in good time. If you are planning to invite people to fly in from other cities, two months is a good time frame. For people who are in the same office as you or people who intend to work remotely, one month is enough. After all, how many of us know now whether we will be free to contribute to a doc sprint in two months' time?

Where possible, send a personal invitation to each sprinter. Let them know why you have chosen them. For example, they may be active in the community, or perhaps they have already contributed to the documentation, or they have a specific skill set that is valuable for this sprint. Personal email invitations are very powerful. One of our external attendees mentioned this specifically during a retrospective session.

It is a good idea to have the wish list (which may still be a draft) and the basic information about the doc sprint available online by the time you send out the invitations. Although nothing is yet finalized, suggest that people put their names down next to the tutorials that they would like to develop. If they wish, they can partner with another developer on the same tutorial.

As the sprint date draws closer, remind people via email. Tweet, blog, send Google+ and Facebook updates. Do whatever it takes to keep the date and the importance of the doc sprint in everyone's mind.

If the sprinters take an active part in refining the wish list, and they put their names down for a specific document early, this will give everyone a good focus on the sprint. Come the first morning, they can dive right in without even waiting for the kickoff meeting. This is particularly helpful if people are working in different time zones. Some may be half way through their day by the time you start the kickoff meeting.

A communication plan

It works well to map out a plan of the blogs, emails, and other communications that you will send before, during, and after the sprint. When things get busy, it is all too easy to forget to send the final reminders or even to kick off the first webinar. A communication plan also helps to make sure you do not spam people at some times and leave them in the dark at other times.

The communication plan can be very simple, just a list of tasks with a planned date and an indication of who is responsible.

Templates, style guide and technical instructions

Templates are extremely useful. They help to get people up and running quickly (an empty page is a fearsome thing) and they ensure a basic framework that is standard across all the tutorials. For some sprinters a style guide is too much information, but technical writers will find it useful when they have a debate about wording or punctuation.

Technical guidelines are essential for using the wiki and for the development environment if people are writing code as well as documents. Include basic information such as URLs, login instructions, and a getting started guide for the wiki editor. On the code side, link to any applicable coding standards. Tell people where to find the source repository and build tools. Point out any plugins or macros that dynamically pull the code from the repository onto the wiki page.

Development environment

If the sprinters will write code as well as documents, make sure that they will be able to set up the development environment required and tell them what they will need. This may include an SDK (software development kit), an IDE (integrated development environment), and the development tool itself (Java, .NET, or whatever is applicable). Recommend to them very strongly that they set all this up before the sprint so they don't waste an hour or two on the first day.

They will also need access to a source control repository and a build tool. They may need a continuous integration server to run their automated tests.

Online facilities

The main business of the sprint happens on the wiki, but other online tools are useful, too, especially when sprinters are working remotely and in different geographical regions.

Webinars are a good way of getting people to talk to each other. I suggest at least one webinar a day during the sprint. A webinar is a web-based meeting. People can participate from different locations around the world using just their computer and a telephone or a headset with earphones and microphone. There are a few companies that provide webinar facilities, such as GoToWebinar at `http://-www.gotomeeting.com/fec/webinar`.

The first webinar in a doc sprint is in effect the kickoff meeting. The organizer will introduce the sprint, ask people to introduce themselves, point out the templates, and explain the format and logistics of the sprint.

A good way of using webinars is as "remote standups," held once a day during the sprint. (See Chapter 15 for a description of standup meetings in an agile environment.) Attendees take turns to talk about their progress to date, what they still need to achieve, and any problems they have encountered. If someone else is working on something similar or has a solution to a problem, they can get together after the meeting to discuss it.

Hint: Schedule a practice webinar before the sprint starts so you can get to know the webinar software. Ask one or two of your colleagues to log in from different locations and experiment with the use of a telephone and the computer's built-in microphone and speakers. Find out what works best in the sprint venue.

If you have the luxury of video conferencing facilities, they work well when most of the sprinters are grouped in two locations. For example, the Atlassian doc sprints are usually based in Sydney and San Francisco, with a few remote sprinters sprinkled around the world too. We set up a daily video conference session between the two cities and run a webinar simultaneously for the remote attendees. It can get a bit tricky juggling all the technology, but it is fun and rewarding to see everyone participate.

An online chat room is essential for keeping the noise down in the sprint room itself and for sharing information quickly and efficiently. We use an IRC (Internet Relay Chat) channel, and it is a hive of activity throughout the sprints.

An email group is useful when you want to send an announcement to all sprinters, whether they are online or offline at the time. A Google group fits the bill.

Catering, venue, and other practicalities

There are a lot of practicalities to consider: mundane things like tables, chairs, network hubs, cables, and signs. Book the rooms. If people are coming to a central location, tell them what they need to bring. In particular, they will probably need to bring their own laptop computers.

Find out what your budget is. Will your organization pay for any meals? Then let the sprinters know so that they can decide what to do about their food. If you are taking people out for dinner, remember to book a table in a restaurant.

Organize someone to take photographs. Pester the remote participants to take photographs of themselves and send them in too.

Hint: Add a list of tasks on the planning wiki page with a person's name next to each task. Include even simple tasks like booking the restaurant. Think of the satisfaction you will feel as you tick off each task!

Fun

A doc sprint *is* all about the people, and people love fun. They will do a better job if they have a temporary diversion every now and then. A bit of laughter, a bit of social contact, and then the nose back to the grindstone.

A haiku competition is a feature of the Atlassian doc sprints.[10] People compose the haikus at any time during the sprint, and then a panel of judges (the technical writing team) chooses the winner. Matt Doar won the competition in the first doc sprint, with his profound piece of poetry based on a children's joke ("What's brown and sticky? A stick!"):

> Brown and then sticky.
> But the answer is not "stick."
> It is chocolate.

Daniel Green won the second haiku competition:

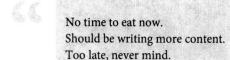

> No time to eat now.
> Should be writing more content.
> Too late, never mind.

Other diversions may include a crossword (chocolate-themed, of course), a collaborative story, and whatever springs to mind.

Demo and retrospective

Towards the end of the last day of the sprint, people gather together for a demonstration and retrospective session.

In the first part of the session, sprinters demonstrate what they have achieved. This is very interesting to the other sprinters. Consider inviting other members of the organization, too. For the sprinters, it is a good motivator to know that they have a deadline and will need to show their goods. It gives them something to work towards and finishes the sprint with a bang.

A theme and fun in a doc sprint

A theme is great for making a doc sprint sound like fun, for giving the sprinters a sense of unity, and even just for prettying up the wiki pages. Choose a theme early in the planning phase. Ask potential sprinters to contribute ideas. Publish the theme along with the schedule and wish list. The very best theme of all? Chocolate of course.

After the demonstrations, sprinters attend a retrospective session[11] to give feedback and suggestions to the doc sprint organizers. A good format is to ask people what they thought went well and what could have been better. Use the wiki to gather comments from remote sprinters in different time zones, then collate all the feedback and ask people to vote on the top good points and the top points for improvement. This will help to make the next sprint even better.

From an organizer's point of view

One really useful thing I have learned, as a doc sprint organizer, is that the organizer's work does not stop when the sprint starts. If you, as organizer, are planning to take part in the sprint, make sure you assign yourself something simple and short. People will come to you with all sorts of "interesting" questions, ranging from bugs in the software development kit to where to go for lunch, or to the location of the fan that should have arrived yesterday. You need to feel free to devote time to things like updating templates, telling people what is happening, taking photos, and writing notes for the post-sprint blog.

Is that a chocolate keyboard?

Just look at the cover of Anne Gentle's book, *Conversation and Community: The Social Web for Documentation.*[12] You can almost taste the chocolate!

Actually, the keyboard pictured on the cover is not made of chocolate, although it looks just like it.

The book is brim full of information and ideas about people, documentation and social technologies.

Another hint is to be flexible and expect change. Some people may want to take part but be unavailable on the chosen dates. Other people may commit to the date but then have to pull out at the last minute. If feasible, let them do the sprint a couple of days earlier or later. If all the instructions and templates are online, these sprinters can work quite happily by themselves, with the occasional contact via email or Skype to keep their spirits up.

People may also change their mind about what they want to do on the first day of the sprint. They may see who else is attending and decide they want to learn something from the other attendees, pair with someone, or just do something simpler.

The best tip I can give, after organizing a few doc sprints, is this: Get other people involved as soon as you can. People are enthusiastic, generous, and inventive when given an idea to play with. Start a wiki page or a Google document, hold a kickoff meeting, and throw around some chocolate. The ideas will start flooding in.

Aftermath

The work is not finished when the sprint ends. The documents will need a technical review. They may need to be moved from the doc sprint wiki into the official documentation. The technical writers will need to update templates, make any other changes arising out of the feedback given during the retrospective, and summarize the other recommendations for use in the next doc sprint. Then they will assess and report the success of the sprint.

It is also important to acknowledge the work the sprinters have done. Whether they are from inside or outside the organization, they have contributed a lot of time and expertise on something that is not their day job. Some of the rewards are inherent in the doc sprint itself such as learning from the other sprinters, having contact with external and internal developers, helping others, and developing a satisfyingly perfect tutorial. Even so, it is enjoyable and nice to recognize their contribution. Mention their names in blog posts. Make sure their names appear as authors of the new documents. Create a "hall of fame."

A doc sprint hall of fame

A hall of fame celebrates the people who took part in the sprint, the documents they developed, and anything special, such as winners of a competition. It is a good way of acknowledging the sprinters' contribution to the documentation.

Photographs are an excellent tool for recording and sharing the activities of a doc sprint. For the people who attended the sprint, photographs bring back fond memories of the event. For people who were not there, the pictures show how much fun it was, what it was like being there, who the sprinters were, and what was achieved. This encourages people to sign up for the next sprint.

If the hall of fame displays photographs from a social site, such as Flickr, the sprinters can contribute photographs, too. In Chapter 20 we saw how to include images from external web services on a wiki page using the Confluence widget macro. One such web service is Flickr (`http://www.flickr.com/`). People can add pictures to the Flickr set at any time, and those pictures will appear on the wiki page the next time someone looks at it. For example, the Atlassian doc sprint hall of fame[13] includes slide shows drawn from Flickr.

Be sure to get permission from all sprinters before putting photographs of them on display. If some people are working remotely, ask them if they would like to send in a photograph. The hall of fame mentioned above has a couple of great shots of people working at home.

Unusual kinds of doc sprints

This chapter has described the usual type of doc sprint. There are others. For example, Chapter 15 describes an innovation sprint. This is a short period of time, such as a day, set aside for the technical writers to try something new. And a documentation blitz test is a much shorter period of time, usually an hour, in which people do concerted and concentrated testing of the documentation. The testers can include developers, technical writers, product managers, and QA engineers.

Chapter 17 describes a documentation blitz. This is similar to a blitz test, but instead of looking for problems, the idea of a documentation blitz is to fix a specific and large-scale problem. The participants may be the technical writers and the development team. The purpose is to tackle something that needs to be done quickly, but the technical writers on their own do not have the resources to do the job in time.

A retrospective on this chapter

Given a focus and a framework, people do high quality work. Given motivation and recognition, people do outstanding work. A well-planned doc sprint gives people that focus, framework, motivation, and recognition. Readers benefit from guidelines written by the subject matter experts, framed and organized by the technical writers.

A doc sprint also brings the work of the technical communicators to the attention of the world. A room full of people working with obvious attention and enjoyment makes a good impression on the rest of the organization. A flurry of activity in the blogosphere and Twittersphere lets everyone know that the documentation is alive.

And while we didn't prove the infinite monkey theorem, we did get a number of good tutorials. And some haikus.

References

These are the tools, websites, blog posts, and other references relevant to the content of this chapter.

Endnotes

[1] *Developing Technical Documentation on Confluence Wiki*, a guide in the Confluence documentation: `http://confluence.atlassian.com/display/DOC/Developing+Technical+Documentation+on+Confluence+Wiki`

[2] *Ogg Theora Book Sprint*, a blog post by Adam Hyde on the *Google Open Source Blog*: `http://google-opensource.blogspot.com/2009/09/ogg-theora-book-sprint.html`

[3] *Challenges from a book sprint: the great things about ignorance and disorder*, a blog post by Andy Oram on *O'Reilly Community*, March 2009: http://broadcast.oreilly.com/2009/03/challenges-from-a-book-sprint.html}}

[4] *Agile Doc Reviews – The Documentation Sprint*, a blog post by Julie Stickler on *HeraTech*, December 2009: `http://heratech.wordpress.com/2009/12/04/agile-doc-reviews-the-documentation-sprint/`

[5] *OpenStack Doc Sprinting*, a blog post by Anne Gentle on *Just Write Click*, November 2010: `http://justwriteclick.com/2010/11/23/openstack-doc-sprinting`

[6] Summary of doc sprints on *Mozilla Hacks*: `http://hacks.mozilla.org/category/sprints/`

[7] *MDN docs get more sprint love*, a blog post by Janet Swisher on *Mozilla Hacks*, August 2011: `http://hacks.mozilla.org/2011/08/mdn-docs-get-more-sprint-love/`

[8] *FLOSS Manuals at Google, October 2011*, a bundle of posts by Andy Oram, October 2011: `https://bitly.com/bundles/praxagora/4`

[9] *Why do people write free documentation?*, a blog post by Andy Oram on O'Reilly Media's *OnLamp.com*, June 2007: `http://onlamp.com/pub/a/onlamp/2007/06/14/why-do-people-write-free-documentation-results-of-a-survey.html`

[10] Chocolate haiku competition on the Atlassian doc sprint wiki: `http://confluence.atlassian.com/display/DOCSPRINT/Chocolate+Haiku+Competition+-+November+2010`

[11] An example of a doc sprint retrospective, on the Atlassian doc sprint wiki: `http://confluence.atlassian.com/display/DOCSPRINT/Doc+Sprint+Retrospective+-+November+2010`

[12] *Conversation and Community: The Social Web for Documentation*, a book by Anne Gentle, published by XML Press in 2009: `http://xmlpress.net/publications/conversation-community/`

[13] Hall of fame on the Atlassian doc sprint wiki: `http://confluence.atlassian.com/display/DOCSPRINT/Doc+Sprint+Hall+of+Fame`

23

Driving wiki development

The wiki world is one of innovation and ideas. People dream, then build their dreams. Technical communicators, too. Confluence is a platform extraordinaire for technical communication, but it's not perfect yet. We can help to make it so. The wiki is extensible. That's what makes it a platform to dream of.

Zooming into this chapter

This chapter describes ongoing wiki innovation, gaps in wiki functionality, and how technical communicators can help to drive wiki development. These are the topics covered:

- **Innovation in the world of wikis.** Let's take a look at some recent innovative developments: a federated wiki, a wiki on a stick, and elearning integration. Then, I will introduce a company whose mission is to build tools that make developing documentation easier – on a wiki.

- **Gaps in wiki functionality.** Not all is rosy. There are some longstanding requests awaiting fulfillment and some problems in the new Confluence 4 releases, too. This sections describes the problems and some workarounds. It also suggests how we can work with the Confluence developers to show them what we need.

- **Ideas for new wiki functionality.** Thinking creatively, what features do we as technical writers lust after?

- **How to drive wiki development.** The way to turn our ideas into reality is to bribe Atlassians (with chocolate, of course) and get in touch with the development community.

- **How to develop new functionality.** We can write user macros or plugins and use the Confluence APIs to interact with the wiki.

- **Technical writers who are already building plugins and add-ons.** Let's check in with the technical writers responsible for a report that shows broken links and macros, a macro that makes a magical glossary, on-demand translations of wiki pages, and more.

- **Zooming out.** Let's go kiss some wiki.

- **References.** These are the plugins, tools, and references relevant to the content of this chapter.

Innovation in the world of wikis

Web technologies are growing rapidly, adding more capabilities, and becoming smarter and faster. The cloud is the place to be. Integration with other services and platforms is another fast-developing area. Wiki developers and the community of people who build add-ons are fully engaged in these technologies. That is a good reason for hitching our wagons to a wiki star. Let's take a look at some recent innovative developments.

Ward Cunningham's federated wiki

Ward Cunningham is the man who invented the world's first wiki (see Chapter 2). Now he is working on an intriguing new project. This is not related to Confluence specifically, but is an example of ongoing innovation in the wiki arena.

The idea of a federated wiki has been around a while. The aim is to link wikis together into a distributed network of wiki servers. A reader should be able to move from page to page in search of information, not really caring which wiki hosts each page. The wikis could even be hosted on different wiki software, such as Confluence and MediaWiki. Taking it even further, people have proposed that any communication platform could be able to participate in the federation. The technologies and standards required include:

- Ways of linking the individual pieces of information to each other.

- Ways of transferring a piece of information from one platform to another, independent of the way the content is stored on each platform.

For more information on the original ideas, take a look at *Federated Wiki*[1] and *Inter Wiki*[2] on *WikiWikiWeb*. The *Inter Wiki Map*[3] suggests one solution for the inter-wiki linking system.

Ward describes his new project as the "smallest federated wiki."[5] One way it could be used is to share content across wikis! At the moment, the content in one wiki is more or less shielded from the rest of the world. You can export and import content, but there is no feature specifically designed to share content across wiki repositories.

Who cares about printing money, let's print chocolate!

A new technology, called 3D printing, creates objects by "printing" them from an electronic blueprint. It works by adding successive layers of material on top of each other to build the desired object. People at the University of Exeter are using the technology to print chocolate.[4]

Compare this to the source repositories that developers use. Developers can take a copy of a project ("fork" the project) and then work on it independently. When ready, they can ask the repository owner to accept the changes back into the main repository ("pull" the changes). The source management system will then compare the two versions of the files and merge them, resolving any conflicts where possible and reporting the conflicts that it cannot resolve. Imagine if we could do that with content on a wiki. It is a whole new way of collaboration.

Ward's federated wiki is a work in progress. He says that the concept will change and grow as people contribute to the project. The federated wiki, designed to allow new ways of collaboration, is in itself under development as a collaborative project on *GitHub*.[6]

Appfire's Firestarter, a wiki on a stick

Appfire has created Firestarter, a portable wiki appliance. It is a Confluence wiki on a USB drive. Why would a technical writer want a wiki on a stick? One scenario is to work on the documentation while offline. Perhaps on a long air trip or in a remote forest cabin. Another scenario is to distribute the product documentation to customers on a USB stick, packaging it along with the product. The USB stick would contain the latest version of the documentation at the time the product is shipped. Firestarter can synchronize automatically with the server-based Confluence documentation whenever the customer is online. That way, they can get the very latest documentation even after the date of shipping.

Gilly, an elearning plugin

Gilly is an elearning plugin for Confluence. It integrates the EduBrite learning management system with the wiki so that tests and assessments can be delivered in Confluence.

K15t Software's focus on wiki documentation tools

K15t Software's mission is to build tools and solutions that make developing documentation easier. The company focuses on tools for wiki-based documentation. The "About Us" page on their website, at `http://www.k15t.com`, says:

> Our Vision: We believe that wiki-based technologies will be the future of documentation. Our tools for wiki-based documentation extend the wiki approach to work for technical, project and process documentation, and to involve traditional office users.
>
> Our tools help everyone on the team to contribute their bit of information, and to keep track of the process.

Reading through this book, you will have seen the K15t plugins mentioned many times. The various *Scroll Wiki Exporter plugins* convert Confluence content to PDF, DocBook XML, EPUB, HTML, EclipseHelp, and JavaHelp. *Scroll Office* exports content to Microsoft Word based on Word templates (see Chapter 11). The *Scroll Wiki Forms plugin* extends the functionality of Confluence's page templates (see Chapter 8).

I take my hat off to the K15t team for their vision, passion, and skill. They have steeped themselves in the requirements of technical documentation, consulted technical writers the world over, and then devised innovative ways of building the solutions into Confluence.

K15t have already made the wiki fly, and they are continuing to build more aerodynamic and flexible wings. *Scroll Versions* is a new plugin, in alpha testing at the time when this book is going to press. Public release is expected early in 2012. When this plugin is installed and activated in a Confluence space, you will be able to define a number of versions of your documentation. You can then schedule updates for a given version, and publish any version at any time, in either the current space or a different space. The plugin supports conditional publication of a page, or part of a page, to a particular space. A light workflow is included (draft, review, done).

The plugin also supplies stable page keys across versions for use in online help solutions, and allows duplicate page names within a space. The chocolate-coated cherry on top is the enhanced content reuse. The plugin handles content reuse across versions, which is tasty. Sweetest of all, it provides cross referencing of reused content. When viewing a page that includes content from somewhere else, you can see where the included content comes from. When viewing a page that is included elsewhere, you can see all the pages (and versions) that include it. Chocolate fest.

Gaps in wiki functionality

While innovation is alive, there are gaps in wiki functionality, too. Looking at Confluence in particular, the wiki lacks some tools that really should be there to make it the perfect platform for technical communication. Of course, we need to acknowledge that technical documentation is not the only use case for the wiki. The developers need to cater to other groups of users, too.

First, from our point of view as technical communicators, let's take a look at some known gaps and some specific problems in the latest Confluence 4 releases. Then let's see how we can get involved, drive wiki development, and even develop add-on functionality ourselves. Some technical writers are already doing it.

From the mouths of wiki newbies

A couple of experienced technical writers joined the Atlassian team recently. I asked them these questions soon after they arrived, while they were still new to the task of developing technical documentation on a wiki:

> As a technical writer who has recently started using Confluence, what would you nominate as the best aspect of using Confluence for technical documentation? What aspect do you think most needs improvement?

From Lingbo Lu:

> Great questions! From my own experience:
>
> - Best: Real-time publishing is the most powerful functionality offered by Confluence for tech writers, as any new info or changes are presented to readers immediately when we hit that Save button.
>
> - Not so great: Version management would be an area that's completely missing and is what a lot of writers need.

From Mary Anthony:

Hmm, that is an interesting question. Ok, well my opinion is purely as a technical writer writing documentation for software.

Best Aspect: People are comfortable with it. Confluence is a tool which many of our users use themselves in their work. This means they are comfortable with it in a way they might not be with other "readers" such as Adobe PDF or even a styled HTML page in a browser. Whether or not a reader actually comments on the page, I do think they enjoy the idea that they can comment or even change the content. Our users are comfortable with Confluence also as an authoring tool. In some ways, seeing documentation in Confluence validates for them their own business sense. For some reason, many people who entered the business world using Confluence or other Wikis, feel like other documentation technologies are out-dated — even when the technology surpasses Confluence in search or presentation. As a writer, I want to take advantage of that "comfortable" draw.

Needs Most Improvement: Round tripping content and comments and history. No doubt about it. A wiki like Confluence is a repository of information. Users should be able to move large amounts of information out of it, process it, and put it back into Confluence with your processing in place. Users should be able to put large amounts of data in easily because the format is open and easy to understand. Once you truly open wikis up with a good structure of the underlying data (content/comment/versioning) and an API layer that can manipulate that data, users will surprise us with how they use the functionality. Confluence, as a commercial Wiki, should **do** this and the fact that it doesn't is just short-sighted.

Longstanding requests

These are some of the requests that people have already made for improvements in Confluence. In the list below, identifiers such as CONF-1234 refer to issues logged on the Confluence issue tracker at `http://jira.atlassian.com`. You can visit the issue tracker to see the progress on these issues.

- **Search and replace.** There is no site-wide search and replace solution in Confluence (CONF-2522). (Confluence 4.1 introduces a search and replace for use within a single page in the editor. This is not the same thing as a site-wide search and replace solution.)

- **Search combined with content reuse.** Confluence wiki provides two macros that you can use to embed content from one page onto another page: the include macro and the excerpt-include macro. Technical writers find these macros very handy for content reuse. The problem is that the Confluence search results pick up the content where it is originally written, but not where it is displayed. Chapter 8 discusses this issue and some workarounds. It would be nice if Confluence itself would offer the option to include the macro contents in the search results (CONF-19054).

- **Support of the Copy Space plugin.** If you want to copy the contents of a space to another space, you need to install the *Copy Space plugin* (see Chapter 8). This functionality is not part of Confluence, and Atlassian support does not cover this plugin (CONF-14198).

- **Comment management.** In Confluence, you can view and manage comments on each page. You can also retrieve them via an RSS feed or an email notification. It would be nice to see a stream of comments for the entire wiki, or at least for a space, and to be able to remove or reply to them without going to the page on which the comment was added (CONF-23442).

Improvements needed in Confluence 4

At the time when I am writing this chapter, Confluence 4 is still pretty new. Confluence 4.0 came out in September 2011, followed by Confluence 4.1 in December. The Confluence 4 series to date has some awkward problems for technical documentation.

- **Image references on copy and paste.** When you copy the content of a page to another page by opening the editor and copying and pasting the content, the image references are absolute. The images in the new page will refer to the original page. In other words, the new page will load its images from the original page, not from its own attachments (CONF-23332).

- **Finding the location of an image used on a page.** The image properties panel in the Confluence 4 editor does not show the location of the image (CONF-23945). By location, I mean the page to which the image is attached or the URL of the image source. (In Confluence 3.5, the location was visible in the wiki markup.) This is a concern for technical writers who put a collection of often-used images in one location and then display them on a number of pages.

There is a workaround, though cumbersome. Right-click on the image in the editor and select "View Selection Source." (That is the name of the option in Firefox. Other browsers may word it differently.) You will see the HTML `<img>` element. Find the `src` attribute. If the image comes from an external source, you will see a URL. That's easy. If the image is attached to a Confluence page, the `<src>` attribute will look something like this: `src="/download/attachments/181535157/1.png"`. The number in the middle, in this case `181535157`, is the identifier of the page to which the image is attached. Now construct a URL to view the page using the page ID instead of the page name. Use that magic number as the `pageID`: `http://MY.CONFLU-`

ENCE.SITE/pages/viewpage.action?pageId=181535157. That URL will take you to the page to which the image is attached.

Note: The Confluence team is planning to add the location to the tooltip that appears when you hover over the image. This will alleviate the problem, but it is not a solution. Many authors do not notice the tooltip, and you cannot select or copy the content of the tooltip.

- **Searching in the editor.** When the editor is open, there is no way to search for the metadata elements of the page content, such as macro names, macro parameters, or image names. They are all wrapped up in the HTML markup and are no longer part of the text (CONF-23560, CONF-23561, CONF-23667). A new plugin for Confluence 4 is currently under development by Arsenale that will allow authors to edit the XHTML source of the page: *Arsenale Invisible Ink.* This plugin will offer an alternative to the default Confluence 4 rich-text editor. At the time that this book is going to press, the plugin is in alpha testing. The initial release is planned for January 2012. The aim is to provide power users with a rapid text editor that includes a quick way of finding and editing macro parameters and other markup elements on the page.

- **Global search, macros and metadata.** In earlier versions of Confluence, it was possible to use the global search to find macro parameters and other markup elements in the wiki markup. This is no longer possible in Confluence 4. This problem is somewhat alleviated by the special search fields available in the global search.[7] You can search for a specific attachment by entering "filename:" in the search box followed by the attachment's file name. Enter "macroName:" and the name of a macro to find occurrences of a given macro.[8]

However, it is not possible to search for macro parameters because they are hidden in the markup and are not part of the page content (CONF-24568). This means, for example, that you cannot discover where a particular excerpt or page is included on another page (CONF-9293, CONF-23775). See the section titled "K15t Software's focus on wiki documentation tools" (p. 436) for information about the *Scroll Versions* plugin by K15t Software – this new plugin will cross reference inclusions very nicely.

How did this happen? How could Atlassian release a new version of the product with these gaps? For the Confluence product team, it is a matter of priorities. They aim to get new features to the customers as soon as possible. The sooner a new feature is released, the sooner customers benefit from the value of the new features, and the sooner the Confluence team can get feedback and continue with the cycle of improvements.

I think the answer lies also in the complexity and size of the design effort involved in building a new editor and a new storage mechanism from scratch. In my view the new editor is smart and innovative. It is already the smartest I have seen in a web application. It has the potential of becoming even smarter now that the development team has laid that solid foundation of XHTML and a single editing environment. However, the editor itself has some bugs and it suffers from the clumsiness characteristic to rich text editors. The Atlassian technical writers did not discover the full implications of hiding the markup until very late in the release cycle. We should have spent more time dogfooding the release in a production-level technical documentation environment. We started pointing out the problems very soon after the documentation wiki was upgraded to Confluence 4.0, but it was too late to get the problems fixed in the current release cycle.

The point about the Confluence team's priorities is important. Technical documentation is just one of the target use cases for Confluence. When deciding whether a product is ready for release, the development team weighs the benefit to all customers, across all use cases, and considers the relative importance of known bugs and functionality gaps.

How can we fix the problems now? By explaining them to the Confluence team and helping them to continue improving the editing experience. They have already fixed a number of bugs and made the fixes available in the Confluence 4.0 and 4.1 series of releases. Watch the issue numbers listed above if you would like to keep up with any improvements.

Ideas for new wiki functionality

Let's move on from the rather dire area of functionality gaps and think more creatively. If you could compile a wiki wish list, what would be in it? Here are some ideas for new functionality that would be useful or just plain cool. I am sure technical writers the world over have more ideas.

Build me a book

Imagine this scenario: Caraque, whom we first met in Chapter 13, would like to make his own quick reference guide by collecting pages from the *ChocTactics* documentation. He logs into the wiki and drags the pages he wants into an "album" in his personal wiki space. He names the album "Caraque's ChocTactics Selection" and saves it. Then he spots a related blog post on another Confluence site and a workaround for a known problem on the issue tracker. He drags them into his album, too.

Eating chocolate while you write?

The type of chocolate that you are eating may affect the words that you write! Oxford University's Department of Experimental Psychology conducted a series of experiments where they asked people to taste different types of chocolate and associate the taste with a range of sounds.[9] Milk chocolate and truffles make people think of rounded, soft and low sounds. Dark chocolate, with high percentages of cocoa, are associated with sharper, more clipped sounds. The results of this study may be useful for manufacturers when thinking up new names for their chocolate products.

Ganache updates a page about *ChocTactics* reporting. Since this is an area that especially interests Caraque, he has already included this page in his album. No problem. The links in the album are relative. Next time Caraque opens the page in his album, he will see the updated content.

A new starter joins Caraque's team. Caraque gives him a link to the album because it contains the most relevant information for their team.

In another area of the wiki, Ganache is building more albums. She has documented a product that people use on different platforms, with different feature sets on each platform. Ganache needs to provide different sets of documentation for the different platforms. In the documentation space, she builds an album for each platform and drags the relevant pages into each one.

The "albums" feature request is logged as CONF-17025.

Content reuse across wiki sites

Confluence already supports content reuse within the same wiki (see Chapter 8). Now let's take it a step further. Would it be useful if Confluence supported content reuse across different wiki sites?

It is fairly common for organizations to take what they need from a set of external technical documentation and provide it to their own employees as part of their internal written procedures. In this way, our documentation forms the basis of procedures and help for our customers. They need to grab bits, mix them up with their own procedures, and put them into a different order. When we update the documentation, they need the update to be reflected in their own copy, automatically and immediately, without manual intervention.

Confluence's page gadget allows content reuse, but setup is tricky and options are limited. Using a page gadget, it is not easy to affect the display of the content. You cannot choose to include just part of a page or a set of related pages unless you have update access on both wikis.

A more comprehensive and flexible solution for cross-wiki content reuse would be interesting, useful, and exciting.

Choice of product version or configuration

A documentation site usually offers different versions of the documentation to support different versions of the product. This is because customers upgrade to the latest release at different times. Let's use Gianduja, our acquaintance from Chapter 13, as an example. She is an administrator of a site running *ChocTactics*. Although *Choc Ink* has already released *ChocTactics 2.0*, Gianduja is still using version 1.1. When she comes to the documentation, she needs the information relevant to that version only.

Our indomitable technical writer, Ganache, has organized the different versions of the *ChocTactics* documentation in different spaces (see Chapter 10). How can Gianduja quickly find the version that she needs?

If Gianduja comes to the documentation by clicking an online help link in the product, it is easy. The product itself can construct a URL that points Gianduja to the right documentation space (see Chapter 12). But if she comes via a search or some other link, it would be nice to ask her which version she needs. This would be useful as part of the Confluence Documentation theme.

This idea can apply to more than just version control. For example, some documentation sites may offer different documentation for different platforms (mobile, web) or different hardware models.

This feature request is CONF-23912.

How to drive wiki development

Technical communicators have plenty of ideas. We know what we need and what would be the icing on the top. How can we convert ideas to reality? One way is to tell the developers what we want. Developers fall into two groups: the in-house developers who work for Atlassian and the community developers who build plugins and other add-ons for the wiki.

Letting Atlassian know what we want

At times it may be difficult to persuade product owners of the usefulness of our request. Technical communication is just one of the target use cases for the wiki. We need to be positive, proactive, maybe even slightly aggressive. Call it the "kiss my wiki" attitude.

To let Atlassian know what you want, add a request to the Confluence issue tracker, which you can find at `http://jira.atlassian.com`. That is the primary means of requesting a feature or improvement or reporting a bug. The development team monitors and responds to these issues regularly. The issue tracker is also the place to go to find existing requests, follow their progress, and express your views by commenting on the issues.

Other options are to join a user group[10] or use the contact form on the website.[11] Appendix 1 has more resources for getting in touch with Atlassian and community members.

There are some notable examples of features already developed with technical documentation in mind, where Confluence developers worked under close consultation with the technical writing team. We have been known to bribe them with chocolate.

- **Copying a space.** The *Copy Space plugin*, described in Chapter 8, makes it possible to copy the content of a space to a new space. Don Willis developed the plugin back in 2007. Technical writer Rosie Jameson shepherded the plugin through its initial tests and made sure it was installed on the Atlassian documentation wiki. Now, this plugin is an indispensable part of the documentation procedures. The technical writing team keeps a careful eye on its compatibility with each new release.

- **Page ordering.** Until April 2008, the order of wiki pages was determined automatically, more or less alphabetically. Thanks largely to the technical writers' powers of persuasion, Confluence 2.8 introduced manual page ordering. Authors can put pages into any order they like, and Confluence will respect that order when displaying, printing, and exporting content. We no longer need to manually number our chapter titles.

- **Documentation theme.** Jens Schumacher developed the Documentation theme for release in March 2010, closely consulting the technical writers at every step. He may even be a bit tired of chocolate by now.

Talking to the plugin developers

Plugin developers are beings of power. Chapter 2 introduces plugins, add-ons and extensions, and the people who build them. Find the developers' contact details on the *Atlassian Plugin Exchange* at `http://plugins.atlassian.com`. Choose the developers whose interests coincide with yours. Talk to them. Buy them chocolate.

How to develop new functionality

Here are a few ways you can build new functionality that plugs into or extends Confluence.

- **Write a user macro.** You can define a simple macro by typing some HTML into a form on the Confluence administration screen – without having to add a plugin. Chapter 7 describes these "user macros" and gets you started by writing a chocolate macro.

- **Develop a plugin.** A plugin is a more complex piece of software. To get a plugin working, you write the code and bundle it up in a JAR file, then install it onto the Confluence site. The *Atlassian Developers* documentation shows you how to get started with the Atlassian Plugin SDK.[12]

- **Interact with the APIs.** Use the Confluence APIs[13] to integrate the wiki with other applications and platforms.

The *Atlassian Developers* site, at `http://developer.atlassian.com`, is the one stop shop for anyone interested in developing with Confluence and the other Atlassian products.

Technical writers who are already building plugins and add-ons

Many technical writers are experimenting with plugins and add-ons. These are a few that I know of.

Busted Stuff Report

Edwin Dawson has developed a tool for finding broken links, images, include macros, and other macros. The tool is a Python script that connects to a Confluence site, retrieves a list of pages via the XML-RPC library, and checks the source of each page for errors. It generates an HTML report which technical writers can use to help repair their documentation. The tool has the whimsical name of *Busted Stuff Report*.[14]

Glossary macro

Jodie Miners writes a blog at `http://jodiem.com.au/` containing many tips about, and experiences with, Confluence. One of her posts describes a user macro that she wrote to make a sophisticated glossary for her wiki-based documentation.[15]

On-demand translations, analytics and more

Ellen Feaheny is a technical writer who has devoted time, energy, and brilliance to championing and extending the power of Confluence. Ellen is now CEO of AppFusions, an engineering company that specializes in building add-ons and integrations for web applications. I have spoken to Ellen a couple of times at conferences. She is still a technical writer at heart.

AppFusions (https://www.appfusions.com) develops and sells a number of innovative extensions for Confluence. For example, their *Speak My Language* plugin provides on-demand translations of Confluence pages. In Chapter 14 we saw their two plugins that add analytical data to the wiki via Google Analytics and Webtrends. Their website has an *AppFusions Labs* page that describes their upcoming products and invites input from interested parties.

Zooming out

Experiment, build, innovate. Drive wiki development. That's a tall order. But technical writers are up to it. In fact, we are the best people to do it. Let's go kiss some wiki!

References

These are the tools, websites, blog posts, and other references relevant to the content of this chapter.

Useful plugins and tools

- *Arsenale Invisible Ink plugin*, a new plugin for Confluence 4 that will allow authors to edit the XHTML source of the page. As this book is going to press, the plugin is in alpha testing. Public release is expected early in 2012: https://plugins.atlassian.com/plugin/details/918877

- *Busted Stuff Report*, a tool for finding broken links, images, include macros, and other macros in Confluence: https://bitbucket.org/edawson/busted-stuff-report

- Firestarter by Appfire, a portable wiki appliance: http://appfire.com/products/fire-starter/

- *Gilly plugin* by EduBrite, an elearning plugin for Confluence: `https://plugins.atlassian.com/-plugin/details/28320`

- *Scroll Versions*, a new plugin in alpha testing at the time when this book is going to press. Public release is expected early in 2012. The plugin will be available on the *Atlassian Plugin Exchange*.

- *Speak My Language* plugin by AppFusions, providing on-demand translations of Confluence pages: `https://plugins.atlassian.com/plugin/details/512647`

Endnotes

[1] *Federated Wiki*, a page on *WikiWikiWeb*: `http://c2.com/cgi/wiki?FederatedWiki`

[2] *Inter Wiki*, on *WikiWikiWeb*: `http://c2.com/cgi/wiki?InterWiki`

[3] *Inter Wiki Map*, a page on *WikiWikiWeb*: `http://c2.com/cgi/wiki?InterWikiMap`

[4] *Photos: Come in Willy Wonka, your time is up - meet the printer that prints in chocolate*, an entry posted on *silicon.com* by Shelley Portet, July 2011: `http://www.silicon.com/technology/-hardware/2011/07/06/photos-come-in-willy-wonka-your-time-is-up-meet-the-printer-that-prints-in-chocolate-39747665/`

[5] *Smallest Federated Wiki*, a page on *WikiWikiWeb*: `http://c2.com/cgi/wiki?SmallestFed-eratedWiki`

[6] Ward Cunningham's *Smallest-Federated-Wiki* project on GitHub: `https://github.com/Ward-Cunningham/Smallest-Federated-Wiki`

[7] Special fields in the Confluence search, described in the *Confluence User's Guide*: `http://con-fluence.atlassian.com/display/DOC/Confluence+Search+Fields`

[8] *How to search Confluence for usage of a macro*, a blog post by Sarah Maddox on *ffeathers*, November 2011: `http://ffeathers.wordpress.com/2011/11/04/how-to-search-con-fluence-for-usage-of-a-macro/`

[9] *Assessing the shapes and speech sounds that people associate with chocolate samples varying in cocoa content*, a study at the Department of Experimental Psychology, University of Oxford, published in April 2011 (`doi:10.1016/j.foodqual.2011.03.009`): `http://www.sciencedirect.com/-science/article/pii/S0950329311000668`

[10] Atlassian user groups: `http://www.atlassian.com/company/community`

[11] Contact form on the Atlassian website: `http://www.atlassian.com/company/contact`

[12] A guide to developing with the Atlassian Plugin SDK: `https://developer.atlassian.com/-display/DOCS/Developing+with+the+Atlassian+Plugin+SDK`

[13] Confluence developer documentation: `https://developer.atlassian.com/display/CON-FDEV/Confluence+Developer+Documentation`

[14] Edwin Dawson's *Busted Stuff Report* for Confluence: `https://bitbucket.org/edawson/bus-`
`ted-stuff-report`

[15] *Building a Glossary in Confluence*, a blog post by Jodie Miners on *JodieM.com.au*, August 2011:
`http://jodiem.com.au/2011/08/13/building-a-glossary-in-confluence/`

So long, and see you on the wiki

It's not over 'til the wiki sings

Technical communication is all about the people. It's all about you and me. Thank you so much for reading this book. It's not over yet.

Here is an extract from the article *Conversation is the New Attention*, by Christopher Fahey and Timothy Meaney on *A List Apart*, 19 April 2011 (http://www.alistapart.com/articles/conversation-is-the-new-attention/):

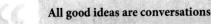

All good ideas are conversations

Conference presentations, like comic books, magazines and movies, are media. And they are all evolving away from a broadcast heritage.

From the printing press through 20th century Big Media, from large-scale commercial print to AOL, information has flowed from central producers to consumers. From Geocities to blogging to The Daily, collectively we still believe that the best way to share ideas is through broadcast publication. Produce and distribute; receive and consume.

This remains the case today—just take a look at your average magazine iPad app. It's essentially a PDF that you can touch. It's like paper. It's one way. It's broadcast.

But the future of publishing, of media in general, isn't broadcast.

As humans, we're on an unwavering path to make it easier to communicate with one another. All media is influenced by our desire to talk to one another, our need to contribute to the conversation. Compare I Love Lucy to Lost: the conversations about Lost were as important as the show itself. Compare old time radio, families gathered around for a fireside chat, to modern sports or political talk radio, where conversation and debate is the show. Compare CNN to Reddit.

Successful media endeavors of the near future will embrace experience, engagement, and conversation.

Reprinted with the permission of A List Apart Magazine (alistapart.com) and the authors.

The article is about conference presentations. I think we can extend the ideas to printed books, and to this book in particular.

You have reached the end of this book, but the conversation continues. Come and join us on the *Confluence, Tech Comm, Chocolate* wiki at `http://wikitechcomm.onconfluence.com`.

Sarah Maddox,
February 2012

Getting help with Confluence

These are some resources that are useful when you have questions about using Confluence or if you run into a problem with your Confluence site. Each entry contains the name of the resource, the location, and a description of what it offers.

Resource	Location and description
Atlassian Answers	http://answers.atlassian.com The *Atlassian Answers* site is a forum where people raise questions and discuss solutions. The participants are customers, community developers, Atlassian developers, support staff, technical writers, product managers, and more. You can ask for advice about plugins, check whether the functionality you need is available, or find out if anyone else has tried to do what you want to do.
Atlassian QuickStart	http://quickstart.atlassian.com/ This site contains quick-start guides for the downloadable version of Confluence as well as Confluence OnDemand (the hosted option).

Resource	Location and description
Confluence documentation	http://confluence.atlassian.com/display/DOC
	In this book, references to the Confluence documentation refer always to the DOC space, as shown in the above URL. The documentation for each version of Confluence is held in a separate space. The DOC space always holds the most recent version, which at time of writing is Confluence 4.1.
	The earlier versions are in spaces with a key of CONFxx, where xx is the version number. For example, the Confluence 3.5 documentation is in CONF35 and the Confluence 4.0 documentation is in CONF40.
	To find the version-specific guides, just replace the key DOC with the version-specific key in the URL. For example, the most recent *Confluence User's Guide* is here: http://confluence.atlassian.com/display/-DOC/Confluence+User's+Guide
	The guide for Confluence 3.5 is here: http://confluence.atlassian.com/-display/CONF35/Confluence+User's+Guide
Confluence Knowledge Base	http://confluence.atlassian.com/display/CONFKB
	This is a collection of trouble-shooting guides and articles maintained by the Confluence support team.
Atlassian Support	http://support.atlassian.com
	If you cannot find an answer to your problem in the documentation or knowledge base, raise a support request for the Atlassian support team to handle.
Atlassian Plugin Exchange	http://plugins.atlassian.com
	Plugins are add-ons you can install into Confluence to extend its functionality. See Chapter 2 for an introduction to plugins, add-ons, and extensions. At the *Atlassian Plugin Exchange* you can find and download plugins, learn about plugin developers, and find links to support and documentation.

Resource	Location and description
Atlassian Experts	http://www.atlassian.com/resources/experts
	The Atlassian Experts are people and companies who supply products and services to complement and/or extend the Atlassian products. If you need some custom development or specialized help, get in touch with one of these experts.
Atlassian Developers	http://developer.atlassian.com/
	This is a site for developers, both inside and outside Atlassian. It contains the documentation and tools that developers need to build on top of the Atlassian products, including Confluence. Here you will find the Atlassian Plugin SDK, the API reference guides, tutorials, information on making your plugin work within the Atlassian frameworks, information about the Confluence architecture, and more.
Atlassian website	http://www.atlassian.com
	The website describes the company and its people. For information about Confluence in particular, go to the "Products" section of the site. There are some good overviews of the product and its features for selected use cases, including technical writing. This site also has information about pricing, free trials, and ordering.

Glossary

The glossary is an alphabetical selection of terms used in the book with an explanation of each term. If you are looking for something that is not in the glossary, try the index. Rest assured, chocolate is in both.

Term	Definition
Agile methodology	A collection of guidelines on how to organize a development team. An agile methodology is based on a set of philosophical principles about the best way to produce software systems. These principles are encapsulated in the *Agile Manifesto*: http://agile-manifesto.org See Chapter 15.
Attachment	Any file that you upload to the wiki and associate with a specific page. If you want to display a screenshot or other image on a page, you upload it as an attachment and then embed it onto the page. You may also want to attach Office documents, multimedia files, and so on.
Caraque	This word has a few different meanings. In the world of chocolate, it can refer to a variety of the cacao bean, a small square of

Term	Definition
	chocolate, or the long thin scrolls of chocolate that are often used to decorate a cake.
	Caraque is also one of the characters in *Confluence, Tech Comm, Chocolate*. He is a community developer and customer of *Choc Ink*. See appendix 3 for a description of the characters in this book.
Chocolate	Ambrosia for technical writers.
Collaboration	The coordinated activity of a group of people working together to achieve a shared objective. In terms of this book, that objective is to produce a document or a set of documents. Collaboration is the strong point of wikis – it is what they are designed to encourage. Collaboration is also what makes good technical writers great. See chapters 17 and 19.
Confluence	A wiki made by a software development company called Atlassian. Wondering how to pronounce the word "Confluence?" Put the emphasis on the first syllable, not the second. Think about "impudence" rather than "imprudence."
	A confluence is also a coming together of rivers, and of other things that flow.
Content	A collective term for the text, images, and other components that make up a page or a set of documentation.
Content reuse	Using the same piece of content in more than one place. When technical writers talk about reusing content, they mean that they will write a piece of content only once, store it in a single location, and use it again and again. See Chapter 8.
Doc sprint	An event in which a group of people collaborate to write a specific set of documents. It usually spans two or three days and focuses on developing a set of technical documents. See Chapter 22.
Documentation blitz	A short period of time in which technical writers and the development team work together to fix a specific, large-scale problem.

Term	Definition
	Their aim is to tackle something that needs to be done quickly, where the technical writers on their own do not have the resources to do the job in time. See Chapter 17.
Documentation blitz test	A very short period of time, usually an hour, in which people do concerted and concentrated testing of the documentation. The testers can include developers, technical writers, product managers, and QA engineers. See Chapter 15.
Dog food and dogfooding	"Eating your own dog food" is a phrase used by software developers. It means using your own products so you can experience them in the same way your customers do. "Dogfooding" is an abbreviation of "eating your own dog food."
Editor	The functionality in the wiki that allows people to change the content of a web page. The editor opens when someone clicks the edit button. See Chapter 6 for a description of the editing modes available in Confluence.
Extreme Programming (XP)	A variant of agile methodology. See Chapter 15.
Ganache	A mixture of chocolate and cream used to fill chocolates or cakes. People often add a flavor to the mixture, such as chopped raspberries for a raspberry ganache.
	Ganache is also a fictitious character, a technical communicator, and the hero of *Confluence, Tech Comm, Chocolate*. See appendix 3 for a description of the characters in this book.
Gianduja	A paste made with chocolate, ground hazelnuts or almonds, and sugar. Sometimes spelled "gianduia."
	Gianduja is also one of the cast in *Confluence, Tech Comm, Chocolate*. She is a wiki administrator and dabbles in development, too. See appendix 3 for a description of the characters in this book.
Innovation sprint	A short period of time, such as a day, set aside for the technical writers to try something new. See Chapter 15.

Term	Definition
Kanban	A variant of agile methodology. See Chapter 15.
Label	A piece of metadata on a page. Other wikis and content management systems may call it a tag. The Confluence search gives a higher weighting to pages where the label matches the search term. Using macros you can pull together all content tagged with a given label. Some plugins add more functionality that makes use of labels.
Macro	A small piece of code that does something. In Confluence, the purpose of a macro may be as simple as displaying a piece of text or drawing a panel around a block of text. Or it may open up a new application within your web browser, such as a drawing tool. In wiki markup, and in the Confluence 3.5 editor, a macro is represented by a word or words in curly brackets. The Confluence 4 editor provides a more graphical representation of a macro using "placeholders" to draw a border around the macro and its contents.
Markup, or markup language	In the world of computers and the World Wide Web, a markup language consists of a set of recognized elements and characters that you can use to add formalized meaning to a section of text. HTML and XML are well-known examples of markup languages. Wiki markup is another example.
Permission	A tool for controlling access to the content on the wiki and to its administrative functions. Confluence permissions determine who can do what on the wiki. The Confluence administrator or space administrator assigns permissions to people and to groups of people. Chapter 5 describes the levels of permissions available in Confluence.
Plugin	A small piece of software that you can install onto your wiki site. It is similar to the add-ons that you can install into your web browser. Once installed, the plugin adds extra features to the core wiki. It may add new options to the menus, add a macro for use on a page, or change the look and feel with a new skin or theme. Some plugins are supplied by the company that develops

Term	Definition
	the wiki itself. Other plugins are written and supported by community developers.
Round tripping	The repeated conversion of content from one format to another and back again. See Chapter 11.
Roundtrip error	An unexpected result that occurs when your content is pushed through a procedure and emerges with unintended changes. Also called a round-tripping error. In the context of a wiki editor, you may enter text in a rich text editor and apply formatting, then save the page and view the result. In the process of saving the content into the backend storage format and then rendering it in view mode, the wiki may change the content's appearance slightly. For example, extra white space may appear.
RSS	A protocol for requesting and receiving updates from a web server. Popular belief has it that RSS stands for "really simple syndication." It originally stood for "RDF site summary." The server sends the response in an XML file conforming to a standard that various RSS readers can interpret. Many online news sites make RSS feeds available for use by other services. Confluence offers some prepackaged RSS feeds and a feed builder that you can use to define the information you want included in an RSS feed.
Screenshot	An image of an application screen, often used in technical documentation to show the reader which area of the application is being described.
Scrum	A variant of agile methodology. See Chapter 15.
Search engine optimization (SEO)	A set of techniques for making your pages appear high up in the search results when people search for a term that is important to you. See Chapter 9.
Single source publishing	A methodology for developing content on a single platform and then publishing it to a number of different mediums and platforms. Also known as single sourcing. See Chapter 11.

Term	Definition
Social media	A loose, growing collection of web-based technologies that focus on making it easy for people to talk to each other. In so doing, they harness the power of the crowd. People bounce ideas off each other and share information across the world, sometimes in just a few hours. Examples of the social media are Twitter, YouTube, FlickR, SlideShare, and many more. Sometimes wikis are classified as social media too. Many businesses now use social media as a way of communicating with customers. The topic of social media is gaining more and more attention in the technical communication field too. See Chapter 17.
Software as a service (SaaS)	Some companies, including Atlassian, offer customers the use of applications that are hosted on remote servers. This is called "software as a service" (SaaS). Customers do not need to install the application on their own servers. Atlassian's SaaS offering of Confluence is called Confluence OnDemand.
Software development kit (SDK)	A set of software tools bundled into a neat package containing everything that a developer needs to get started with creating a specific type of application or add-on. The Atlassian Plugin SDK is such a kit, providing the tools needed to build a plugin for Confluence or any other Atlassian application.
Space	A collection of pages and blog posts in Confluence. A space offers a way of grouping logically related information. The way you define your spaces depends on how you are using the wiki. If your wiki is an intranet, you may give each department its own space. In a technical documentation wiki, you might choose to have a space per product or per version. You might even have separate spaces for your user's guide, administrator's guide, and installation guide.
Tech comm	A common abbreviation for technical communication. Sometimes spelled in one word: "techcomm."
Technical writer and technical communicator	A person who writes technical documentation and communicates technical information in other ways. The terms "technical writer"

Term	Definition
	and "technical communicator" are used interchangeably in this book unless otherwise stated.
Template	A skeleton structure of a page or of some other documentation component. People use templates when writing a page from scratch, or when creating a new space, to ensure that the page or space is consistent with similar pages and spaces. Elements of consistency in a page, for example, are the layout of the page, the existence of introductory and closing sections, the vocabulary used, the heading styles, and more. See Chapter 8.
Theme	A collection of presentation styles that can be applied independently to a wiki space or wiki site. A theme defines the color schemes, fonts, and layout of the pages in a space. Another name for a theme is a skin. Some themes offer additional functionality such as a smart search. Within a Confluence site, each space can have its own theme. When you first install Confluence there is a choice of themes available, and you can add more themes as plugins. See Chapter 7.
Twitter	A web service that people use to send short messages to anyone who is interested in reading them. A message sent via Twitter is called a Tweet. See Chapter 21.
Uniform resource locator (URL)	The address of a website, web page, or other object on the World Wide Web. For example: http://www.google.com
User macro	A macro that you can define yourself via the Confluence administration screens without having to add a plugin. See Chapter 7.
Wiki	Ah, now there's a question. Start at Chapter 1... Just joking. A wiki adds an edit button to a web page. It is a piece of server-based software that makes it possible for people to edit a web page and save their changes to the server. When other people request the page, the wiki serves them the updated content.
Wiki markup	A markup language designed specifically for a wiki. You can use it to apply styles and other smarts to your document by adding

Term	Definition
	recognized codes to the text. Another name for wiki markup is wikitext language. See Chapter 6 for a lengthier introduction to the wiki markup syntaxes used by Confluence and other wikis.
Workflow	The set of steps needed to move from one state to another and the rules that govern that progression. In terms of technical documentation, a common workflow is to create a page as a draft, put it through an iterative review process, have it approved, and then publish it. See Chapter 10.
XHTML	An XML markup language based on HTML 4. XHTML can be parsed by an XML parser and processed using XML tools. The most obvious difference between XHTML and HTML is that in XHTML all elements need to be closed. For example, in HTML the element does not need a closing tag. In XHTML, it must either have a closing tag (</br>) or be an "empty element tag" ( ). Another difference is that in XHTML the element and attribute names are case sensitive, but they are not in HTML.
XML	An acronym for Extensible Markup Language. XML is a standard for creating markup languages. DITA, DocBook, and XHTML are the most common XML-based languages used by technical communicators. XML resembles HTML – for example, elements are set off using angle brackets (<element>) – but it has a stricter syntax.

Characters in this book

Here is a roundup of the fictitious characters in this book. They are not based on real people. Not even Mick.

Ganache is the hero of *Confluence, Tech Comm, Chocolate*. She is a technical writer at *Choc Ink*, a fictitious company specializing in software tools for the chocolate industry. Ganache handles the technical communication for the company's chocolate assessment tool called *ChocTactics* and for a new product called *ChocSpice*. Readers of the book follow in the steps of Ganache as she develops the documentation on the wiki.

One of *Choc Ink*'s customers is **Gianduja**. She is an administrator of her company's *ChocTactics* application. She also dabbles in development. Gianduja first came to the documentation wiki by clicking an online help link on a *ChocTactics screen*. (See Chapter 12.) She has become quite closely involved with the documentation wiki and the community of people who use it.

Caraque is another customer and a community developer. In Chapter 13, he exchanges tips with Gianduja about customizing the *ChocTactics* reports.

The *Choc Ink* employees form a merry band of wiki users. See them come to life in Chapter 13, which describes a day in the life of wiki inhabitants. Ganache is there, of course. So is **Roald**, a support engineer at *Choc Ink*. His main focus is the support queue, but he spends quite a bit of time on the wiki, too, building a knowledge base and responding to customers.

Charlie is the *ChocTactics* product manager. He has many channels for getting feedback from *Choc Ink*'s customers. The wiki is one of those channels. Charlie watches for comments on the documentation and responds when a customer asks something about the direction of the product. In particular, he keeps a close eye on the release notes, the supported platforms page, and the security advisories. He is also working closely with Ganache on the change management pages for the upcoming release of *ChocTactics 2.0*.

Then there is a developer called **Mick**. His name does not have anything to do with chocolate. Sometimes that worries him. But usually not. (Does this last comment tease your curiosity about the relationship of the other names to chocolate? Try looking up the names in the glossary. And think of a famous fictitious chocolate factory, too.)

Index

CPSIA information can be obtained at www.ICGtesting.com
Printed in the USA
LVOW11s0848031013

355255LV00007B/302/P